Designing Human–Machine Cooperation Systems

Series Editor
Jean-Charles Pomerol

Designing Human–Machine Cooperation Systems

Edited by

Patrick Millot

iSTE

WILEY

First published 2014 in Great Britain and the United States by ISTE Ltd and John Wiley & Sons, Inc.

ISTE Ltd
27-37 St George's Road
London SW19 4EU
UK

www.iste.co.uk

John Wiley & Sons, Inc.
111 River Street
Hoboken, NJ 07030
USA

www.wiley.com

Library of Congress Control Number: 2014939767

British Library Cataloguing-in-Publication Data
A CIP record for this book is available from the British Library
ISBN 978-1-84821-685-3

MIX
Paper from responsible sources
FSC www.fsc.org FSC® C013604

Printed and bound in Great Britain by CPI Group (UK) Ltd., Croydon, Surrey CR0 4YY

Contents

Foreword

There are three central words to this book: "ergonomics", "human being" and "machine". This book is not a book on ergonomics, although the topic is duly covered; neither is it a book on human science, even though human beings play a key role in it. This book is what I like to call a book of interfaces, a book that is the result of research at the junctions of various topics. Topics encompass the scientific domains that research – in France, and elsewhere – consists of. Classically, one belongs to a single topic, and one conducts research *within* this single topic. However, this categorization does not enable a true rendition of all the problems of research, including those found at the boundary between several topics.

Separation into different topics satisfies a certain element of Cartesianism in its presentation, but can cause confusion with regard to the work of researchers focusing on these interfaces. Indeed, they are not focused on the difficult problems of topic A or topic B; they are rather focused on the difficult problems that involve both topics A and B. This structure therefore presents shortcomings in appreciating the quality of much important research. Simply by looking at the history of scientific progress, it would appear that a lot of breakthroughs happen at these interfaces.

For this reason, among others, some countries (the number of which is always on the increase) have called upon the help of research organized into projects through agencies (such as the National Science Foundation (NSF), the Defense Advanced Research Projects Agency (DARPA), etc., in the United States, and the National Funding Agency for Research (ANR) in France, etc.). A project is presented as a scientific objective and, therefore, is not restricted by topic. It is run by a consortium consisting of members of the different topics required to reach this research objective over a certain timescale. The evaluation of a project allows us to determine the quality of the research conducted, and this evaluation is no longer limited to one topic. This organization into projects allows for a greater focus on various difficult problems, which the one-topic limit did not.

To conclude this point, this does not mean that research must only be conducted through various projects: actually, depending on the goal of the planned research, two types of presentation may be necessary, and these must therefore coexist. However, this also means that research at interfaces is just as important as purely topic-based research and that both must be evaluated using relevant scientific criteria.

The work put together by Patrick Millot sits at the junction of several topics, since it covers work on human–machine systems and their conception. It can therefore be defined as a project whose objective is to assemble the results of the most recent research in this field; this project is run by a consortium of acclaimed scientists from a variety of different backgrounds. Let us not mention the timescale of this project, as it equates to the maturing time of a book: a time that is usually underestimated.

Human–machine systems are as present in our working world and everyday life as they are in the technological world. These systems are therefore very important, as poor choices during conception can have very big consequences, especially in terms of safety, as recent examples have certainly shown. I have chosen one example in particular, as it is universally known. One system, the automobile, is itself linked to another, the road system and the other drivers and one pilot: the driver of the vehicle itself. To this collective, a significant limitation must be added: the driver is not a professional. The human–machine systems must therefore be simple (the driver has not received any specialized training other than that required for the obtainment of a driving license) but very informative, without being overbearing (there is also a lot of information coming from outside the vehicle), with the goal of making driving as safe and enjoyable as possible. We can see the difficulty and complexity of such a human–machine system and therefore the necessity of permanent research on this topic. The proposed work will offer solid lines of thought and solutions, especially as it comes at the topic from an original angle: putting the person at the center of human–machine systems.

From this point of view, the book is organized into three complementary parts that enable the different aspects of the problem to be addressed: part 1 focuses on the methods of conception, part 2 focuses on the methods of evaluation and, finally, part 3 focuses on human–machine cooperation. Undoubtedly, the readers will find in this book an idea of the state of research in this area, and hopefully the answers to many of their questions.

Finally, this book introduces us to a selection of authors from very different disciplines: specialists in "human engineering", cognitive psychology, artificial intelligence, etc. This fits in well with the requirement of uniting acclaimed specialists from different topics so as to conduct or even understand research at the interfaces.

To conclude, I will reiterate something that I often say, which is that one cannot be precise at the interface of topics: in other words, one cannot be a specialist in inter-disciplinarity. On the contrary we can be excellent in our discipline and know how to cooperate with other specialists, and from this cooperation new advances arise from knowledge. However, this cooperation is only fruitful if the different players are excellent in their respective domains.

This book is the perfect illustration of this concept, and I am convinced that the readers will take great pleasure and interest in reading a book that offers a complete vision of the conception of a human–machine system that is centered on the "human fully involved in the loop".

Bernard DUBUISSON
Professor Emeritus, UMR Heudiasyc
University of Technology of Compiègne

Introduction: Human–Machine Systems and Ergonomics

I.1. What has ergonomics got to do with human–machine systems?

This book on the ergonomics of *human*–machine[1] systems is aimed at engineers specializing in informatics, automation, production or robotics, and who are confronted with an important dilemma during the conception of human–machine systems:

– on the one hand, the human operator guarantees the reliability of the system and he has been known to salvage numerous critical situations through reasoning abilities in unplanned, imprecise and uncertain situations: the Apollo 13 space mission is a mythical example of this[2], where the three astronauts owed their survival to their own genius, their innovative capabilities, as well as to those of the engineers on the ground;

Introduction written by Patrick MILLOT.

1 The word *human* used here without prejudice as a synonym for a human being, or a human operator. For this reason, the masculine form *he* is used throughout the text to avoid weighing down the syntax of the text with the form *he/she*.

2 Apollo 13 (April 11, 1970, 13.13 CST – April 17, 1970) was a manned moon mission of the Apollo program that was cut short following the explosion of an oxygen tank in the Apollo service module during the flight to the Moon. As the vessel could not be turned around, the crew was forced to pursue their trajectory toward the Moon, and harness its gravitational pull during orbit so as to return to Earth. As the service module had become uninhabitable, the crew took refuge in the lunar module, Aquarius. Occupation of this module by the entire crew for an extended period of time had obviously not been anticipated. The astronauts and the control center on Earth had to find ways of recuperating energy, saving enough oxygen and getting rid of carbon dioxide. The crew eventually made it safely back to Earth. See http://fr.wikipedia.org/wiki/Apollo_13.

– on the other hand, the human operator can be unpredictable and create disturbances in the automated system; the nuclear industry is an "interesting"[3] example of this in that it gave three dramatic examples in a little over 30 years: Three Mile Island in 1979, Chernobyl in 1986 and Fukushima in 2011. The Mont Sainte Odile accident is another significant example, from the aeronautic field.

At the beginning of the 1990s, a well-known researcher in the French control community said to me: "human-machine systems are interesting, but I don't see what they've got to do with automation!" On the contrary, the three nuclear incidents mentioned show what the consequences of badly designed human–machine interaction can be. Kara Schmitt accurately summarizes the problems that can be encountered with human–machine interaction. The Three Mile Island accident was the result of *automation misunderstanding* in that the operators did not understand the function of the automatic safety system that would have avoided the accident, and unplugged it. The major Chernobyl accident was characterized by a *lack of confidence in automation* associated with poor understanding of nuclear physics, and the lack of a culture of automated safety in Eastern European countries at the time. These combined factors caused the operators to conduct tests that pushed the reactors to their limits after having turned off the safety systems. Finally, the Fukushima accident, which took place after a tsunami that damaged the nuclear station, was the result of a *lack of appropriate automation* associated with an under-estimate of the risks during conception: the height of the anti-tsunami chamber was only 5.7 m while the waves reached 10 m, the emergency generators in the underground were flooded, their batteries no longer having enough power to feed the cooling systems and to secure the reactors after they had stopped. Moreover, the several emergency stop systems were not automated and the security principles were passive and not active, and therefore required energy to operate [SCH 12].

As a matter of fact, automation does not rival a human remaining in the control loops and supervising the systems, but the human operator must not be reduced to an emergency device to control non-automated activities. On the contrary, the teams of human operators must be fully integrated in the

3 More than the terrible impact of the accidents, importance lies in the lessons learned that could lead to increased safety levels.

command, control and supervisory loops of human–machine systems, so as to get as much as possible out of their capabilities, without suffering from the disadvantages. So, this book focuses on these problems of human-centered *automation* and the factors it addresses.

The approaches of different solutions lie in models, in the sense of a "greater understanding" of human operators, as much as for the systems themselves and their environment. Human modeling has united the *human factors* community over the last 70 years, since the end of World War II. Considering the limitations of the systems of that time, which had relatively low levels of automation, and therefore required a human presence in piloting, control and regulating tasks, researchers tried out unique approaches between the modeling (for the command) of the technical component of the human–machine system and the modeling of the human component. These approaches were inspired by theories of "engineering", first of all the information theory, and then the control theory [SHE 74]. Human engineering research belongs to this movement and was mainly brought to France by Noel Malvache [MAL 73]. The reader can find a history of the approaches used in the human factors research field in [MIL 03] and [SHE 85].

Since the end of the 1990s, the application domains studied have strongly evolved toward large, complex systems, whether they are discrete continuous or hybrid. These are designated as systems of systems, network systems and multi-agents. The *automation level* has greatly increased, which has brought about an increase in the *performance* of the production or service system.

Nevertheless, other objectives must be taken into account, particularly *safety* and *security*. The interest in *life critical systems has increased increasingly*. At the beginning of the 2000s, Amalberti came up with the following categories of *risky systems* [AMA 05]:

– the riskiest systems involve amateur individuals, alpine mountaineering for example, with a risk level around 10^{-2};

– next, he places systems available to the public in which the safety culture is poorly developed (or not consistent) and the choice of operators is not very discriminative, such as car driving, with a level of 10^{-3};

– the chemical industry is next, with a risk level of 10^{-4};

– charter flights with a level of 10^{-5};

– finally come systems that are said to be ultra-safe, such as commercial aviation, the nuclear industry and transport by train, with a risk level of 10^{-6}.

In these systems, the human is seen as an unreliable factor: in 1950, for a hundred accidents, seventy were due to a technical problem, and humans caused thirty. Since 2000, globally this proportion has been reversed, with seventy human causes for thirty technical causes. This is particularly due to a great increase in technical reliability, while human causes have not changed. This explains a natural reflex in the designer to minimize the role of the human in systems by increasing the level of automation. Aside from the technical difficulties surrounding complete automation, increasing automation levels is not actually that simple, in that it involves aspects other than ergonomics, i.e. contextual and organizational. This book attempts to show all the dimensions relating to this problem.

I.2. Increasing level of automation?

The level of automation determines the role and the involvement of human operators to guarantee these objectives: *performance, safety and security*. In highly automated systems, operators have migrated toward control rooms to carry out *supervisory* functions, i.e. *monitoring* and managing failures: *diagnosis* for *re-use*, the *accommodation* or the *reconfiguration* of the automated system. Human tasks become decision based, at the expense of action tasks (reactive), the outcomes of which can be very important for the integrity of the system, but also for its security. In these systems, operators are usually professionals, trained and supervised in an *organization* that is often hierarchical, where they must follow *procedures* to respond to known situations, whether they are normal or part of an incident. However, the main difficulties relate to unexpected and new situations, for which the operators are not prepared and where they must "invent" a solution. The *designer's dilemma* can here be summarized as follows: on the one hand, he can be tempted to aid, or even limit, human activity in the known situations so as to avoid possible mistakes; on the other hand, he can only rely on human inventiveness to deal with the unexpected. However, to try to understand these unexpected situations, the human operator needs information of the system operation in known situations, the very information that is being taken away from him!

These problem-solving tasks are cognitive in nature, and theories that support their modeling can mainly be found in the vast spectrum of cognitive sciences, which include artificial intelligence, cognitive psychology, sociology and ergonomics. These approaches are multi-disciplinary and participative where each discipline contributes to the model and to the proposing of solutions. This book develops these different multi-disciplinary approaches of analysis and modeling for the design of modern human–machine systems and attempts to give an answer to the *designer's dilemma* mentioned above.

In the large transport systems (airplanes, high-speed trains, metro), the operators can still remain directly involved in the driving or piloting loop, all the while carrying out a supervisory role. However, the domain of automobile driving is atypical. It is currently the object of considerable effort to increase its safety, but the problem is difficult because the population of car drivers possesses very heterogeneous capabilities, practice and training, and the organization is hardly controlled, except in an open manner by traffic laws, and with some "sampling" in a closed manner through police controls. Its level of automation is low and efforts are focused on the automation of certain security features, rather than driving in general [INA 06]. Several chapters of this book are focused on this field of research.

Organization itself plays an important role. From an informatics point of view, Guy Boy provides a diagram for a human–machine system using a pyramid, made up of five summits and of their relations which he names AUTOS: A for artifact, U for user, T for task, O for organization and S for situation (see Figure I.1) [BOY 11]. Transposed onto the dynamic systems, the artifact becomes the system and the user becomes the operator.

This figure therefore shows the well-known classic triangle of the *human engineer* O–S–T, as the operator has been formed in the system, carrying out tasks depending on the needs of the system by applying *procedures* (or by trying to innovate in the case of a new problem), these tasks needing to be helped by the *ergonomic quality* of human–machine interaction but also of the interface.

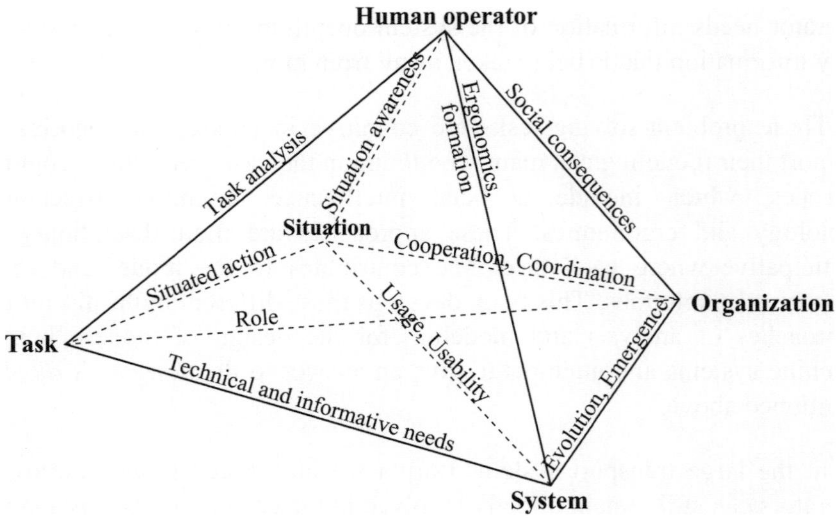

Figure I.1. *Human–machine system environment (adapted from [BOY 11])*

The fourth summit, organization, introduces the *level of automation* involving the *role of the operator* and *task sharing* (and function sharing) between humans within the control or supervision team, but also between humans and automatic systems (of control or of decision). *Task sharing* (or function sharing) between humans and machines gives humans a *level of responsibility* regarding the management of performance and of the risks and a *level of authority* that determines this responsibility. The socio-organizational context of the systems must then make compatible these two levels of authority and responsibility: this is part of the *designer's dilemma* mentioned above. To this effect, human–human and human–machine task sharing cannot be static and defined from the earliest design, but instead must evolve dynamically according to criteria that integrate the performance of the global system and/or the human workload [MIL 88]. It can go even further by establishing a cooperation between human and machine. These advanced aspects will be covered in the last part of this book.

The fifth summit concerns the situation of the task that can introduce new limitations requiring an evolution of the *situation awareness* of the human operator to detect an unusual situation, an evolution of decisions, of the competences being used of even a dynamic evolution of the organization as mentioned previously.

The connection of these five dimensions shows that the successful automation of a system goes well beyond the problem of making it automatic, and that it needs to be a part of a process of *human-centered design of the human–machine system.* Indeed, this approach is developed in this book, in three parts.

Part 1 is dedicated to the methods of human-centered design, from three points of view:

– Chapter 1, written by Patrick Millot, presents the models developed by human engineers and bases itself on functional models to explain human behavior in his environment. It looks at the approaches for positioning levels of automation, notably through principles of task and/or function distribution between human and machine, and extends these to the sharing of authority and responsibility. To attempt to resolve the apparent ambiguity of the role of the operator, this chapter also introduces the mastering of the *situation awareness* of operators, widely studied today.

– Chapter 2 by Christine Chauvin and Jean-Michel Hoc develops models of cognitive psychology and proposes a methodology of design derived from the works of Rasmussen and Vicente called *Cognitive Work Analysis* [VIC 99].

– Chapter 3, written by Gilles Malaterre, Hélène Fontaine and Marine Millot, can be situated in the domain of automobile driving, which unfortunately is the victim of numerous real accidents. The approach the authors use is to analyze these cases to deduce the need for adjustments or assistance tools for the design of new vehicles and the improvement of infrastructure.

Part 2 develops the methods of evaluation of human–machine systems:

– Chapter 4, by Jean-Christophe Popieul, Pierre Loslever and Philippe Simon, evaluates the activity of the human operator at work using methods of automatic classification to define different classes of behavior. The data come from sensors that give the parameters of the task and of its environment, but also from sensors placed on the human body which record characteristic signals of the human state such as the electroencephalograph (EEG) or characteristics of the person's decision and action strategies through eye movements. The methods are illustrated by experimental examples obtained in an automobile driving simulator during studies on the detection of hypo-vigilance.

– Chapter 5, written by Frédéric Vanderhaegen, Pietro Carlo Cacciabue and Peter Wieringa, presents human error analysis methods that are inspired by and adapted from technical reliability analysis methods and which in a sense form the dual approach of modeling methods based on "normal" human behavior. This chapter concludes with the results of the integration of such methods in the design process of human–machine systems.

Finally, Part 3 is dedicated to *human–machine cooperation* through four complementary (between themselves) chapters. We shall see that a cooperative agent comprises a *know-how* and a so-called *know-how-to-cooperate*. The organization of the *cooperative system* is defined according to a structure in which the inputs and the outputs of each of the agents are connected to their environment and to the system that they must control or manage. The functioning of cooperation is related to more functional aspects. Finally, operational aspects such as parameters called cooperation catalysts play a role:

– Chapter 6 by Jacky Montmain contributes to the *know-how* of the cooperative agent. It develops the *causal reasoning* that permits a human–machine cooperation by creating tools founded on artificial intelligence (AI), to support the operator in the control room confronted with situations requiring complex decisions. The author moves from the observation that human reasoning is neither based on a mathematical model of the process, nor on the detailing of the numerical data that are presented, but on the symbolic interpretation of these, which is the key to the explanations that a support system should give. The principal quality expected of the models is no longer precision, but pertinence and compatibility between the representation in use and the cognitive modes of the operator. Examples from the supervision of a chemical process in a nuclear reprocessing plant illustrate these principles.

– Chapter 7, written by Jean-Michel Hoc, contributes to the functional aspects of cooperation. In particular, it presents the models of cooperative activity, the concept of the *COmmon Frame Of Reference* (COFOR) and draws up the lessons for the design of the cooperative human–machine systems. It then describes cooperative activities according to the three levels of abstraction corresponding to the three temporal horizons, by deriving some implications for the design: *cooperation in action*, where the agents manage the interferences between their goals, *cooperation in planning*, where the agents negotiate to come up with a common plan or to maintain a common reference and *meta-cooperation*, which establishes the structures of knowledge of cooperation, such as models of partners or models of oneself.

– Chapter 8 by Serge Debenard, Bernard Riera and Thierry Poulain, describes the development of the human–machine cooperation through the definition of the cooperative structures and through the definition of the *cooperative forms* between human and machine and the implication that they have on human activities. They introduce the concept of "common work space" (CWS), which is very import to encourage cooperation between the agents. Two examples of application processes are detailed, each having different levels of automation, the first application process is low and is concerned with air traffic control (ATC), the second application process is high and concerns a nuclear waste reprocessing plant.

– Finally, Chapter 9, by Patrick Millot and Marie-Pierre Pacaux-Lemoine, widens the notion of the dynamic sharing of tasks or functions between the human and machine toward human–machine cooperation by integrating two dimensions: the structural and organizational dimension and the functional dimension linked to the know-how of the human and automated agents, but also (and especially) their know-how-to-cooperate. The CWS is shown as a way to make the COFOR concrete. Indeed, COFOR is mandatory for any cooperation. Three examples are given to illustrate these ideas: human–machine cooperation in the cockpit of a fighter aircraft, cooperation between a human and a robot in a recognition task and human machine cooperation in the ATC. Finally, we show that more than just being a useful tool facilitating cooperation, the CWS improves the situation awareness of the team. This is of major interest for holding humans in the loop.

I.3. Bibliography

[AMA 05] AMALBERTI R., AUROY Y., BERWICK D., *et al.*, "Five system barriers to achieving ultrasafe health care", *Annals of Internal Medicine*, vol. 142, no. 9, pp. 756–764, 2005.

[BOY 11] BOY G., "A human-centered design approach", in BOY G., (ed.), *The Handbook of Human Machine Interaction: A Human-Centered Design Approach*, Ashgate, Farnham, pp. 1–20, 2011.

[INA 06] INAGAKI T., "Design of human-machine interactions in light of domain-dependence of human-centered automation", *Cognition, Technology and Work*, vol. 8, no. 3, pp. 161–167, 2006.

[MAL 73] MALVACHE N., Analyse et identification des systèmes visuel et manuel en vision frontale et périphérique chez l'Homme, State Doctorate Thesis, University of Lille, April 1973.

[MIL 88] MILLOT P., *Supervision des procédés automatisés et ergonomie*, Hermès, Paris, 1988.

[MIL 03] MILLOT P., "Supervision et Coopération Homme-Machine: approche système" in BOY G., (ed.), *Ingénierie Cognitive IHM et Cognition*, Hermès, Lavoisier, Paris, Chapter 6, pp. 191–221, 2003.

[SCH 12] SCHMITT K., "Automations influence on nuclear power plants: a look at the accidents and how automation played a role", *International Ergonomics Association World Conference*, Recife, Brazil, February 2012.

[SHE 74] SHERIDAN T., FERREL R., *Man-Machine Systems*, MIT, Cambridge, 1974.

[SHE 85] SHERIDAN T., "Forty-five years of man-machine systems: history and trends", *2nd IFAC/IFIP/IFORS/IEA Conference Analysis, Design and Evaluation of Man-Machine Systems*, Varese, Italy, September 1985.

[VIC 99] VICENTE K.J., *Cognitive Work Analysis: Toward Safe, Productive, and Healthy Computer-based Work*, Erlbaum, Mahwah, 1999.

PART 1

Design of Human–Machine Systems

1

Human-centered Design

1.1. Introduction

The theme covered in this chapter is the design of dynamic systems, of production, transport or services, that integrate both human operators and decision or command algorithms. The main question during the designing of a human–machine system concerns the ways of integrating human operators into the system.

As mentioned in the Introduction, human-centered design of human–machine systems must take into account five dimensions and the relations between them: not only the operator, the system and the tasks to be carried out, but also the organization and the situation of the work. These five dimensions are tightly linked; the tasks are different depending on the type of system, particularly its level of automation, but also depending on the potential situation and the expected safety level, the organization of the agents in charge of operating it (operators and/or automatic operating systems). The manner in which the tasks are carried out depends on the human operators themselves, who have different profiles depending on their training, aptitudes, etc.

In section 1.2, we cover the diversity of the tasks human operators are faced with, and the difficulties that they encounter in various situations. The models that explain the mechanisms of reasoning, error management and maintaining situation awareness (SA) are then explored. The creation of tools to support either action or decision in difficult situations leads to a modification of the level of automation, and as a result the global

Chapter written by Patrick Millot.

organization and the *task* or the *functions sharing* between humans or between humans and machines. The concepts of authority and of responsibility are then introduced. All of these points are the topics of section 1.3. Section 1.4 draws these different concepts together into a method of design-evaluation of the human–machine systems.

1.2. The task–system–operator triangle

1.2.1. *Controlling the diversity of the tasks depending on the situation*

First of all, we must make the distinction between the *task*, which corresponds to the work that is "to be done", and the *activity*, which corresponds to the work carried out by a given operator, who has his own aptitudes and resources. Thus, to carry out the same task, the activity of operator 1 can be different from the activity of operator 2.

The tasks themselves depend on the system and the *situation*, as the latter can be either normal or abnormal, or even dangerous. The *level of automation* determines the level of human involvement in the interaction with the system: often in highly automated systems, humans rarely intervene during normal operation. However, they are often called upon during abnormal situations and for difficult tasks. The example of the supervision of nuclear power plants is given hereafter.

In systems with low levels of automation, such as the automobile, the operators are involved both in normal situations (driving on clear roads in normal weather) and in difficult situations such as during the sudden appearance of an object at night in snowy weather. The involvement of the driver is then different.

To be able to deal with the difficulty of a task, we can attempt to decompose it. For example, the task of driving an automobile can be functionally decomposed into three sub-tasks according to three objectives (see Figure 1.1):

– *strategic,* to determine the directions between the start point and the destination;

– *tactical*, to define the trajectory and the speed on the chosen road; and

– *operational,* to control the speed and the trajectory of the vehicle on the road.

Functionally, this can be decomposed in a hierarchic manner, the three subtasks performed by the driver having different having different temporal horizons, and the functions and the resources necessary to execute each of these also being different. Assistance tools are also added which can be applied to specific sub-tasks: the speed regulator, ABS brakes (the wheel anti-blocking system) and automated cruise control (ACC) are applied to the operational sub-task, and GPS to the strategic sub-task. These additions are only assistance tools, i.e. they do not increase the level of automation since the human operator remains the sole actor.

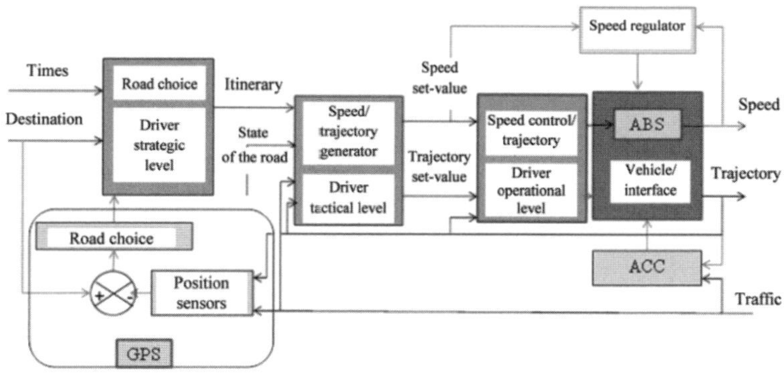

Figure 1.1. *Diagram of the task of automobile driving according to three objectives*

An increase in the level of automation could, however, be applied to one of the sub-tasks, for example the ABV project (*Automatisation à Basse Vitesse*, or low speed automation), which aims to make driving in peri-urban areas completely autonomous through automation, for speeds that are below 50 km/h [SEN 10], or the "Horse Mode project", which is inspired from the horse metaphor, in which the horse can guide itself autonomously along a road, and the horse rider deals with the tactical and strategic tasks. A corollary project is looking into sharing the tasks between the human pilot and the autopilot [FLE 12], and we will look into it further later on.

At the other end of the spectrum, a nuclear power plant is a highly automated system that is very big (around 5,000 instrumented variables), complex (lots of interconnection between the variables) and potentially very risky. The tasks of the operators in the control room have shifted from the direct command level to the *supervision* level (see Figure 1.2). These are therefore decision tasks for *monitoring*, i.e. *fault detection diagnosis* to determine the causes and the faulty elements involved and *decision-making* to define the solutions. These can be part of three types: a *maintenance* operation to replace or repair the faulty element; *accommodation* (adaptations) of the parameters to change the operation point; or, finally, *reconfiguration* of the objectives, for example, to favor fallback objectives when the mission is abandoned or shortened. Planning consists, for example, of decomposing the solutions by giving them hierarchy according to the strategic, tactical or operational objectives like the ones shown above[1].

Supervision loop

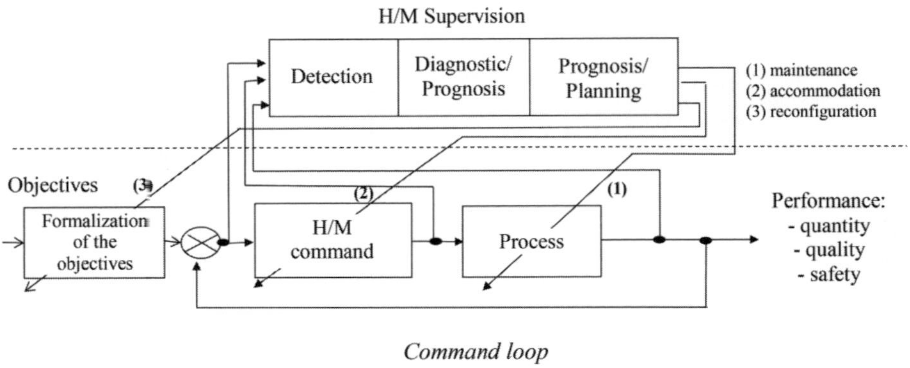

Figure 1.2. *Principles of supervision*

The highly automated systems are also characterized by differences in the difficulty of the tasks, i.e. the difficulty of the problems to be solved during supervision, depending on whether the situation is normal or abnormal, for example during the supervision of critical systems where time-related pressure further increases stress: nuclear industry, civil aviation, automatic

1 For example, in a more public area, such as the automobile industry, the accommodation of function consists of changing a more sensitive style of driving into a more flexible style when meteorological conditions deteriorate, in snow for example. In automation terminology, this is referred to as the *accommodation* of the laws of command to adapt to the evolution of the system's operation point. The new laws of command are defined by the tactical level of the drive and are applied by the operational level (see Figure 1.1). However, if the itinerary must be changed so as to take a clearer and safer route, this is then called *reconfiguration*, and involves the strategic level.

metro/underground systems. To deal with these difficulties, the operators require resources, which, in these cases, is the knowledge required to be able to analyze and deal with the operation of the system, with the goal of coming up with a diagnosis. This knowledge can then be used to write up procedure guides or diagnostic support systems. A distinction can be made between the following:

– knowledge available to the designers: on the one hand, about the *function* of the components, and on the other hand, *topological*, i.e. related to the positioning and the interconnections between the components [CHI 93]; and

– knowledge acquired during usage by the supervision and/or maintenance teams during the resolution of the successive problems. These can be *functional*, i.e. related to the modes of functioning or of malfunctioning of these components, and *behavioral* or specific to a particular situation or context [JOU 01].

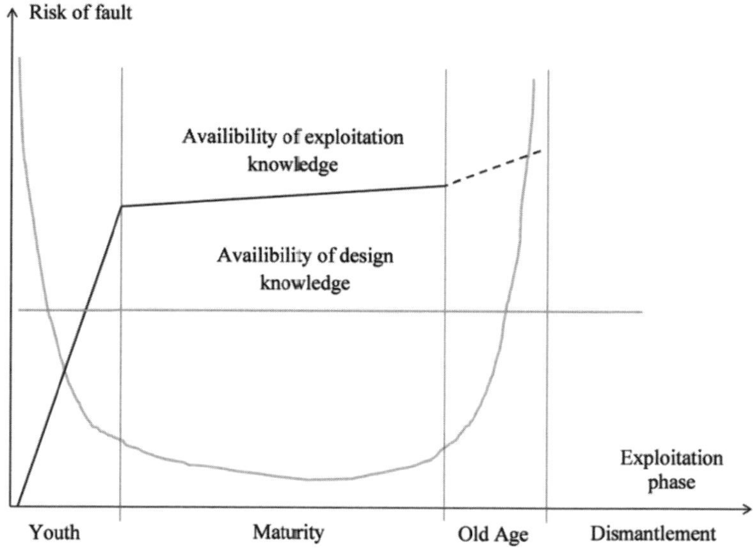

Figure 1.3. *Knowledge requirements throughout the life cycle of a system*

The difficulties of the tasks in the different situations of operation are modulated by the degree of *maturity* of the system (see Figure 1.3), which has an influence on the availability of this knowledge:

– The early period of youth is the one which requires the most adjustments and updates and where the system is the most vulnerable to faults: the problem being that the operators have not yet accumulated enough experience to control and manage the system at this stage (REX "retour d'expérience": feedback experience) to effectively deal with system malfunctions. However, knowledge of the process design is available, but not always very clear or well modeled, describing the structure and the topology of the system, i.e. its components, their operation modes and the relations between them. This knowledge can make up a strong basis to help during the process exploitation phases. In Chapter 6, Jacky Montmain describes a method of modeling that is based on the relations of causality between the variables. This is important for the composition of the diagnostic support systems which are based on a model of normal operation of the process when expert knowledge of possible malfunction is not yet available [MAR 86].

– In the period of maturity, both exploitation knowledge and knowledge of design are available; often they are transcribed in the form of exploitation and/or maintenance procedures. Moreover, the risks of fault due to youth imperfections are reduced, both for the system and the operators.

– Finally, during the period of old age, the process presents an increasing number of faults due to wearing of the components, but the operators have all the necessary knowledge to deal with this, or to apply a better maintenance policy.

Air traffic control is another example of a risky system with a low level of automation and a high number of variables, namely the airplanes and their flight information. Problems, called *aerial conflicts*, take place when two or more planes head toward each other and risk a collision. It is then up to air traffic controllers to preventatively detect these conflicts and to resolve them before they take place, by ordering the pilot(s) to change their trajectory. Considering the expertise of the controllers, the difficulty does not lie so much in the complexity of the problems to be solved, but rather in their sheer number, especially during periods of heavy traffic[2]. Thus, tens of minutes can pass between the moment of detection of a conflict by a controller and the adequate moment when the resolving order is transmitted to the relevant pilot(s). In this way, the controller risks forgetting the conflict and sending

2 The morning and the evening of weekdays when most internal flights take place, and at the start of vacations.

the order too late. Several practical cases in this book involve air traffic control.

After this overview of the diversity of the tasks, depending on the types of systems and the situations, we now move on to look at the approaches to modeling the system itself, and the methods to be developed so as to attempt to make their understanding easier.

1.2.2. *Managing the complexity of the system*

The large dimension of the technical system makes classical modeling and identification techniques extremely time consuming and leads to models that are not suited to real-time simulations. This has been the basis for work on hierarchical modeling in the systems trend led by Lemoigne [LEM 94], producing several methods of analysis and of modeling, such as SAGACE [PEN 94]. SADT follows the same idea, which relies on a decomposition of the global system. More recently, the multilevel flow modeling (MFM) method by Lind decomposes the system according to two axes: the means/ends axis and the all/part axis (see Figure 1.4) [LIN 10, LIN 11a, LIN 11b, LIN 11c].

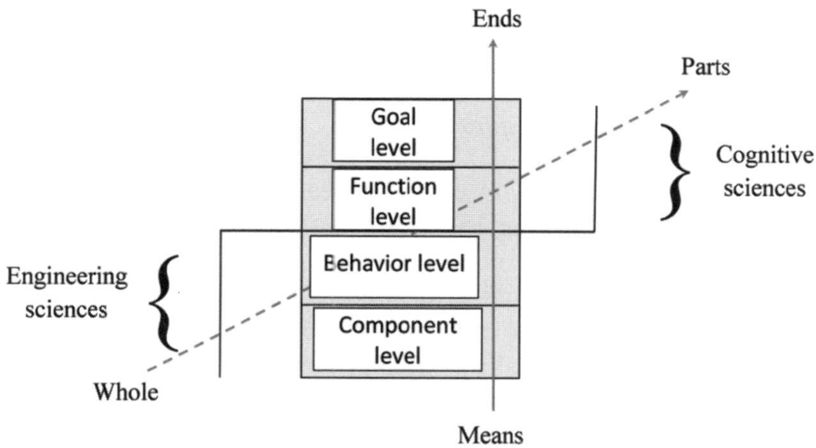

Figure 1.4. *MFM multilevel decomposition of a large system according to Lind*

According to the means/ends axis, there are four levels of model, from the most global (and least detailed) to the grainiest: the goals, the functions, the behaviors and the components. The models of one level are thus the goals of the models of the lower level and the means of the higher level. Let us note that the models of control theory are found at the level of behavior and very technical models, for example electronic and mechanical models found at the level of components. These two levels are part of *engineering sciences*. The two higher levels are themselves part of the *cognitive sciences* and concern the nature and the realization of the more global functions (and their sequencing), ensured by the behaviors of the physical level. Among the possible methods of modeling, we can cite the qualitative models [GEN 04], Petri networks, etc. The making of decisions related to the putting in place of the functions is often the result of optimization algorithms, or even human expertise, which is therefore symbolic, and which can be put into place through certain rules. This starts to be part of the domain of artificial intelligence (AI).

Decomposition according to the whole/part axis is the corollary of the decomposition imposed by the means/ends axis: the closer we are to the ends or goals, the more the entirety of the system is taken into account, the closer we are to the means, the more the model involves the different parts. This method of modeling puts new light on the subjects concerned and shows their complementarity. It most importantly shows that disciplines other than the physical sciences are involved in this vast issue. The method of modeling of the human operator has also followed a similar evolution.

1.2.3. *Managing human complexity*

A lot of multi-disciplinary research has been conducted on human factors since World War II, and a significant amount of methodological know-how has resulted. The goal here is not to produce a comprehensive review of this, but to introduce the designer with well-established and understandable models that help bring constructive and accurate results, even if, in the eyes of the most meticulous specialist, they may appear incomplete or simplified. Three points appear to be most important, knowing that in reality human behavior is a lot more complex; we will discuss other sociological aspects later in the chapter:

– the operator's adaptive behavior to regulate his workload during the execution of a task;

– reasoning mechanisms and decision-making mechanisms that the operator uses during complex decision tasks, such as in the supervision of the large, risky automated systems (nuclear power plants, chemical plants, etc.), but also during reactive tasks (which a short-response time) such as piloting an airplane or during the driving of an automobile; and

– mechanisms of errors and suggestions of solutions to deal with them.

1.2.3.1. *The regulation of human activity*

The system of the "human operator carrying out a task" can be considered to be a complex, adaptive system, made up of interconnected sub-systems, partially observable and partially commandable. A summary produced by Millot [MIL 88a] is presented in Figure 1.5. It brings together the inputs of the system, disturbances and the internal state parameters affecting the outputs of the system. Human functioning is modeled by three regulation loops with three objectives: in the short term, the regulation of performance; in the medium term, the regulation of the workload caused by the task; and in the long term, regulation of the global load, due to the global environment but also to the internal state of the operator.

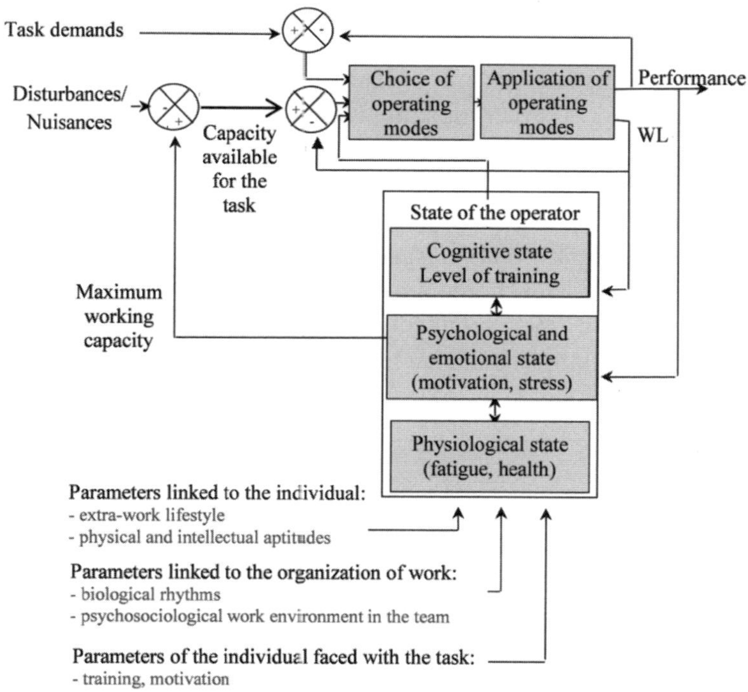

Figure 1.5. *Model of the regulation of human activity (from a detailed summary in [MIL 88])*

1.2.3.1.1. Inputs and outputs of the "human operator carrying out a task" system

The inputs are the *demands* of the task, i.e. the characteristics of the work to be accomplished, gathering the objectives to be reached. These are translated as timeframes to be respected and specific difficulties due to the task and/or the interface.

Nuisances that are not linked to the task can disturb the system. They are induced by the physical environment, and come in the form of vibrations, sound, light, heat, etc. Some of these disturbances increase the difficulties of the task, for example the reflection of lights on a screen, or vibrations of the work surface during manual control. One of the objectives of ergonomics is first to arrange the environment of workstations to reduce or even eliminate these nuisances.

The *output* is the performance obtained during the execution of the task. Observation of the performance is one of the methods possible to evaluate the ergonomic characteristics of a human–machine system. It is obvious that one of the big methodological difficulties relates to the choice of performance evaluation criteria. Generally, it is defined in terms of production indices, whether quantitative or qualitative, that are directly relative either to the procedures applied by the operator (response time, error rate, strategies, etc.) or to the output of the human–machine system, for example a product in the case of a production system. It also integrates criteria linked to *safety* and to *security*, particularly for critical systems [MIL 88]. Ten years later, ergonomists have joined this idea by underlining the necessity to take into account the performance of the human–machine system as an evaluation criterion during conception (see Chapter 2 by C. Chauvin and J.-M Hoc).

1.2.3.1.2. Workload

To carry out the task, the operator chooses the *operating modes*, which, once applied, produce a certain performance. If the operator has some knowledge of his performance, he can refine it by modifying his operating modes. But the performance alone is not enough to characterize the state of mobilization of the operator induced by the task and thus to evaluate the difficulties really encountered during its execution. For this, ergonomists use a state variable, called *workload*, which corresponds to the fraction of *work capacity* that the operator invests in the task. Sperandio defines this as the "level of mental, sensorimotor and physiological activity required to carry out the task" [SPE 72].

The operator carrying out the task has a certain amount of work capacity, which is limited and different depending on the individual and susceptible to vary in function of the state of the individual. The notion of a limited *maximum work capacity* corresponds, in the case of mental tasks, to the old notion of a canal with limited capacity. The treatment and/or the filtering of the disturbances uses some of the work capacity, thus reducing the *capacity available for the task.*

Seeing as the human operator is adaptable, he regulates his workload, as limited by his *available capacity*, but modifying his operating modes so as to satisfy the task demands. The modification of the operating modes is done through a different organization of work, in which the operator dynamically hierarchizes the assigned objectives of the system. This behavior is made evident by Sperandio in a study on the tasks of air traffic control, the levels of the demands of the tasks being defined by the number of airplanes to be simultaneously dealt with by the controller. When this number increases, the controller successively employs different strategies that are more economical with regard to his workload by reducing the number of variables to be considered for each plane [SPE 78].

1.2.3.1.3. Parameters that characterize the state of the operator

The maximum work capacity depends on the state of the operator: physiological, psychological and cognitive. This state is influenced by the task carried out, notably by the performance obtained and the induced workload: a state of permanent stress can, for example, cause physiological problems in the operator (overworking, insomnia) as well as psychological ones (lack of motivation, depression, etc.).

This state is equally affected by other parameters linked to the individual himself, such as his physical and intellectual aptitudes, his lifestyle outside of work (quality of sleep, hobbies, trips, psychological issues, etc.), his training and motivation, these being both influenced by the individual himself and by the organization of work (circadian rhythms, psycho-sociological environment in the team or the company, level of automation, etc.).

1.2.3.1.4. Application to the design evaluation of human–machine systems

The workload is therefore a variable that is characteristic of the state of the working operator, and its evaluation must be able to estimate the appropriateness of the tasks with the capabilities and human resources. Many methods of Work Load evaluation were studied in the 1980s [MIL 88a], based on the creation of an observer of the working operator, physiologically (by electrocardiogram, electroencephalogram and pupil dilation) or psychologically (by the double-task method and methods based on questionnaires). Among the latter methods, two questionnaire methods were found to give the best results: the subjective workload assessment technique (SWAT) [RED 87]) and the task load index (TLX) developed by NASA [HAR 88].

The first one establishes a load index from the operator's answers to a series of questions asked online, related to three indicators: *temporal demands*, *functional demands* and stress. The questions are asked with relatively low sampling frequency (a few minutes); the triplets of indicators obtained are accumulated to give the evolution of the load during the task.

The second one gives a global index by conducting the questionnaire after the end of the task according to six indicators: *temporal demands*, *mental demands*, *physical demands*, *satisfaction with regard to the performance*, *effort and stress*. The indicators are combined into a weighted sum whose weightings are also determined by the operator.

An extension targeting the real-time assessment of the load [MIL 88a] provided encouraging results in simulated car driving tasks [RIE 90]. This load indicator has also been used as a criterion of the allocation of the tasks between human and machine in supervision tasks of a simulated continuous process [MIL 88b] and in an air traffic control simulator [CRE 93, MIL 93]. We will look at this again in Chapter 9, which focuses on the dynamic allocation of tasks between human and machine and further on the human–machine cooperation.

Today, these methods benefit from resurging interest due to the fact that the ergonomic evaluation of the human–machine systems needs measurements to be able to compare several positions, and they have become references, notably when testing new methods [MIT 05, PIC 10, RUB 04].

1.2.3.2. *Human decision-making*

The human operator integrated in the control or the supervision of a big system is no longer only considered as reacting, but also as a problem-solver. A model is therefore needed so as to describe these different behaviors. The Rasmussen model [RAS 83, 86] has been revolutionary in this effect; a more recent revision by Hoc [HOC 96] is presented in Figure 1.6.

The operator detects an abnormal event and evaluates the situation by observing the information available, by identifying the state of the system (diagnosis) or by anticipating its evolution (prognosis). He then elaborates a solution in function of the limitations and the risks run. This solution is planned into goals, sub-goals and implementation procedures, which makes up the task to be conducted. If this task results in an action, the task is executed. This can be compared with the hierarchical decomposition into strategic, tactical and operational objectives mentioned earlier.

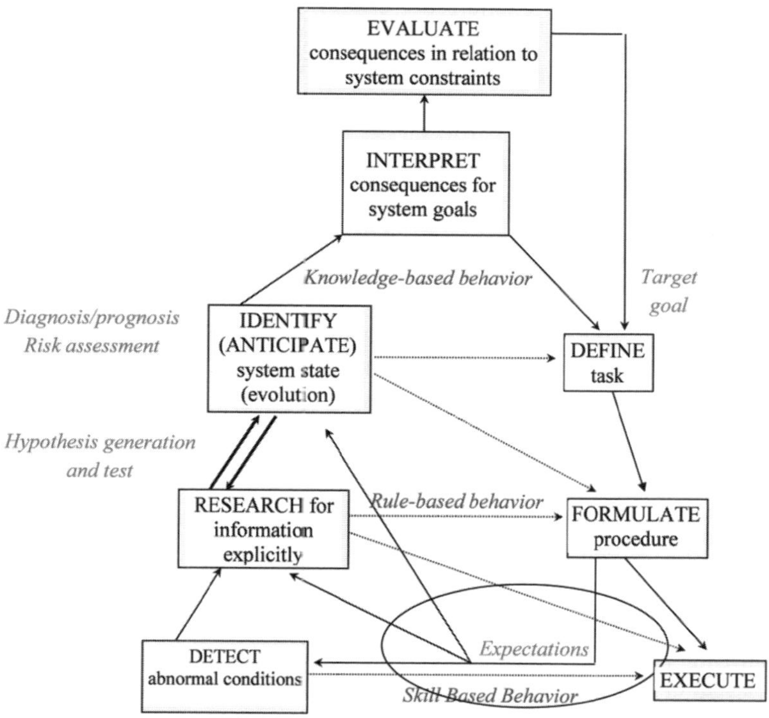

Figure 1.6. *Problem-solving model (Rasmussen revised by Hoc and Reason)*

Hoc's revisions complete Rasmussen's initial model, by detailing the cognitive mechanisms of the situation evaluation, inspired by Reason [REA 90]: diagnosis and/or prognosis by a method of hypothesis generation (data-driven reasoning), followed by tests of these hypotheses (goal-driven reasoning). It also introduces a temporal dimension (diagnosis: current state, prognosis: future state, expectations of the evolution of the system leading to a new evaluation of the situation).

The second strength of this model lies in the three levels of behavior it is comprised of:

– the lower level is reactive; the well-trained operator spontaneously carries out the appropriate action as soon as he detects abnormal conditions: it is called a *skill-based behavior* and comes back to the control theory models.

The two higher levels, on the other hand, are cognitive:

– *Rule-based behavior*, where the expert operator, after having identified the state of the system, directly applies a pre-defined task that he has learned: the corresponding models were very important in the knowledge-based AI systems in the 1980s and the 1990s; and

– *Knowledge-based behavior*, where the operator is faced with a problem he has never encountered before and must invent a solution.

We can therefore remark, just as in modeling the technical system, that modeling the human operator requires the cooperation of several sciences, notably the cognitive sciences: cognitive psychology to propose concepts that describe human reasoning, AI to put them into place, control theory to model the reactive behaviors, for example the guiding of a vehicle.

1.2.3.3. *Dealing with errors*

The human operator attempts to compensate for his errors, either in the short term, correcting them as they occur, or in the longer term to learn from them. *Human error* and *human reliability* have been studied over the last three decades. Rasmussen's model presented in the chapter has also served as a starting point for Reason in understanding the mechanisms of human error [REA 90] and to define barriers that would help prevent and/or deal with these errors. For example, an erroneous action can be a result of the incorrect application of a good decision, or the correct application of an inappropriate action. This erroneous decision can itself result in a bad solution, even when it is based on a correct evaluation of the situation.

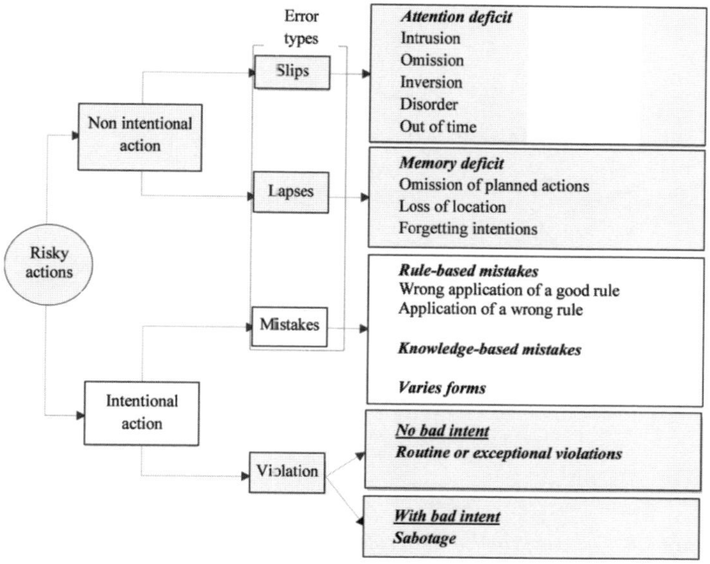

Figure 1.7. *Taxonomy of human errors according to Reason [REA 90]*

Reason divides human error into two categories: non-intentional and intentional (see Figure 1.7). These categories are themselves sub-divided into *slips* and *lapses* for the non-intentional actions, and into *mistakes* and *violations* for actions/decisions that are intentional. The violations comprise two categories: one is without bad intent, for example preventing an accident through procedure, while the other is malicious (sabotage).

Rasmussen [97] proposes an explanation for the appearance of certain errors through the necessity for the operator to achieve a compromise between three conjoined objectives that are sometimes contradictory (see Figure 1.8):

– the performance objectives imposed either by the management of the company, either by the operator himself, and which equate to extra efforts;

– cognitive or physiological costs (workload, stress) to achieve the objectives indicated previously and that the operator attempts to regulate;

– the efforts that result from the precautions to be taken to uphold the safety of the system operation of the environment and of the operators themselves.

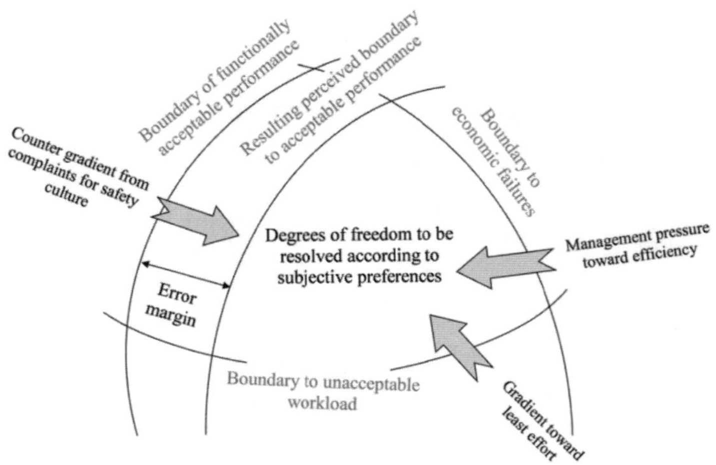

Figure 1.8. *The compromises that rule human action (adapted from [RAS 97])*

If the pressure from management and the cognitive costs caused by the task increase, the operator will tend to "push back" the fixed safety limit, and eventually pass the "error margin" and even the ultimate limit, which can lead to a loss of control, and as a result, an incident or an accident.

Several methods of risk analysis have been defined to detect risky situations and propose methods of countering them [FAD 94, HOL 99, HOL 03, POL 03, VAN 03]. Barriers can then be designed to stop these crossings of the limits [POL 02]. Technical, organizational or procedural *defenses* can aim to prevent or correct erroneous actions or decisions [ZHA 04].

Generally, risk management involves three complementary steps during the conception of a system:

– *Prevention*: this involves anticipating risky behavior during conception by putting technical and organization defenses in place to avoid this behavior (using standards, procedures, maintenance policies, efficient supervision, etc.).

– *Recovery*: if prevention is not sufficient, the second step consists of attempting to detect these unexpected behaviors (by alarm systems like in nuclear power plants or through a mutual control system with a second operator, like in commercial airplanes) and to correct them; this method can quite simply mean informing the operator of his error and allowing him to correct it himself.

– *Management of consequences*: if recovery through correcting actions is not efficient, an accident can happen. The third step is therefore to anticipate the occurrence of the accident so as to minimize the negative consequences, for example through the establishment of quick emergency aid response systems for road accidents, or the building of an isolation chamber around a nuclear reactor to block any possible leak.

1.2.3.4. *Situation awareness*

Situation awareness (SA), according to Endsley, refers to the human capacity to develop an internal representation of the current situation, of the environment, and to predict likely future states of this environment [END 95a].

Formally, SA is defined by three components, "the perception of elements in space time (SA1), the understanding of their significance (SA2) and the projection of their state into the near future (SA3)" (see Figure 1.9).

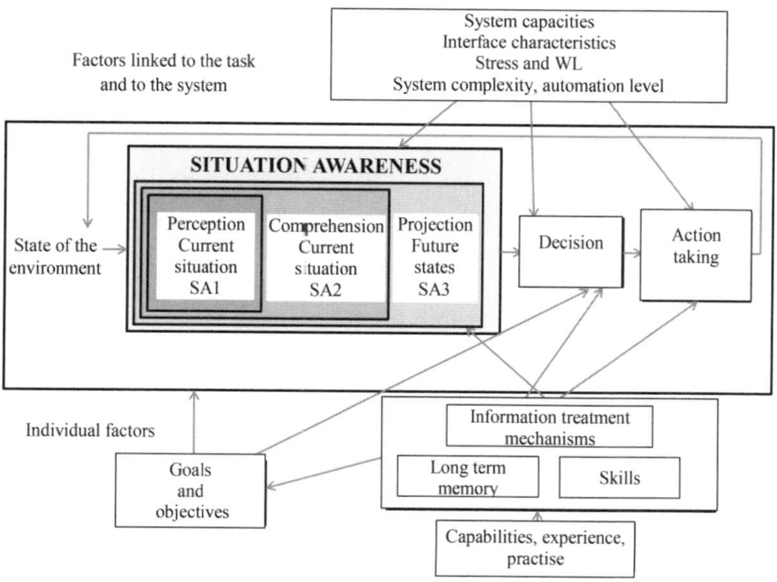

Figure 1.9. *Situation awareness model during a dynamic decision according to [END 95a]*

This is therefore of major importance in the control and/or the supervision of human–machine systems. Also, Endsley proposed methods of measurement of the three SA components [END 95b].

The most well known is *situation awareness global assessment technique* (SAGAT), but it only works in a simulated environment: the simulation is frozen at moments randomly chosen and the screens of human–machine interface are hidden while the human operator is solicited to quickly answer questions on his understanding of the running situation. The operator's perception is then compared with the real situation on the basis of the information provided by an expert in the subject (*subject matter expert*, SME) who answers the same questions but can see the screens. The method has been validated in a context of air combat simulation, the temporary pausing of the simulation for five to six minutes apparently having no effect on the operator's memory of the situation. The questions raised are very specific to the work situation and the operator cannot answer if he is not conscious of the situation: for example, for SA1, "which routes present an urgent situation?" According to Jones and Endsley, this method constitutes an objective and "unbiased" evaluation of SA [JON 04]. However, this method can only be applied in a simulated context.

Another well-known method is *situational awareness rating scale* (SART) for tasking in real situations [TAY 90]. It consists of a 10-point scale where the operator can indicate the amount of SA he had during the task, after the task is finished. The 10 items are then combined to form a measurement of the main factors: *attention resources*, *attention requirement* and *understanding*. This method is criticized by [JON 04] because of its application at the end of the task, which can introduce bias due to changes in the operator's memories. In our opinion, this criticism is valid, and situation awareness must be estimated during the execution of the task. However, it is true that SAGAT is hard to put into practice. In its place, Jones and Endsley have proposed and tested the *real time probes* method, which is a variation of *situation present assessment method* (SPAM) by Durso [DUR 98], which consists of periodically asking well-targeted questions related to the three levels of SA (and without blanking the screens), but also, by measuring the delay between question and answer as an additional indicator of the quality of SA. This method has been partially validated experimentally in comparison with SAGAT.

Other methods can be cited which rely on other theoretical bases, such as Neisser's theory of perception [NEI 76], or on very empirical principles that use metrics for each of the operator's tasks like in the *man–machine integration design and analysis system* (MIDAS) simulator in the Next Gen project study of "new generation" American air traffic control. However, Endsley's definition based on three levels of SA remains the most commonly retained because it is the easiest to understand [SAL 08].

It acts again as a support for an extension of the studies for the evaluation of SA in teams or during cooperative work [END 00, SAL 08] and its interrelation with the level of automation of the human–machine system. We will return to this in Chapter 9.

This axis of research is far from being completely drained: studies are being carried out in many application domains such as automobile driving, but also in diverse organizations such as road traffic, the deployment of ground troops or air traffic control. New theories are also being put forward and tested in an attempt to understand it better [POP 11]. This is explained by the importance of, on the one hand, increasingly current automation to avoid giving too much importance to the human operator, and, on the other hand, the necessity of keeping the operator in the loop to deal with new problems. Evaluating and maintaining one's level of SA through astute levels of automation can be a promising route to compromise.

1.3. Organization of the human–machine system

1.3.1. *The ambiguous role of the operator in automated systems*

Technological progress has greatly increased the decision-making capabilities of machines and has facilitated the creation of decision support systems (DSSs), for example diagnostic assistance tools. The capabilities of these tools allow them today to have a certain degree of autonomy in highly automated systems such as commercial jets or nuclear power plants. We refer to human agents and automatic or artificial agents. Human–machine systems become more and more complex and the interaction between human and artificial agents increases as the systems create tasks that are more and more cognitive and increasingly complex [HOL 03].

One of the essential problems concerns *task and function allocation* to agents (human or artificial) involved in the system and as such, the definition of their respective *responsibilities*. The allocation of responsibilities between humans and the DSS highlights the general problem of the sharing of *authority* [BOY 11, MIL 11] and involves human factors such as *self-confidence* and *trust*. Moray and Muir explain the *need to use* the DSS for a function depending on *self-confidence* and on *trust in the DSS* [MOR 95], each of the two having influence in opposite ways:

– the operator's *need to use* the DSS decreases if his self-confidence increases and if his trust in the DSS decreases;

– the *need to use* the DSS increases if his trust in the DSS increases and if his self-confidence decreases.

Rajaonah describes in detail the mechanisms that build confidence and their impact on the relationship between humans and the artificial agent, in particular in automobile driving [RAJ 08].

These behaviors are corroborated by observations made in other fields:

– On the one hand, human operators who are responsible for the operation of a complete system can have doubts with regard to their interaction with the machine if they do not feel that they are able to control it fully. For example, it has been observed in hospitals that people do not change the control on a machine out of fear of not being able to and in case they are not able to change it back properly [WEA 05].

– On the other hand, even if the machine has the responsibility of making decisions and carrying out actions according to the choices of the designer, the operators still continue to intervene when they notice a safety problem. This is the case, for example, in aeronautics with the onboard anti-collision systems [ROM 06].

An evaluation has highlighted the effects, both positive and negative, of the use of DSS by human operators [BAR 97]:

– The use of a DSS (1) increases the human understanding of the problems to be solved, (2) improves the human performance in information processing and (3) stimulates the operator's self-confidence by allowing him to focus on the strategic aspects of the problems to be solved.

– However, the use of DSS can render the operator passive and facilitate his *complacency*, particularly since he does not understand that the system is suggesting what it is suggesting. The human operator, if he does not understand the technical system and the DSS, can tend toward *passiveness* and accept the decision made by the DSS that is worse than one he would have made himself [SMI 89].

Thus, as previously underlined in the introduction, the roles of humans can be antagonistic: (1) human competences must guarantee performance, integrity and safety of the system, and (2) human decision can be erroneous and place the human–machine system as a whole in danger, in particular in the case of critical systems.

1.3.2. *Allocating humans with their proper role*

Performance and safety are directly linked to the complexity of the human–machine system and to the human capacity to understand it and deal with it. Technical failures become more likely with the size and the complexity of the system, and depending on its lifecycle. The supervision operator must be able to understand the behavior of the system, so as to deal with the complexity that results, i.e. being able to deal with the *right information*, at the *right level of abstraction*, at the *right moment* and in *the time that corresponds to the dynamics of the system*. We mentioned the modeling approaches that aim to decompose the system [LIN 11a, LIN 11b] and/or order the tasks according to the hierarchy of *strategic, tactical* and *operational* abstraction [RAS 91].

In an automated industrial system, organization is hierarchical: the tasks of the *operational level*, which is the lowest, are related to the command of the system which is split into sub-systems, each with its own local command unit. The second level concerns the *tactical tasks* of coordination of the local units in accordance with a production plan, and can contain DSSs. The higher level deals with *strategic objectives* by establishing this plan, and it contains the human teams [MIL 99, SHE 92].

As we go up from the lowest level toward the highest:

– the *nature of the information* evolves from precise digital information at the lowest level toward information that is more symbolic and global at the highest level;

– the *levels of abstraction* evolve from the means toward the goals; and

– the *temporal horizons* go from real-time at the bottom of the hierarchy (the sub-system's command tasks) toward the long term (planning or strategic decision) at the top.

This helps in defining the nature of the tasks that come from human capabilities considering the performances expected at each level. The classification of the tasks hereafter furthers the initial idea of Fitts' laws [FIT 51], which even 60 years ago aimed to attribute tasks to either humans or machines depending on who had the best aptitudes to execute them, MABA MABA[3]:

– human capabilities are more adapted to dealing with symbolic information, with decision tasks to do with planning and anticipation, according to more global objectives rather than specific ones, and this in the medium to long term;

– for these reasons, activity located at the bottom of the hierarchy does not suit human capabilities and can lead the human operator to make mistakes. Unfortunately, these criteria are not systematically applied during the conception of human–machine system.

Today however, the allocation of tasks between humans and machines goes a lot further than Fitts' laws.

1.3.3. *Sharing tasks and functions between humans and machines*

1.3.3.1. *Levels of automation*

1.3.3.1.1. Sheridan's scale

The influence of the role of human operators and its degree of implication on the global performance of the human–machine system, related to production and safety, have been studied since the 1980s, Sheridan [SHE 92] having defined the famous *levels of automation* and their consequences (see Figure 1.10).

In a system whose command is completely manual (level 1), safety lies entirely in the reliability of the human operator. Conversely, in a completely

3 *Men Are Better At – Machines Are Better At.*

automatic system (level 10), the human operator is not present in the command loop, nor is he present in the supervisory loop (see Figure 1.2). This level of automation can lead to a decrease in vigilance and to a decrease in the skill of the operator involved in supervision, making him less able to fulfill his responsibilities in the system. As a result, safety relies entirely on the system's technical reliability.

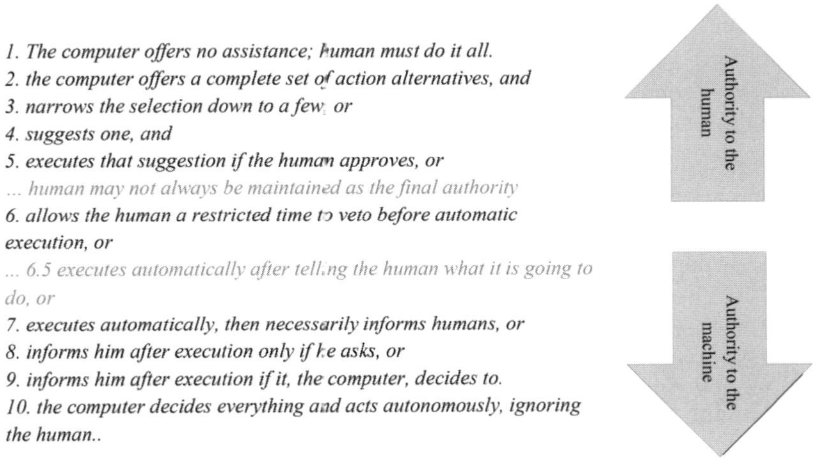

Figure 1.10. *Sheridan's 10 levels of automation [SHE 92] adapted by [INA 06] and [MOR 00]*

Between these two extremes, the intermediate levels consist of introducing dedicated DSSs and establishing supervision procedures that facilitate the sharing of authority between the humans and the automatic agents. Levels 2–4 correspond to static allocation according to which the human operator controls the system but a DSS provides him with the solutions. The operator has the *authority* to control the system (including the DSS) and can either put his own solution in place or choose the DSS's solution. The two agents interact on a tactical level according to an appropriate model of human–machine cooperation (which we will cover in detail in Chapter 9). At levels 5 and 6, the approaches are different due to the integration of strategic aspects that relate to the allocation of authority to execute the task. Finally, levels 7–9 correspond to a static allocation according to which the machine is given the *authority* to carry out the solutions. The levels are different depending on the type of return that they provide the human operator with.

This scale with 10 levels mixes the tactical and strategic aspects of allocating functions or tasks. Intermediary levels of automation can be added to take specific contexts into consideration. For example, Inagaki defines level 6.5, where the calculator carries out the action automatically in the case of an emergency, by telling the operator what it is about to do so as to minimize the surprise effect and to maintain the operator's SA [INA 06].

Parasuraman, Sheridan and Wickens [PAR 00] extend this approach from the simplified version of the Rasmussen model in four steps: *information perception*, *identification of the system state*, *decision-making* and *executing the decision*. The question of the level of automation is thus asked at each step, allowing in this way, a better representation of the interactions between the agents, allocation of the sub-tasks (static or dynamic) and the sharing of authority between the agents, whether to execute a task or to hand out the sub-tasks or the functions of each step.

1.3.3.1.2. Kaber and Endsley scale

Kaber and Endsley [KAB 03] go even further by proposing a finer division of the four types of functions: *supervision of the process, strategy production, selection of a strategy* among the proposed list and, finally, *the implementation of the action*. Each function is allocated either to the human agent (H) or to the artificial agent (M), or to both, with the different alternative allocation being spread over 10 levels (Table 1.1):

– Manual control: the human agent carries out all the tasks from the supervision of the state of the process to the production of alternatives and to the selection of an alternative and its implementation.

– Action assistance: the artificial agent helps the human agent implement the selected action, although some human actions of control are still asked for. This level is typical of tele-operation.

– Batch processing: the human operator produces and selects the alternatives to be implemented and then transmits them to the artificial agent that implements them. The artificial agent operates at a relatively low level of automation, very close to the physical implementation. This level is representative of the way of dealing lots used by manufacturing systems or of the speed control in automobile driving.

– Shared control: the human agent and the artificial agent generate alternatives. The human agent chooses which one is implemented either by him or by the artificial agent.

– Decision assistance: the artificial agent and/or the human agent propose a list of alternatives. The human agent selects one that he transmits to the artificial agent to implement. The expert systems are an example of this type of assistance. Contrary to the shared control, the artificial agent is here able to propose alternatives and then implement them alone.

– Mixed decision-making: the artificial agent generates a list of alternatives and then chooses one that is proposes to the human agent. The latter accepts it or chooses another from the proposed list or from another list generated by a human agent. The artificial agent then implements the alternative.

– Rigid system: only the artificial agent generates a list of alternatives. The human agent must select one from this limited list which he transfers to the artificial agent for implementation.

– Automated decision-making: that artificial agent and the human agent generate a list of alternatives, but only the artificial agent selects the best which is then implemented.

– Supervised control: the artificial agent controls most of the functions, but is supervised by the human agent, who can regain control, notably to propose another alternative and thus change the level of automation (going from level 6 with decision help).

– Fully automated: the artificial agent controls all of the functions. The human agent is completely absent from the control of the process that is entirely automated.

It is important to point out here that the length and frequency of function allocation must be taken into consideration. The choice of the level of automation then depends on the performance obtained during the execution of the task and on the need to maintain the human operator's SA. The frequency of change of the level of automation depends on the human workload [KAB 03].

Contrary to the levels proposed by Sheridan, the agents, whether human or artificial, have the possibility of confronting their list of alternatives, and, depending on the level of automation, it is either the human or the machine that chooses the best one.

		Supervision	Strategy production	Strategy selection	Implementation
1	**Manual control**	H	H	H	H
2	**Action assistance**	H/M	H	H	H/M
3	**Batch processing**	H/M	H	H	M
4	**Shared control**	H/M	H/M	H	H/M
5	**Decision assistance**	H/M	H/M	H	M
6	**Mixed decision-making**	H/M	H/M	H/M	M
7	**Rigid system**	H/M	M	H	M
8	**Automated decision-making**	H/M	H/M	M	M
9	**Supervised control**	H/M	M	M	M
10	**Fully automated**	M	M	M	M

Table 1.1. *Levels of automation, Kaber and Endsley [KAB 03]*

1.3.3.2. *Autonomy, authority, responsibility*

These 10 levels of automation implicitly situate the notions of *autonomy, authority* and *responsibility*.

Autonomy can be seen as the capacity of an agent to define its own objectives and to execute them. The human is implicitly autonomous in level 1, as is the machine in level 10. The human progressively loses his autonomy as soon as the calculator progressively becomes a part of the decision process. The autonomous agent, placed in an organization with *decisional independence*, can make decisions and act on the system without asking other systems to validate its decisions or its action.

Authority is the capacity to take responsibility for the final decision, whether this concerns a task carried out in an autonomous manner or orders transmitted to one or several agents, as part of a coordination for example. Authority takes place as part of an organization: on Sheridan scale, the human possesses authority at levels 1–5, and from level 7 onward this authority is transferred to the machine.

Obviously, these two definitions do not take into account the performance of the agent who possesses the autonomy and/or the authority. However, usually this agent is expected to have an adequate amount of *know-how*, i.e. *knowledge*, *reasoning capabilities* and *methods of communication* with his environment and the other agents, whether it is to receive information or to carry out actions or transmit commands or orders related to the task.

Responsibility is the duty of the agent to answer for his actions or decisions in front of a body (agent or group) that possesses the authority. This is also part of an organization and performance is a criterion of evaluating this duty.

During the design of a human–machine system, the level of automation must respect the competence hierarchy when it comes to attributing authority, but it must also ensure coherence between the authoritative agent and the responsible agent so as to avoid conflict. For example, some assistance tools in automobile driving perform better than any human driver, notably in the avoidance of obstacles, as they are quicker and more efficient. Logic would therefore require that they be given the authority that would allow them to take decisions instead of the human driver in the case of a risk of an accident. However, for legal reasons, responsibility must always remain fully in the hands of the human. Also, in the case of automobile driving, the onboard tools are only aids, under the responsibility of the human driver, who can turn them off at any time.

We will see in Chapter 9 that during a study on the dynamic task allocation in air traffic control, a certain level of automation was turned down by the human controllers, because it removed their authority (and gave it to an automatic control tool which performs better) while leaving them with legal responsibility in the case of an accident. Being obliged to accept responsibility without possessing the authority to exercise it is difficult, even impossible, and these controllers refuse this quite rightly.

1.3.3.3. *Task allocation or function allocation*

In section 1.1, we have presented examples of the diversity of the tasks to be carried out depending on the domains. Let us recall that the task is different from the activity in the sense that task is to prescribed work what activity is to real work. This distinction is commonly made in ergonomics.

One of the pillars of human-centered design is the allocation of tasks or functions between the human and the machine, so as to give each the role that fits each other's capacities the most. This allocation can be static and defined as early as the designing of the human–machine system. If not, it can be flexible and evolve dynamically according to criteria involving the decision-related needs of the system and the human's need for assistance. In this second approach, we define the initial conditions, i.e. the initial allocation of the tasks or the functions and criteria of dynamic evolution during conception [MIL 93].

The literature mentions two approaches, one allocating the tasks [MIL 88], the other the functions [BOY 98]. Guy Boy makes a clear distinction between tasks and function by introducing *cognitive functions* which he defines as processes that transform the task into activity and that have three attributes: a *role*, a *validity context* and *a set of resources.*

In the *task allocation*, we start with the *cognitive analysis of the task* and we decompose it into sub-tasks, according to different criteria. These can be linked to the nature of the task (similar sub-tasks or different complementary sub-tasks) or can be hierarchical (strategic, tactical, operational). Cognitive analysis of the tasks determines the demands of these sub-tasks and their feasibility by the humans and by the machines. *Tasks allocation* (or allocation of the sub-tasks) then consists of finding the agents that will carry them out based on their capabilities [MIL 88a]. If the agent is human, we can determine the resources that the (sub-)task requires, but also the organization in which he is placed, his responsibilities, his level of authority, the interfaces, and, if necessary, the assistance he needs to furnish with DSS [MIL 11]. If the agent is artificial, the task demands make up the specifications for its implementation [MIL 99]. We will see in Chapters 8 and 9 the example of air traffic control tasks, already mentioned at the start of this chapter, which involve detecting conflict in airplanes (which are going to cross at too small a distance with a risk of collision), and then calculate a new trajectory for one of the airplanes, and then order the plane pilot to follow it. Air conflicts can be different in nature, more or less complex: each conflict is a task. Depending on their complexity, these tasks can be allocated: the simplest ones to an artificial agent and the more complicated ones to the human controller. These tasks are then completely handed over to the agent, the criterion of allocation obviously taking into consideration the capacities of each agent to which it is being allocated.

The *function allocation* is part of the AI trend of cognitive science. The function being a process, it needs to be put to practice through an agent that will control it, with this agent being able to control several functions, as long as he possesses the resources, as much cognitive as sensorimotor. Boy defines the *cognitive function analysis* to determine the different functions that are necessary to complete the task. The *allocation of functions* thus distributes the processors and consists of finding capable (and willing) agents to ensure the functions [BOY 11b]. Let us take the example of a task of resolving an aerial conflict as described above. It is composed of five functions: *information perception, problem comprehension, schematic decision-making, precise decision-making* and *execution of the decision* according to Pacaux-Lemoine and Debernard [PAC 02]. If the function is to be controlled by an artificial agent, it must be created. If the function is to be controlled by a human, the conceptor must build the interfaces (and the resources) to achieve it. In Chapters 8 and 9, we will see how the functions of *perception, problem understanding, precise decision-making* and *action execution* can be indifferently allocated to human or machine. However, *schematic decision-making* is only ever allocated to the human.

As a summary, the allocation of functions decomposes the process of resolution of the tasks into functions and allocates its steps (the functions) to the agents. The task allocation allocates the tasks directly (and therefore in blocks, the functions corresponding to the steps), but takes into account the capabilities of each agent to carry out these functions to allocate the tasks. The two approaches are equally interesting in what they bring for multi-agent human–machine systems, for the sharing of authority and for human–machine cooperation [BOY 11a, MIL 11]. These points will be dealt with in the last three chapters dedicated to human–machine cooperation.

1.3.3.4. *Criteria for task allocation*

The tasks[4] are first listed and defined in terms of objectives, functionality, means and limitations by cognitive task analysis. This raises the problem of their allocation between humans and automatic system. For this reason, the criterion has two mandates: a technical feasibility and an ergonomic feasibility.

4 The approach would be the same for the allocation of functions.

The criterion of the technical feasibility leads to two classes of tasks, among the initial group (see Figure 1.11):

– the TA tasks that can be automated;

– the non-automatable tasks (TH) that must be given to humans. The reasons can be due to problems of availability of information for example, or difficulties in automating the decision, whether technical or theoretical.

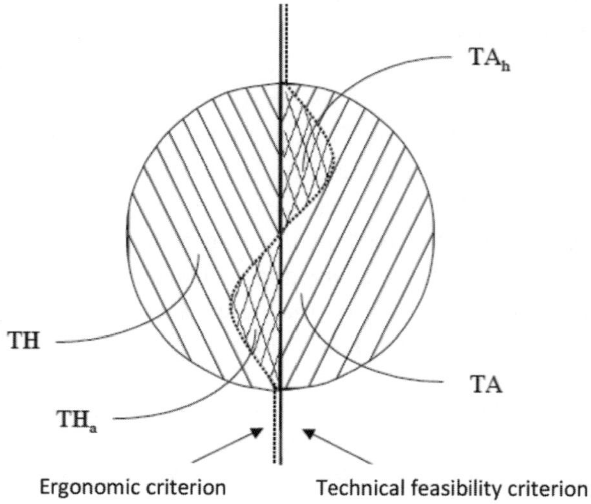

Figure 1.11. *Task sharing between humans and automats: TA: automatable tasks, TH: non-automatable tasks, TA_h: tasks that can be carried out by both automatic system and by humans, TH_a: tasks that cannot be automated and outside of human capabilities*

The second criterion must be applied to the two groups TA and TH to guarantee the ergonomic quality on which the reliability and the security of the global system depend:

– Among TA, some TA_h tasks are also compatible with human capabilities and pairing them with human operators can allow them to understand and better supervise the global system and the automations. It is the whole TA_h group that makes up the whole of the distributable tasks for a dynamic allocation that we will define in Chapters 8 and 9.

– Among TH, some TH_a tasks are extremely complex (or their complexity is increased by a response time that is necessarily short) and can be assisted for example by a diagnosis support tool. These TH_a tasks can be the object of cooperation between human and machine, also covered in detail in Chapters 8 and 9.

The ergonomic criterion of task sharing is based on models of human operators that define the human resources available, and, equally, the intrinsic limits of the operators, as much perceptual, for the acquisition and treatment of information, as physical, for the execution of the resulting actions. The cognitive resources are to be identified taking into account the application context, i.e. the situation. The limits of physical resources can be defined using ergonomic guides [KOL 91].

In Chapter 2, the reader can find a point of view that is centered on ergonomics.

1.4. Human-centered design methodology

We propose hereafter to combine these different points into a method for human-centered design (see Figure 1.12).

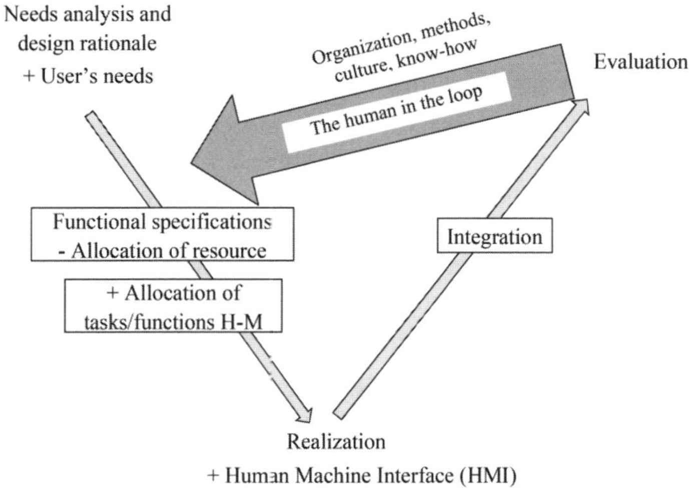

Figure 1.12. *Outline of human-centered design*

The process is well illustrated by the famous V cycle of the design of systems, revisited in [BOY 13] to introduce the human dimension. It is detailed in [MIL 99], and we summarize below:

– The process starts with the *needs analysis*. The needs of future users and of the operators of the system must not be ignored. According to the different situations of function that can be predicted in the system, the technical limitations (dynamic, security) must first be highlighted. In systems that are highly automated, we can rely on regular analysis methods of functioning or malfunctioning of systems [FAD 90, VIL 88] to deduce (or induce) the prevention methods to be put in place to "manage" the system. These make up the tasks to be carried out. In other systems like road traffic, the number of accidents is high enough that analyzing them provides a database that is statistically valid. Chapter 3 illustrates these methods. Generally speaking, previous feedback experience (REX) is used to define procedures. This is where the dilemma of the constructor with regard to the human appears: must he be limited by procedures to prevent possible errors? Or is it better to use more of his competencies in inventing solutions when faced with unknown problems? This second option is the best, even if it seems more difficult. The previous section on levels of automation, the allocation of authority and responsibility has already detailed the arguments and the approaches.

– Once the tasks are listed, they must be *specified* in terms of objectives, of means available (or necessary, for example sensors and actuators) and of functionalities. Knowing these functionalities and their contexts, the problem of their allocation between humans and automatic system arises, as was seen previously. The integration predicted in the increasing step below must also be specified at this point, seeing as the role of humans is established there, considering their *know-how*, their *culture* (in particular for safety and security) and the *organization* which we will specify.

– the following step concerns the execution of automatic processors to carry out automated tasks and of human–machine interfaces for carrying out human tasks. The forms of information and the means of action must follow the ergonomic recommendations to respect the human limits [KOL 91].

– Once the human–machine system is established, the upward step of its integration and evaluation begins; and in addition to the technical tests, whether unitary or of integration, and considering the non-totally deterministic nature of humans, the *ergonomic quality* and the *usability* of the human–machine system must be evaluated. The human–machine system is placed in situations of functioning or malfunctioning, and through well-chosen scenarios, the performance of the system and the induced human cost (workload, SA, etc.) are tested. If they are shown to be unacceptable, changes are asked for. Examples are given in Chapters 8 and 9.

The acceptability of the system can also be evaluated and under different forms: *ergonomic acceptability* (or individual) that joins usability is followed by *economic, social* and *legal acceptabilities* [DUQ 09] according to different criteria provided by experts. For example, legal acceptability was mentioned previously in the area of automobile assistance, and according to current French law, a vehicle must be driven by a human holding a valid driving license. This limits the level of automation and the sharing of authority and responsibility.

In the rest of this book, extensions to multi-agent teams involving several human operators and several tools will be progressively introduced.

1.5. Conclusion

This chapter asks the question of the role of humans in the human–machine system, and, more generally, the principle of human-centered design to answer this. The success of this design makes it imperative that humans, with all their characteristics, must be taken into account at the very beginning, and throughout the design process.

The human must be considered an element that not only interacts with the system by carrying out tasks (or more precisely activities), but which is also placed in an organization and deals with work situations. The organizational aspects are of high importance when we want to define the level of automation of the human–machine system, in particular the allocation of functions and tasks between human and machine. The principles of authority and responsibility must therefore be respected inexorably.

In fact, human-centered design aims to reconcile two apparently antagonist behaviors: the imperfect human, who can correct and learn from his errors, and the attentive, and inventive human capable of detecting problems and bringing solutions even if they are difficult and new. The desire for control of the imperfect human tends to lead to procedures, the result of returns on experience, i.e. a form of automation of the human activity. However, this "proceduralization" of human activity does not favor vigilance and inventiveness and therefore goes against the human qualities that justify keeping the operators active in the loop.

At the opposite end, favoring human SA maintains attention, vigilance and inventiveness. With this goal, human resources can be amplified by tools to support with decision and action. The integration of such tools leads again to the question of the level of automation, since these tools could become real decision partners and even real cooperators for humans, as we will see in the last chapters.

1.6. Bibliography

[BAR 97] BARR S.H., SHARDA R., "Effectiveness of decision support systems: development or reliance effect?" *Decision Support Systems*, vol. 21, no. 2, pp. 133–146, October 1997.

[BOY 98] BOY G.A., *Cognitive Function Analysis*, Ablex Publishing, Greenwood, 1998.

[BOY 11a] BOY G.A., GROTE G., "The authority issue in organizational automation", in BOY G. (dir.), *Handbook for Human-Machine Interaction*, Ashgate, Wey Court East, pp. 131–150, 2011.

[BOY 11b] BOY G.A., "Cognitive function Analysis in the design of human and multi-agent systems", in BOY G. (dir.), *Handbook for Human-Machine Interaction*, Ashgate, Wey Court East, pp. 189–206, 2011.

[BOY 13] BOY G.A., *Orchestrating Human Centered Design*, Springer, London, 2013.

[CHI 93] CHITTARO L., GUIDA G., TASSO C., *et al.*, "Functional and teleological knowledge in the multimodeling approach or reasoning about physical systems: a case study in diagnosis", *IEEE Transactions on System, Man and Cybernetics*, vol. 23, no. 6, November–December 1993.

[CRE 93] CREVITS I., DEBERNARD S., VANDERHAEGEN F., *et al.*, "Multi level cooperation in air-traffic control", *4th International Conference on Human–Machine Interaction and Artificial Intelligence in Aerospace*, Toulouse, France, 26–28 September 1993.

[DUQ 09] DUQUESNE L., MILLOT P., PACAUX-LEMOINE M.P., *et al.*, "Approach of multicriteria decision aid for the acceptability of dynamic road signs", *EAM09, European Annual Conference on Human decision making and Manual Control*, Reims, France, September 2009.

[DUR 98] DURSO F.T., HACKWORTH C.A., TRUIT T.R., *et al.*, "Situation awareness as a predictor of performance for en route air traffic controllers", *Air Traffic Control Quarterly*, vol. 6, pp. 1–20, 1998.

[END 95a] ENDSLEY M.R., "Toward a theory of situation awareness in dynamic systems", *Human Factors*, vol. 37, pp. 32–64, 1995.

[END 95b] ENDSLEY M.R., "Measurement of situation awareness in dynamic systems", *Human Factors*, vol. 37, pp. 65–84, 1995.

[END 00] ENDSEY M., ROBERTSON M., "Situation awareness in aircraft maintenance teams", *International Journal of Industrial Ergonomics*, vol. 26, pp. 301–325, 2000.

[FAD 94] FADIER E., ACTIGNY B., *Etat de l'art dans le domaine de la fiabilité humaine*, Octarès, Toulouse, 1994.

[FIT 51] FITTS P.M. (ed.), *Human Engineering for an Effective Air Navigation and Traffic Control System*, National Research Council, Washington, 1951.

[FLE 12] FLEMISCH F., HEESEN M., HESSE T., *et al.*, "Towards a dynamic balance between humans and automation: authority, ability, responsibility and control in shared and cooperative control situations", *Cognition Technology and Work*, vol. 14, pp. 3–18, 2012.

[GEN 04] GENTIL S., MONTMAIN J., "Hierarchical representation of complex systems for supporting human decision making", *Advanced Engineering Informatics*, vol. 18, pp. 143–159, 2004.

[HAR 88] HART S., STAVELAND L., "Development of NASA TLX (Task Load Index): results of empirical and theoretical research", *Human Mental Workload*, vol. 1, pp. 139–183, 1988.

[HOC 96] HOC J.-M., *Supervision et contrôle de processus, la cognition en situation dynamique*, Presses Universitaires de Grenoble, Grenoble, 1996.

[HOL 99] HOLLNAGEL E., *Cognitive Reliability and Errors Analysis Method: CREAM*, Elsevier, Amsterdam, 1999.

[HOL 03] HOLLNAGEL E., "Prolegomenon to cognitive task design", HOLLNAGEL DANS E. (ed.), *Handbook of Cognitive Task Design*, Lawrence Erlbaum Associates, London, pp. 3–15, 2003.

[HOO 10] HOOEY B., GORE B., WICKENS C., *et al.*, "Modeling pilot situation awareness", *Human Modeling in Assisted Transportation Workshop*, Belgirate, Lake Maggiore, Italy, 30 June–2 July 2010.

[INA 06] INAGAKI T., "Design of human-machine interactions in light of domain-dependence of human-centered automation", *Cognition, Technology and Work*, vol. 8, no. 3, pp. 161–167, 2006.

[JOU 01] JOUGLET D., MILLOT P., "Performance improvement of technical diagnosis provided by human-machine cooperation", *Proceedings of the IFAC/IFIP/IFORS/IEA Conference: Analysis Design and Evaluation of Human Machine Systems*, Kassel, Germany, pp. 379–384, September 2001.

[JON 04] JONES D.G., ENDSLEY M.R., "Use of real-time probes for measuring situation awareness", *The International Journal of Aviation Psychology*, vol. 14, no. 4, pp. 343–367, 2004.

[KAB 03] KABER D.B., ENDSLEY M.R., "The effects of level of automation and adaptive automation on human performance, situation awareness and workload in a dynamic control task", *Theoretical Issues in Ergonomics Science*, 2003.

[KOL 91] KOLSKI C., MILLOT P., "A rule-based approach to the ergonomic static evaluation of man-machine graphic interface in industrial processes", *International Journal of Man-Machine Studies*, vol. 35, pp. 657–674, 1991. Available at http://www.tandf.co.uk/journals.

[LEM 84] LEMOIGNE J.-L., *La théorie du système général: théorie de la modélisation,* Presses Universitaires de France, Paris, 1984 (re-edited in 1994).

[LIN 10] LIND M., "Knowledge representation for integrated plant operation and maintenance", *Proceedings of 7th American Nuclear Society International Topical Meeting on Nuclear Plant Instrumentation, Control and Human Machine Interface Technologies*, Las Vegas, NV, 7–11 November 2010.

[LIN 11a] LIND M., "An introduction to multilevel flow modeling", *International Journal of Nuclear Safety and Simulation*, vol. 2, no. 1, March 2011.

[LIN 11b] LIND M., "Control functions in MFM: Basic principles", *International Journal of Nuclear Safety and Simulation*, vol. 2, no. 2, June 2011.

[LIN 11c] LIND M., "Reasoning about causes and consequences in multilevel flow models", *Proceeding of ESREL2011*, Troyes, France, September 2011.

[MAR 86] MARRAKCHI M., Représentation des connaissances pour l'aide au diagnostic industriel: application au système expert SEDIAG, Doctoral thesis, University of Valenciennes, 1986.

[MIL 88a] MILLOT P., *Supervision des procédés automatisés et ergonomie*, Hermès, Paris, 1988.

[MIL 88b] MILLOT P., KAMOUN A., "An implicit method for dynamic task allocation between man and computer in supervision posts of automated processes", *Proceedings of IFAC/IFIP/IEA/IFORS Conference on Analysis Design and Evaluation of Man Machine Systems*, Oulu, Finland, pp. 77–82, June 1988.

[MIL 93] MILLOT P., DEBERNARD S., "Men-machines cooperative organizations: methodological and practical attempts in air traffic control", *IEEE/SMC'93 Conference*, Le Touquet, France, 17–20 October 1993.

[MIL 99] MILLOT P., "Systèmes homme-machine et automatique", *Journées Doctorales d'Automatique JDA'99*, Conférence Plénière, Nancy, France, September 1999.

[MIL 11] MILLOT P., DEBERNARD S., VANDERHAEGEN F., "Authority and cooperation between humans and machines", in BOY G. (ed.), *Handbook for Human-Machine Interaction*, Ashgate, Wey Court East, pp. 207–234, 2011.

[MIT 05] MITCHELL L., PICKUP L., NORRIS S., *et al.*, "Assessment of new workload tools for the rail industry", in WILSON J., NORRIS B., CLARKE T., A. Mills (ed.), *Rail Human Factors: Supporting the Integrated Railway*, Ashgate, Aldershot, pp. 330–340, 2005.

[MOR 95] MORAY N., LEE J.D., MUIR B.M., "Trust and human intervention in automated systems", in HOC J.-M., CACCIABUE P.C., HOLLNAGEL E. (ed.), *Expertise and Technology cognition and Human Computer Interaction*, Lawrence Erlbaum Associates, Hillsdale, pp. 183–194, 1995.

[MOR 00] MORAY N., INAGAKI T , ITOH M., "Situation adaptive automation, trust, and self-confidence in fault management of time-critical tasks", *Journal of Experimental Psychology: Applied*, vol. 6, no. 1, pp. 44–58, 2000.

[PAC 02] PACAUX-LEMOINE M.-P., DEBERNARD S., "Common work space for human-machine cooperation in air traffic control", *Control Engineering Practice*, vol. 10, no. 5, pp. 571–576, 2002.

[PAR 00] PARASURAMAN R., SHERIDAN T.B., WICKENS C.D., "A model for types and levels of human interaction with automation", *IEEE Transactions on System, Man and Cybernetics*, vol. 30, no. 3, pp. 286–297, May 2000.

[PEN 94] PENALVA J.M., PAGE E., "La modélisation des systèmes dont la maîtrise est complexe", *ILCE'94*, Montpellier, February 1994.

[PIC 10] PICHON M., MILLOT P., ENJALBERT S., "Toward an on-line and unobtrusive workload assessment method", *IFAC/IFIC/IFORS/IEA Conference on Analysis Design and Evaluation of Human-machine systems*, Valenciennes, France, August 2010.

[POL 02] POLET P., VANDERHAEGEN F., WIERINGA P.A., "Theory of safety-related violations of a system barriers", *Cognition Technology and Work*, vol. 4, pp. 171–179, 2002.

[POL 03] POLET P., VANDERHAEGEN F., AMALBERTI R., "Modelling border-line tolerated conditions of use (BTCUs) and associated risks", *Safety Science*, vol. 41, pp. 111–136, 2003.

[POP 11] POPKEN A., KREMS J., "Automation and situation awareness", BOY G. (ed.), *Handbook for Human-Machine Interaction*, Ashgate, Wey Court East, pp. 75–90, 2011.

[RAJ 08] RAJAONAH B., TRICOT N., ANCEAUX F., *et al.*, "Role of intervening variables in driver-ACC cooperation", *International Journal of Human Computer Studies*, vol. 66, pp. 185–197, 2008.

[RAS 83] RASMUSSEN J., "Skills, rules and knowledge: signals, signs and symbols and others distinctions in human performance models", *IEEE Transactions on System, Man and Cybernetics*, vol. 13, no. 3, 1983.

[RAS 86] RASMUSSEN J., *Information Processing and Human-Machine Interaction: An Approach to Cognitive Engineering*, North Holland, New York, 1986.

[RAS 91] RASMUSSEN J., "Modeling distributed decision making", in RASMUSSEN J., BREHMER B., LEPLAT J. (ed.), *Distributed Decision-making: Cognitive Models for Cooperative Work*, John Wiley and Sons, Chichester, pp. 111–142, 1991.

[RAS 97] RASMUSSEN J., "Risk management in a dynamic society: a modeling problem", *Safety Sciences*, vol. 27, no. 2/3, pp. 183–213, 1997.

[REA 90] REASON J., *Human error*, (version française traduite par HOC J.M., *L'erreur humaine*, Presses Universitaires de France, Paris, 1993), Cambridge University Press, Cambridge, 1990.

[RED 87] REID G., POTTER S., BRESSLER J., "Subjective workload assessment technique (SWAT): a user's guide", *Harry G. Armstrong Aerospace Medical Research Laboratory*, 1987.

[RIE 90] RIERA B., SIMON P., VANDWALE P., *et al.*, "Activity modelling and workload assessment of a car driver", *Proceedings of the 9th European Annual Conference on Human Decision Making and Manual Control*, Ispra, Italy, September 1990.

[ROM 06] ROME F., CABON P., FAVRESSE A., *et al.*, "Human factors issues of TCAS: a simulation study", *International Conference on Human-Computer Interaction in Aeronautics HCI – Aero 2006*, Seattle, September 2006.

[RUB 04] RUBIO S., DÍAZ E., MARTÍN J., *et al.*, "Evaluation of subjective mental workload: A comparison of SWAT, NASA-TLX, and workload profile methods", *Applied Psychology: An International Review*, vol. 53, no. 1, pp. 61–86, 2004.

[SAL 08] SALMON P., STANTON N., WALKER G., *et al.*, "What really is going on? Review of situation awareness models for individuals and teams", *Theoretical Issues in Ergonomics Science*, vol. 9, no. 4, pp. 297–323, 2008.

[SEN 10] SENTOUH C., DEBERNARD S., POPIEUL J.C., *et al.*, "Toward a shared lateral control between driver and steering assist controller", *11th IFAC/IFIP/IFORS/IEA Conference Analysis Design and Evaluation of Human Machine Systems*, Valenciennes, France, September 2010.

[SHE 92] SHERIDAN T.B., *Telerobotics, Automation, and Human Supervisory Control*, MIT Press, Cambridge, 1992.

[SMI 89] SMITH G.F., "Representational effects on the solving of an unstructured decision problem", *IEEE Transactions on Systems, Man and Cybernetics*, vol. 19, no. 5, pp. 1083–1090, 1989.

[SPE 72] SPERANDIO J.C., "Charge de travail et régulation des processus opératoires", *Le Travail Humain*, vol. 35, no. 1, 1972.

[SPE 78] SPERANDIO J.C., "The regulation of working methods as a function of workload among air traffic controllers", *Ergonomics*, no. 21, pp. 195–202, 1978.

[TAY 90] TAYLOR R.M., "Situational awareness rating technique (SART): the development of a tool for aircrew systems design", *Situational awareness in aerospace operations (AGARD-CP-478)*, NATOAGARD, Neuilly-sur-Seine, France, pp. 3/1–3/17, 1990.

[VAN 03] VAN DER VLUGT M., WIERINGA P.A, "Searching for ways to recover from fixation: proposal for a different view-point", *Cognitive Science Approach for Process Control CSAPC'03*, Amsterdam, Netherlands, September 2003.

[WEA 05] WEARS R.L., BERG M., "Computer technology and clinical work: still waiting for Godot", *The Journal of American Medical Association*, vol. 293, no. 10, pp. 1261–1263, 2005.

[ZHA 04] ZHANG Z., POLET P., VANDERHAEGEN F., *et al.*, "Artificial neural network for violation analysis", *Reliability Engineering and System Safety*, vol. 84, no. 1, pp. 3–18, 2004.

2

Integration of Ergonomics in the Design of Human–Machine Systems

2.1. Introduction

Human–machine systems have become, in various sectors (nuclear, chemical engineering, transport), complex systems where multiple tasks are automated. To be able to ensure their reliability and their performance, it is necessary to take into account all the agents that make them up, as well as their interactions, during the design process.

A human–machine system is made of machines and human agents that interact to monitor an environment or to act on this environment.

In Figure 2.1, the arrows point to transfers of information (or of actions). The information comes from the environment and can relate to the human alone, the machine alone or to both. They can integrate retroactions. The actions can be carried out by the human alone, the machine alone or by both (in cooperation).

The complexity of such a system is related, according to Hoc [HOC 96], to the more or less immediate supervisory field (the spatio-temporal expanse of the possible information and actions) of control (is the effect of a command on the process direct or must the causal chain be known to anticipate the effects of the chosen action?). It is also related to the accessibility of relevant information, to temporal constraints (does the

Chapter written by Christine CHAUVIN and Jean-Michel HOC.

process to be controlled require rapid interventions?) and to the level of automation.

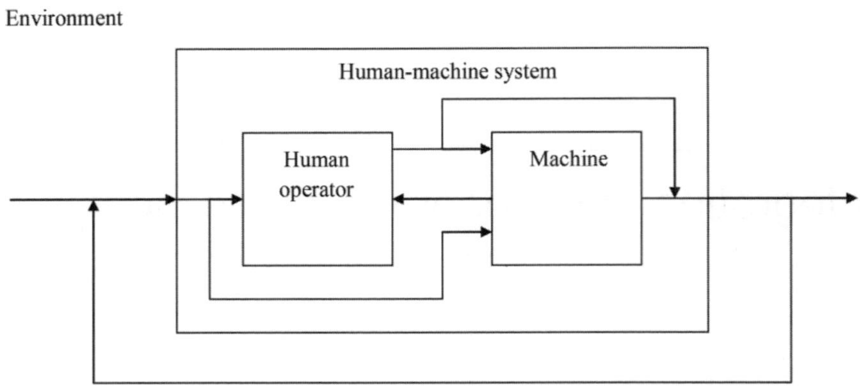

Figure 2.1. *The human–machine system*

Automation is the mode in which the production processes and the operational tasks are carried out and monitored using machines or electronic devices. It often limits the human to a position of supervision. However, it is expected of him to be able to deal with unexpected situations for which no automation is planned [HOC 03]. His contribution, in such cases, is related to his knowledge of the procedures, of the goals and of the methods of supervision, to the topographical layout of implementation and to his behavior [CAC 95].

Today, there is no consensus, either in engineering sciences or in ergonomics, on the best way of introducing ergonomics – taking into consideration the human operator – into the design cycle of human–machine systems. Ergonomics is now more often involved in processes of correction or improvement in the existing systems than in truly revolutionary creations. Works in ergonomics have more often followed than preceded these creations: the radar, the Macintosh "desktop", the Internet, etc. Moreover, ergonomics would avoid the creation of too inflexible norms, which would be in the way of creativity. If ergonomics had existed at the time, would we have abandoned the Roman numbering system (with which all educated people were familiar at the time) in favor of the Arab system? It is hard to imagine, these days, using Roman numerals to perform complex calculations. It is too complicated for humans to predict exactly how they will adapt to an environment that is very different from the one in which they are used to working. The greater the novelty, the more empirical methods are required to validate and sculpt the hypotheses that might have been proposed.

Several authors [CAC 95, VIC 99, HOL 99, HOL 05] agree in saying that the analysis and the design of complex, modern systems requires an update of the methodological tools in relation to:

– the notions of "system" and "complexity"; and

– the effects of automation on the supervision of installations by the human operator.

This chapter supports this notion. After having presented classic design approaches, which are only partial in terms of the design of the "system" (either because they focus exclusively on the machine, or because they are focus exclusively on the human), we will look at two methods that seek to integrate the different components of the system during design: the method by Long, Dowell and Timmer [DOW 98, LON 96, LON 01, TIM 02], which aims to define the performance criteria of the system that could orient the design choices, and the method developed by Rasmussen [RAS 86], then by Rasmussen, Pejtersen and Goodstein [RAS 94] and by Vicente [VIC 99] in a didactical form.

Such methods should be able to help the designer of a human–machine system in answering the following questions:

1) What information must be taken from the environment and, as a result, what are the sensors that are required?

2) How must the information system presented to the operator be organized (what will its structure and content be)?

3) What must the functionalities of the automated systems be?

4) How must the tasks be distributed between computers and humans?

5) How must the responsibilities be distributed between the individuals and the groups?

6) What must the characteristics of the interfaces be in terms of data presentation and of structure?

We will see that, nevertheless, these methods do leave some of these questions unanswered. They must not be considered as procedures but as strategic guides.

2.2. Classic and partial approaches of the system

2.2.1. *Machine-centered approach*

Since designers more often than not have a background in engineering, more so than in ergonomics, the design process that is most often come across is centered on the machine. The functional analysis that is then adopted consists of decomposing the function that the machine must ensure into sub-functions of different types.

In Figure 2.2[1], we can see that this approach favors information and commands received from the human operator, information sent to the human, information received from the environment and the actions of the environment. The distribution of the functions between the human and the machine is already presumed to be fixed. The information (and the commands) that the machine must receive from the human as well as those that it received from the environment are necessarily pre-determined, as much as the actions the machine must carry out on the environment. The activity of the human–machine system is not really considered. Inasmuch as the human may be able to take information in the environment and act on it without going through the machine, the machine is designed in a way as an autonomous agent. There is a possibility for commands or information coming from the human, and there is information that goes specifically to the human. But these possibilities are often envisaged as "ergonomical" limitations, rather than as the best way of having the task carried out by the human–machine system.

This engineer's point of view is obviously very operational, since all analyses come up with results that are directly relevant in terms of the design of the machine. It provides answers to the following questions: 1) what

1 Figure 2.2 is similar to Figure 2.1 except that most thick lines are essentially used in the analysis.

information is directly taken from the environment? 2) What actions are directly carried out on the environment? 3) What information and commands are received from the human? 4) What information is transmitted to the human? Only the first of these questions is common to the questions previously listed. The answer to the third question usually depends on the limits of autonomy of the machine (linked to the impossibility of the total autonomy). The answer to the fourth question is meant to come from ergonomics. A more complete design of the human–machine system in its entirety would lead to asking richer questions. The question of the distribution of functions between the human and the machine would presume an attempt, as far as possible, to define functions without thinking about who/what will carry them out. This is all the more necessary when the distribution must take place with criteria whose values are only measurable in real time (dynamic distribution).

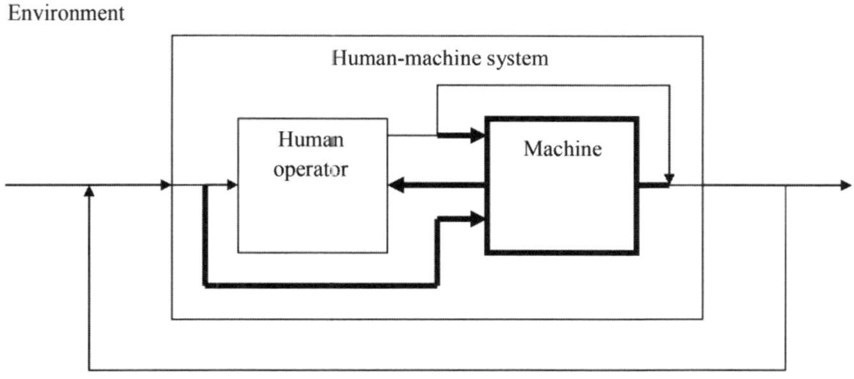

Figure 2.2. *Machine-centered design*

The question of the best format for transferring information to the human is strongly limited to general ergonomic criteria. By focusing on the performance of the human–machine system, information is often transmitted to the human with the goal of getting an answer from him. The choice of format and timescale of the communication obviously depends on the satisfying of this goal. For example, if an automated device for increasing the safety of the trajectory of an automobile is able to send information to the driver, the best question to ask is: what response are we trying to get from the driver? If this is related to adherence, aiming to reduce speed, it is worth asking if it wouldn't be better to transmit a numerical value or a proprioceptive sensation of this adherence (a slight tap of the brake, for example). In the first case, the target is a supervisor who deals with calculation. In the second case, the target is driver, in a control loop, who must integrate this parameter in the motor control of the trajectory.

In summary, an approach too exclusively centered on the machine does not allow for a satisfying performance of the human–machine system. However, it does guarantee the optimal performance of the machine, but only in some conditions of its operation. The fact that the human operator is not fully taken into account can lead to an implicit choice of conditions of validity that will never be fulfilled. In any case, this approach does not get the most out of a human–machine cooperation.

This approach centered on the machine is justified by the fact that apparently there is little more than the machine to be designed, the human being considered as a natural device, equating to data that needs to be taken into account or a condition to be met. However, this is not the case, as humans are always susceptible to learning and developing. A human–machine couple that is successful in reaching satisfactory performance must leave open some degree of freedom for the human, whose expertise can largely be planned.

This design strategy can be qualified as "residual". Everything that can be automated is automated and the human is left to do the rest. On the one hand, this can be quite appealing: it is a good way of giving free rein to technological innovation. Alphonse Chapanis [CHA 70], a famous ergonomist, was a supporter of this idea that let the ergonomist determine whether "the rest" was indeed feasible by a human. Beyond this question of feasibility, on the other hand, it can be disastrous. "The rest" can be meaningless, requiring excessive amounts of work to be carried out in unfavorable conditions. In this regard, James Reason identified the pitfall that needs to be avoided: "the designers of systems have involuntarily created a work situation where a lot of the properties of human cognition that are normally adaptive have become dangerous handicaps" [REA 88, p. 7]. For example, this could involve leaving the human the task of taking control when the machine is lost, of turning it off in a situation that had not been anticipated, because of a lack of preceding warnings. For example, this is the case when the autopilot makes up for turbulences, handing back the commands to the pilot as the plane is about to descend (1985 San Francisco Boeing 747 incident).

2.2.2. *Activity and human-based approaches*

Different methods for designing socio-technical systems explicitly consider human, social and organizational factors. Baxter and Sommerville [BAX 11] describe seven of them and remark that most of them essentially consist of a critical analysis of the existing systems and often do not point out solutions that could be brought to the problems that have been identified.

Several approaches involve describing the human at work in this way. In the methods of design, their main attraction is the ability to make the complexity of human activities apparent. They are based on the analysis of the activity of operators in real-work situations or in simulated situations. They involve the description, followed by the interpretation of this activity, based on theoretical backgrounds that can be very different from each other. Without claiming to be all-encompassing, from Green, Davis and Gilmore [GRE 96] we can identify two main approaches that cover very distinct visions of cognition and of action: cognitive ergonomics is a part of the cognitive psychology, and the theoretical outline of "situated action" is part of social science, and more specifically a part of ethno-methodology and cognitive anthropology.

2.2.2.1. *Cognitive ergonomics*

Cognitive psychology looks at the manner in which a human acquires information, deals with it, stores it and exploits it. It deals notably with the properties of short-term and long-term memory (in terms of the capacities and structure of memories), with the nature of mental representations, forms of reasoning and strategies for processing information (directed by the goals that the operator has set or by the data). These results are used in cognitive ergonomics to provide useful recommendations for the design of human–machine systems (see, for example, the recommendations from Bastien and Scapin [BAS 94]). They are also used in the interpretation of the cognitive activity of expert operators involved in carrying out tasks with a large mental component, and are enriched by such analyses, without neglecting their emotional aspects [HOC 04].

Cognitive task analysis (CTA) methods are derived from methods of *hierarchical task analysis* (HTA). HTA allows the decomposition of complex tasks into a hierarchical structure of operations and sub-operations. The unit of analysis is the operation; any operation is specified by a goal, activated by an input, carried out by an action and finished by feedback [ANN 05]. CTA covers a selection of tools and techniques that describe tasks and sub-tasks in terms of the goals targeted but also in terms of the required knowledge and mental processes [CHI 00]. It is often accepted that such results must be used to design artifacts in conjunction with the strategies and the mental models of the operators.

2.2.2.2. *Ethno-methodology and cognitive anthropology*

The outline of ethno-methodology is defined by Suchman [SUC 87]. It stipulates that the actions observed in a work situation are "situated" actions, i.e. that they are not mainly determined by cognitive processes but by a process based on interaction and circumstance, located in the relations between actors and between actors and the situation. In this approach, the action gains its significance in the physical and social world, and it is analyzed based on specific, local and contingent elements of its significance.

Thus, Theureau and Jeffroy [THE 94] consider the working human as an autonomous, intentional, social agent, motivated by psycho-affective factors, who gives himself goals that make sense to him and who produces actions by interacting with his environment. For both the authors, the analysis of work situations deals with the "course of the action", which is defined as: "the activity of a determined agent, actively engaged in a determined, physical and social environment that belongs to a determined culture. The activity is significant for the latter, meaning that it is demonstrable and describable by him to an observer/listener at any moment of its duration" [THE 94, p. 19]. The description and the analysis of the course of the action consists of reconstructing the process of giving the action meaning. Their approach has been used in the design of human–machine systems. Gaillard [GAI 94] details a model of the course of action of an air traffic controller that can be used in the design of new systems.

More recently, Martin and Sommerville [MAR 04] sought to represent the regularities observed in different work situations in the form a collection of *patterns*; each *pattern* describes both a given socio-technical configuration and the activities carried out in this configuration (in particular the actions and the interactions of individuals with the artifacts). These *patterns* contain knowledge on prototypical work situations that can be used again during similar situations. They can also be used as a support in discussions and exchanges that are established between the concerned parties in the design project. They help in the identification of the good and bad aspects of the studied situation, and the possible modifications that might need to be made.

2.2.2.3. *The limits of work analysis*

In its participation in the design stage, the method of work analysis (in the generic sense of the term, as the techniques are varied) brings up three types of problems.

First, if the design focuses on a relatively new situation, the analysis target of the work is a future situation, which requires a certain extrapolation, as much in terms of tools as organization, training or culture. Could ergonomics have predicted the use of the Internet before its popularization?

Second, work analysis, even when it leads to a certain model, cannot directly deduce criteria for an "ergonomic" design. For example, it is not true that work analysis is able to reliably provide justified support suggestions. It is often admitted that the artifacts must be consistent with the strategies and the mental models of the operators. However, as explained by Miller and Woods [MIL 97] or Vicente [VIC 99], the operators' mental models can be incomplete and their strategies can be less than optimal. It is true that the human operator is most often an expert in his activity, but it would be wrong to presume that only his expertise is relevant. In other experimental conditions, different expertise can be developed for better performance (efficiency/costs ratio: see section 2.3). The same calculation expertise cannot be developed in an environment that only uses Roman numerals as in one that only uses Arab numerals, with skills in mental arithmetic or in using calculators.

Third, it must be noted that the human-centered approach favors information on the environment or on the machine that reaches the human, actions on the environment or on the machine that come from the human (see Figure 2.3). The human-centered approach ends up neglecting the machine less often than the approach centered on the machine ends up neglecting the human. This fault is less common as the human usually builds himself a model of the machine (even if this model is not completely accurate), and therefore takes it into consideration during the development of his activity.

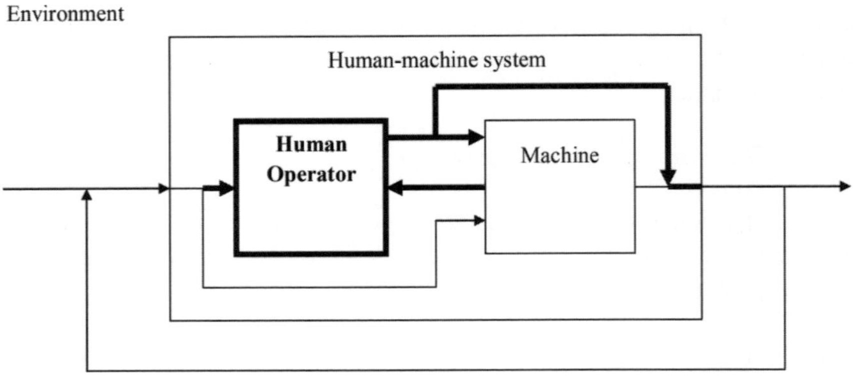

Figure 2.3. *Human-centered design*

All the same, this approach considers the machine as only an assistant to the human activity, which considerably reduces the possible scope of technological innovation toward the design of the autonomous machines that we know today. For example, automats on board in modern cars were designed not only to assist the driver, but also to increase safety during the drive and, soon, to reduce energy consumption. In this way, the human-centered approach still does not entirely satisfy the questions relating to human–machine cooperation.

2.3. The central notion of performance (Long, Dowell and Timmer)

John Long's approach is a reaction against an ergonomic design that does not justify its decisions by noting underperformance. *Performance* is considered to be the ratio between efficiency (*task quality)* and *costs*. The costs (mainly cognitive) relate to the activities of supervision and of control. They are distributed between those that are carried out by the human operator and by the computer. The efficiency obtained by human–machine *work system* in the *work domain* is evaluated in comparison with an expected performance. Task quality is often evaluated in relation to explicit project specifications. The cognitive costs for the human operator are evaluated in relation to a certain form of modeling of cognitive activities, but there is no explicit basis for comparison. Performance is evaluated in real time, allowing the observation of both the added value of the human–machine system's action and of any drops in performance, allowing identification of any design problems. The diagnosis of these problems is obviously in the hope of finding solutions.

This approach has been particularly refined in a micro-world simulating an air traffic control with the goal of leading to a diagnosis of any potential design problems. To illustrate the application of the method, we give some indications on the way in which the concepts have been made particular in the domain of air traffic control, and more precisely in the context of a micro-world, by Dowell [DOW 98], Long and Timmer [LON 01] and Timmer and Long [TIM 02]. In this micro-world, the environment (Figure 2.4) is composed of airspace objects (for example beacons, sector, and the like) and airplane objects (defined by their speeds and climbing rate). The meeting of these two types of object (an airplane and a beacon) results in a traffic event that facilitates the evaluation of a vector for the considered plane, whose coordinates are the position (a beacon), altitude, speed, heading and time.

Such vectors help in the evaluation of the task quality, mainly in reference to two variables: security and promptness (in the sense of the speed of flow of the traffic). Security is expressed in terms of the time or distance that separates the airplanes. The promptness of an airplane can be measured by the state of advancement of the flight.

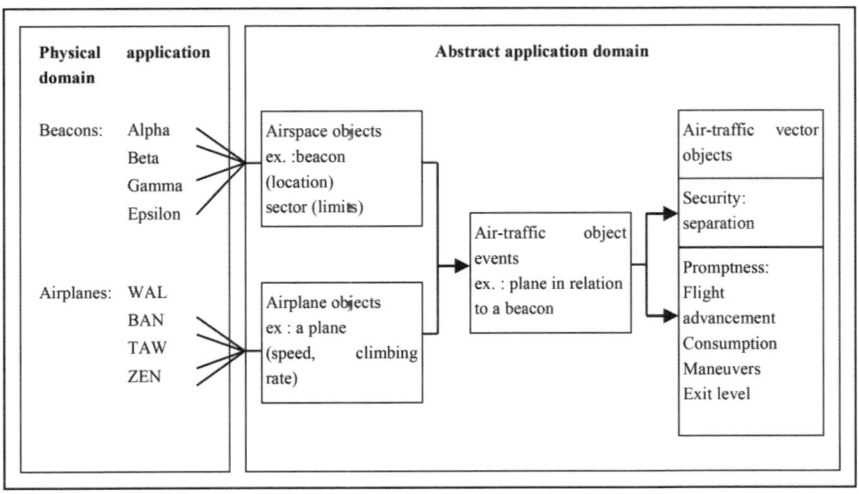

Figure 2.4. *Environment of the micro-world of air traffic control (from [LON 01])*

Other variables are also considered to evaluate task quality: the number of maneuvers asked of the airplane during its flight, possible changes in the cruising altitude and of the exit velocity (exit state) from what was initially planned, and fuel consumption. The cost for the operator is evaluated as a cognitive cost. In this same micro-world, the human–machine system (Figure 2.5) is formed by two types of conjoined devices: the human air controller (with a cognitive architecture and a physical architecture) and the technical devices (screen, radar, *strips*: summaries of the information on each flight). Figure 2.5 details the elements of the models that will not be developed in further detail here.

The human–machine system produces effects on the environment which can be evaluated in terms of performance, by comparing the task quality (according to the criteria relating to the environment: security, promptness, maneuvers, exit state and consumption) to the costs (for the controller and for the devices). Task quality, for example, concerns fuel consumption or the security of airplanes. Cognitive costs are measured, for example, by looking at the abandoned plans (whose elaboration was apparently of no use) or forgotten ones (which needed to be elaborated again).

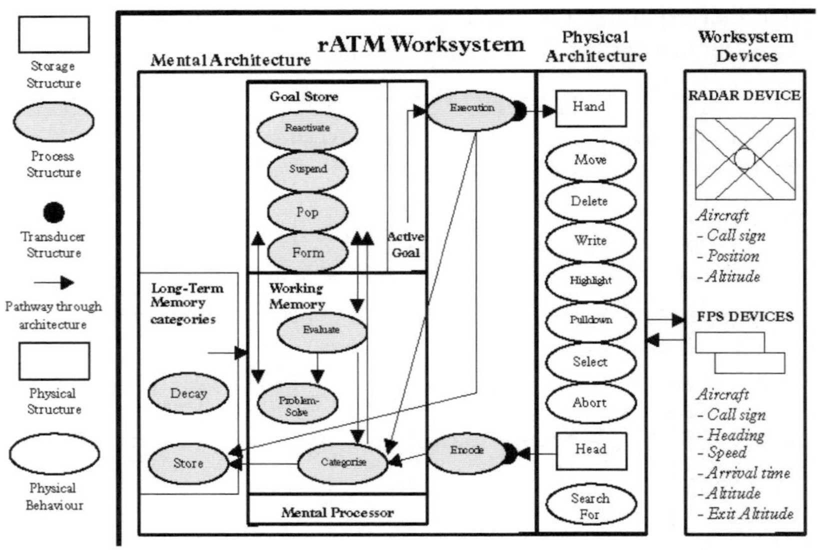

Figure 2.5. *The human–machine system of the micro-world of air traffic control (from [LON 01])*

Figure 2.6 gives an example of the identification of a design problem from the processing of one airplane in particular called ZEN. It is the summary of a much more complex protocol. The column on the left presents the base information that is useful for understanding. The central column indicates the representation that the controller has of the state of the airplane (inferred from what he says). The one on the right highlights the modifications of the "task quality" vector of the airplane. Its *strip* appears and its status becomes one of "entering airplane". Its trace appears on the radar screen. Its status becomes "active plane". Its vector, evaluating the task quality, shows that all its parameters are good (see target parameters in brackets), with the exception of security since a separation (from another airplane) of 1470 is less than the permitted separation (FALSE means that the separation condition is not realized, meaning that there is a conflict with another plane). Some time is needed for the controller to identify the airplane as active, in conflict and prompt. At this point, the controller increases its speed (to resolve the conflict), making the plane active, secure, but not prompt (it will exit earlier). As the controller has not updated the *strip* where he should have written the new speed, he neglects the information on the promptness and his representation of the airplane is limited to considering it as active. The plane will exit this state of bad promptness. The interpretation of the design problem is an oversight with regard to the category "bad promptness" in memory and not updating the *strip*. Other examples presented by Timmer and Long [TIM 02] deal with bad management of the plans that can lead to oversights and to new elaborations. These difficulties can lead to the correction of the design of interfaces, for example to help memory.

The advantage of this process lies mainly in the analysis of the performance dynamics during the running of the activity. At any moment, it is possible to evaluate the added value of the human–machine system to the task quality (in relation to what would have occurred without the system) and the cognitive costs encountered. This facilitates the identification of drops in performance during running, and then the categorization of the situations in which they appear and orientation toward somewhat generic diagnoses. Figure 2.7 illustrates the evolution of performance according to the progression of airplanes, their fuel consumption and security management. Figure 2.8 illustrates the evolution of cognitive costs (evaluated by listing occurrence of categories of activity: surveillance, planning, control). These two figures are taken from the work by Dowell [DOW 98].

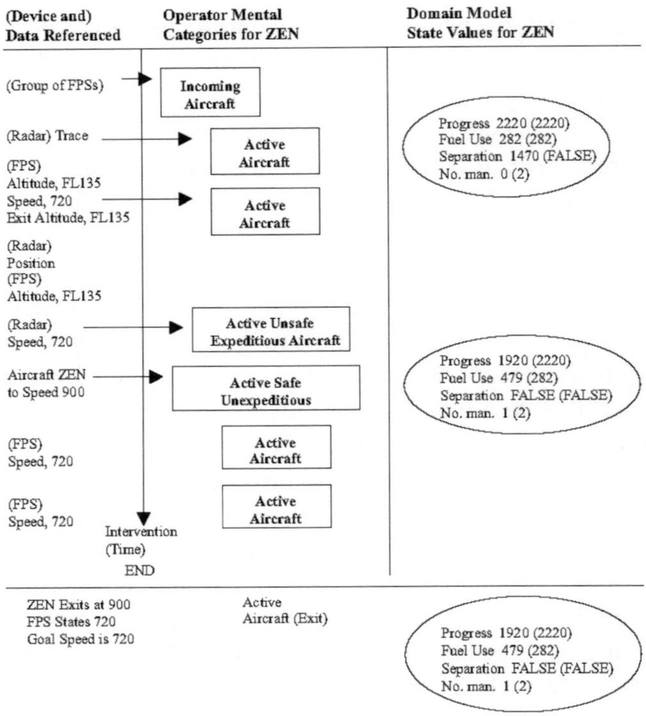

Figure 2.6. *Identification of a design problem (from [LON 01])*

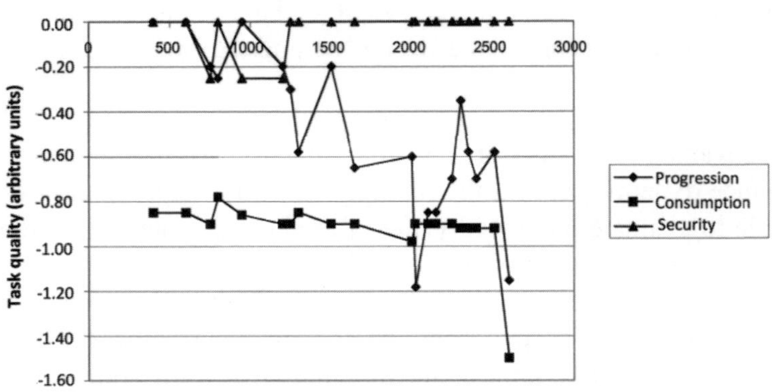

Figure 2.7. *Evolution of the task quality of air traffic control (from [DOW 98])*

This approach provides an outline for guiding a design process, but it presents several difficulties in being put into place.

In the case of air traffic control, it has never been applied to a real-work situation (although it has been applied to medical reception [HIL 95] or to crisis management [HIL 98]. By limiting itself to the micro-world, the method can be explicated, but passing to the complexity of a real situation such as air traffic control has not yet been done. Such a passage presents two types of problem. First of all, the evaluation of the task quality is more often the object of a consensus than a sort of authoritarian decision. The management of a company does not always know how far it is possible to go in terms of task quality and what this will cost from a cognitive point of view. Furthermore, macroscopic evaluation (for example, evaluating the performance of air traffic control systems in France) is not based on the same criteria as microscopic evaluation (and the expert operators are sometimes alone in possessing these evaluation criteria; for example to evaluate the performance of the control of one particular sector). A conflict management strategy can prove to be ineffective in the context of the sector of the controller who resolves problems, but on the contrary very effective if the adjacent sector is taken into account, where resolution will avoid other potential problems with the airplanes involved. Next, we can imagine that the evaluation of the costs of a complex activity can quickly become difficult to accomplish, at least in the details suggested by the Long approach. However, it is clear that the global evaluations of the mental charge are insufficient to point toward a precise diagnosis of the design problems.

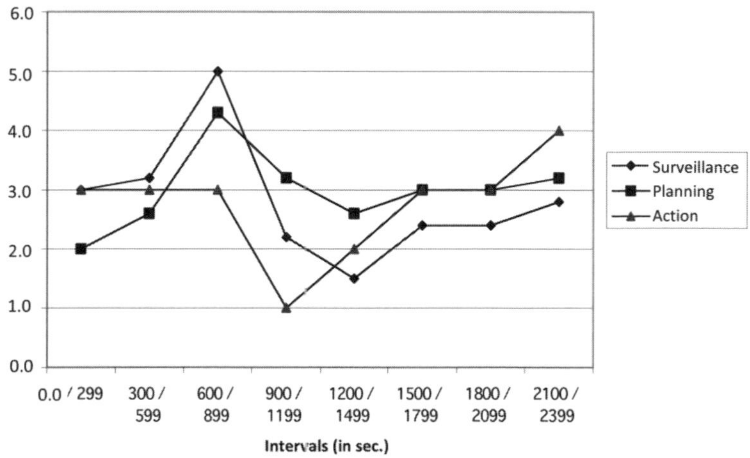

Figure 2.8. *Evolution of the cognitive costs of air traffic control (from [DOW 98])*

It does not take into account risk management. Uncertainty, danger and the necessity of controlling the situation are essential characteristics of dynamic situations. The human operator, in any case, rather seeks an acceptable performance (notably with regard to social norms) than a technically optimal performance [HOC 07]. The reason for this is to leave a margin so as to conserve resources to be able to deal with uncertain and dangerous situations that could threaten the control of the situation. Thus, the operator not only deals with external risks (for the environment) but also internal risks (for himself, indirectly, for the environment). To evaluate a real-task quality in comparison with a desired task quality (optimal), it is easy to forget that comparing one with the other can lead to bad risk management [AMA 96]. For example, by absolutely seeking to satisfy local optima, global optima can be lost. A good measurement of performance must integrate an evaluation of failure risks.

It is more adapted to corrective ergonomics than to ergonomics for design. It presumes that there exists a human–machine system that is similar to the one being designed and whose activity can be observed to carry out inefficiency diagnoses to orientate design. This question is very general: the gap from a model can only be relevant if two conditions are met. First of all, the model must be precise enough. If the reference being observed is inexistent, this will be even less the case. Second, the gap from the model must not be too big if it is to be sufficiently informative of the corrections to be made. When designing a truly new system, the current references can be too far from the new system to be really relevant.

If it developed the phase of identification of inefficiencies, it is a lot more vague on the matter of a diagnosis as the basis of a design decision. If a reference on the matter of low task quality is available, it is relatively easy to detect unacceptable drops in task quality. However, it can be sufficient to use methods for task quality and to carry out relative evaluations rather than using a rigid norm. The reference in the matter of cognitive costs is harder to find. The occurrence of errors (slips, mistakes, etc.) could be used as proof of cognitive cost overloads. However, errors are a part of normal activity. Experts make a lot of them, and some of the errors play an important regulatory role in the activity.

To summarize, John Long's approach is probably not applicable as it stands in too complex situations. However, its simplifying nature clearly highlights the level of expectation that is required, as a guide to a method can only get close to this. This method brings to light the unfounded character of an accepted idea according to which the description of activity (through work analysis) would automatically lead to the justification of design choices. In particular, it puts forward task quality and cost criteria that involve the human as much as the machine. If the design process remains an art rather than an exact science, a likening of this type can only be profitable.

2.4. An integrated approach: cognitive work analysis

An integrated approach is different to normative and descriptive approaches. It stresses the manner in which the work could be carried out and on the limits that shape the work. This approach is already well described in the book by Rasmussen [RAS 86] and in the one by Rasmussen, Pejtersen and Goodstein [RAS 94]. It is presented in a more didactic manner in the work by Vicente [VIC 99], as cognitive work analysis (CWA). It is considered to be the main design method taking human factors into account [SAL 10].

The central philosophy of the vision is that the operator must "finish the design". In other terms, the designer must impose inevitable limitations from the environment on the operator, but non-necessary limitations are not justified. The operator must be left with the most degrees of freedom possible that are useful in the carrying out of adaptive processes. Obviously, the approach is affected when the operators are beginners as some non-necessary limitations could be used as a guide in the elaboration of an activity that is not yet familiar.

As with the previous one, this approach is critical in a dominant point of view in psychology and in the field of human factors studies, according to which priority must be given to the analysis of cognitive limitations on the analysis of environmental limitations. However, the debate remains open on the taking into consideration (even implicitly) of cognitive limitations in the analysis of environmental limitations. According to the device that will carry out the work (human or artificial, expert or beginner), some environmental limitations become essential, others become secondary. For example, taking into consideration the level of radioactivity of an action zone is essential if the work is carried out by a human, but very secondary if it is being carried out by a robot.

The design process proposed can be organized into five successive stages:

– the analysis of work domain aims to define the object of the action and its limitations;

– the analysis of the task aims to identify what needs to be done; it answers the question *what*;

– the analysis of the strategies highlights the manner in which this is done; it answers the questions *how*;

– the socio-organizational analysis aims to determine the distribution of the demands of the task between agents; it therefore answers the question *who*;

– the analysis of the competencies helps in the identification of the competences of an ideal operator.

2.4.1. *Domain analysis*

Analysis of the work domain is the most important stage of the method. The notion of "domain" covers the notion introduced by Long. It is the object of the action. The analysis of the work domain aims to describe its functional properties and its intentional structures, independently of the characteristics of the operators, of the level of automation of the tasks to be carried out, of the goals to be reached and of the characteristics of the interfaces. The description of the domain is essential since it enables us to know the limitations that are inherent and that limit or determine the capabilities of action of the human and of the machine. It must be able to identify the information that the operators will need to be able to deal with unexpected situations. It also helps us understand that the actions required implement different functions and there are different initial conditions in the domain to be controlled. It therefore highlights possibilities of action.

Analysis of the work domain calls upon a modeling tool developed by Rasmussen [RAS 86]: abstraction hierarchy. The conjoined use of abstraction hierarchy with the structural decomposition of the domain leads to its representation in the form of a space of abstraction decomposition.

After having explained these formalisms and the principles of their use, we will present their applications, followed by a discussion on their limits.

2.4.1.1. *Abstraction hierarchy*

Rasmussen uses the notion of abstraction hierarchy to describe the functional properties of a system in the form of means–ends relations. The abstraction hierarchy describes a work domain in different layers going, for the highest layer, from the goals (i.e. the ends) to be reached to a detailed description of the means used to reach these goals, for the lowest layers. It takes into account the observed variations in the verbalizations of human agents acting on a work domain and expressing the level that they are dealing with.

Rasmussen [RAS 86] makes the distinction between five levels representing different classes of limitation. Each level is linked to the higher level by the *why* question and to the lower level with the *how* question:

– The highest level of the hierarchy – the level of *functional objectives* – represents the objectives for which the system was designed and limitations that govern the interactions between the work system and its environment. For a company, this is the quality of the product, productivity, but also the limitations imposed by laws and regulations in terms of financial operations, work conditions, protecting the environment, etc.

– The level of *abstract functions* represents fundamental laws that facilitate measurements to be made, and, consequently, priorities to be established and resources to be distributed between the different general functions of the domain by taking into account its functional objectives. The measure carried out at this level does not depend on the domain or its environment: they follow general laws that provide certain regularity to the functional behavior of the domain. These are the stable laws of nature in the case of technical work domains (laws of conservation of mass and energy, for example; however, a large variety of domains are governed by formal or legal limitations, or by institutional and social practices).

– At the level of *general functions*, the fundamental functions and the work activities are carried out. These functions aim to manage energy, material, information source and human resource usage. In the domain of industry for example, these are design, production, scheduling, maintenance, etc.

– Each general function is put to work using one or several *physical processes*. The characteristics and behaviors of the domain's components are written at the level of physical processes or functions.

– The lowest level of the abstraction hierarchy is the level of *physical forms*. It represents the whole of the material resources of the domain (tools, equipment, staff), as well as their characteristics in terms of appearance and spatial localization.

2.4.1.2. *The abstraction-decomposition space*

It is usual to cross abstraction hierarchy with a structural decomposition. The latter aims to decompose a domain into sub-systems, and then decompose each of these sub-systems into functional units, and to decompose the units into subsets, and finally the subsets into components. It takes into consideration changes in the attention span of the human actors operating on the domain.

To conduct the analysis of the domain, Vicente [VIC 99] suggests following 10 principles:

1) determine the limits of the work domain in such a way that there remain only a few relations between the defined domain and its environment;

2) use a matrix that has in the rows the abstraction hierarchy (decomposition into means and ends) and in the columns the structural decomposition (decomposition into whole and parts);

3) start with structural decomposition, by identifying the smallest entities that are relevant to the model (without going over the level where actions are possible), and then determine the manner in which they can be aggregated;

4) for each knot, ask if the knot connected to the lower level does make up a part, and inversely it must be asked whether a knot does make up a whole with regard to knots that are lower;

5) to develop the abstraction hierarchy, it is easier to start by identifying the elements of the higher level (the functional objectives) and of the two lower levels (functions and physical forms). Next, an attempt must be made to identify the intermediate levels. In the case of a work domain designed to

fulfill several objectives, it can be useful to first develop an abstraction hierarchy for each objective, and to combine them after;

6) keep in mind the distinction between a means–end hierarchy that is applied to actions and a means–end hierarchy that is structural; only the latter must be represented by abstraction hierarchy;

7) make sure that all the knots located on the same level are described using the same formalism and that the description of knots located on different levels uses different formalisms;

8) use the questions why, what, how to verify that all the knots have the proper links;

9) make sure that the connections between the cells of the matrix are links of the means-end or all-part type, and none other;

10) be ready to correct the constructed representation several times. It is an iterative process.

Vicente [VIC 99] presents an example of modeling an experimental work domain that is the DURESS[2] II micro-world (Figure 2.9). This example is also described by Hajdukiewicz and Vicente [HAJ 04]. DURESS II reproduces the supervision of a thermo-hydraulic process composed of two water reservoirs that must be maintained at the respective temperatures of 40 and 20°C, by satisfying a flow requirement. For this, the operator can act on eight gates, two pumps and two resistances. All these components are modeled by first order dynamics, with a delay and with a time constant of 15 seconds for the resistances and 5 seconds for the other components; the other important variables are the temperature of the water that feeds the system and the volumes of the two reservoirs.

In such a domain, the use of structural decomposition leads to the identification of three levels the whole of the system (DURESS), the sub-systems (transport, storage, heating) and the components (pumps, gates, resistances and reservoirs).

2 *DUal REservoir System Simulation.*

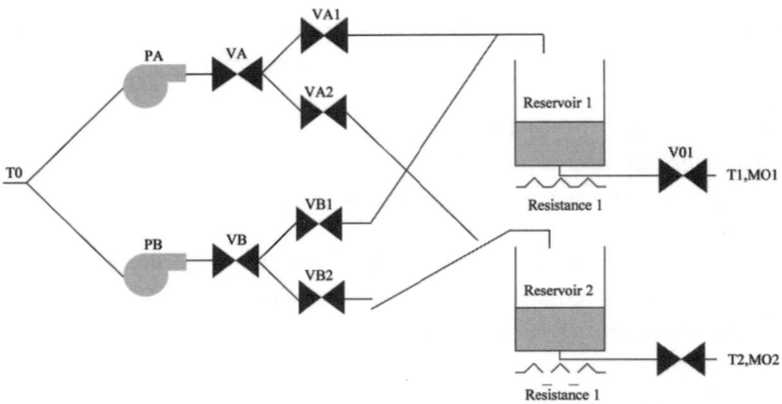

Figure 2.9. *Schematic representation of the DURESS micro-world (from [VIC 99])*

Regarding abstraction hierarchy (Figure 2.10):

– *the function objective* consists of keeping the water at the instructed level for each of the reservoirs and to maintain enough water in each reservoir to satisfy the variable demands of flow;

– in the case being dealt with, the level of *abstract functions* can be described in terms of the mass and energy conservation applied to each of the storage sub-systems. Taking into account this level of abstraction means considering a new level of decomposition: the level of the sub-systems. Each storage sub-system has a mass and a storage energy (the reservoir), a source of mass (the water upon entry), two sources of energy (the entering water and the resistance) and a loss of mass and of energy (the exit gate);

– the flow and the storage of heat and of water are described at the level of the *general functions*. At the level of decomposition, sub-systems are described: the mean water and heat flows coming from the entering flow, the heat flow coming from the heating system, the storage of heat and the storage water in the reservoirs, as well as the heat output flow and water output flow through the output gate. The level of the components shows the description of the heat and water flows through each gate and pump, as well as for the heat flow coming from the resistances, the storage of heat and of water in the reservoir and the flow of water and of heat linked to the demand at the output;

– the states of the components are described at the level of the *physical functions*. This is the tuning of the gates, pumps and resistances and of the volume and temperature of the reservoir;

– at the level of the *physical form*, the appearance and the localization of each of the components are described.

As this example demonstrates, the crossed use of the abstraction hierarchy and of structural decomposition of the system leads to only filling the cells located on the diagonal of the table: the functional objectives are described at the level of the system, the abstract functions at the level of the sub-systems, the general functions at the level of sub-systems and components, the physical functions and the physical forms are only described at the level of the components.

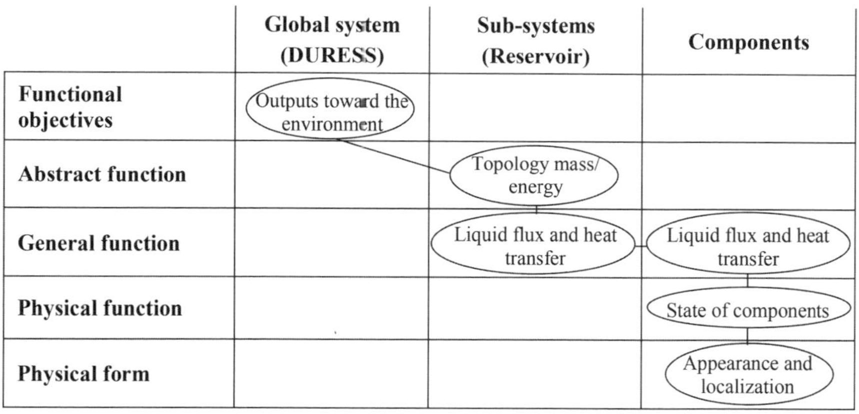

Figure 2.10. *Abstraction-decomposition space describing the work domain for the DURESS II micro-world (from [VIC 99])*

2.4.1.3. *Applications*

Abstraction hierarchy is very useful to analyze a complex system. It is also completely relevant from a psychological point of view, since it helps the operator "manage the complexity" by segmenting the domains of work that would be unmanageable if the operators had to observe them fully. It is relevant because it stresses the goals and because in this way it can provide a support for the problem resolution strategies that are orientated by the goals. Starting at a higher level of abstraction can thus circumscribe problem resolution, and then deciding which of the lower levels is relevant, and, finally, by focusing on the part of the hierarchy that is linked to the function that is of interest to the operator. It can therefore help an operator plan, since the development of the planning relies on the schematic activities that involve the exploitation of hierarchical relations between representations [HOC 92].

Moreover, faced with approaches of human–machine interaction that only aim to guide the operator during the resolution of known problems (by explicating relations between variables), the abstraction hierarchy offers an alternative to help with the resolution of unknown problems.

Beyond analysis, abstraction hierarchy acts as a support in the design of "ecological" interfaces. These interfaces show the limitations of the system as well as the space of the possible interactions between the operator and the machine within the limits of the system. They have been developed in many domains: the medical domain (e.g. modeling of the human body in the form of an interface that can help both the surgeon and the anesthetist [HAJ 01]), process control (e.g. interface for control and supervision of a hydraulic center [MEM 05]; interface dedicated to the operations of electricity distribution [DRI 09]) and aviation (e.g. interface dedicated to the detection and the avoidance of collisions [HO 03]). Vicente [VIC 02] also cites applications in the military domain (control–demand system), or even in the domains of the management of computer networks, of researching information and of software engineering.

In 2004, Burns and Hajdukiewicz [BUR 04] proposed a methodology that would allow the translation of abstraction hierarchy into the form of an ecological interface.

2.4.1.4. *Limits and criticisms of the method*

Several authors have established the limits of the method of analysis of the work domain. By Lind [LIN 99], there is a distinction between problems or limits that are methodological in nature and problems or limits that are conceptual, to which we can add the problems that are specific to the application of the model as part of the design of an ecological interface.

Methodologically, Lind notes that there is no procedure that facilitates the construction of a model of the work domain, to modify it or validate it and be able to answer the questions that a designer might ask himself with regard to the splitting of the units of the system. This problem is all the more critical as the definition of the different levels of the abstraction hierarchy depends on examples coming from the selection of a few domains. It would therefore appear that the method is not generic enough.

Conceptually, Lind notes that the heterogeneity of the concepts used to characterize the contents of the five levels of abstraction hierarchy is a source of confusion. Thus, the levels of forms and physical functions go back to the behaviors and to the localization of physical objects (pumps, heat exchangers, etc.) allowing action on a plant; however, can we also describe, at these levels, the entities that are water or heat? Furthermore, it does not seem to be easy to model substances that are at the same time both agents and objects of an action, since the distinction between objects and agents does not exist in the abstraction hierarchy.

Moreover, it appears that the concept of "function" has several meanings, since some functions are attributed to entities by social convention, while others are dispositional properties of the entity.

Finally, the splitting into whole and parts takes different meanings depending on the level of abstraction. At the level of forms and physical functions, the hierarchy is defined by spatial inclusions; whereas at the level of abstract and general functions, it is defined by temporal relations (the parts are temporal intervals of a period taken as a whole).

It would therefore be invaluable to have a methodology, principles or rules that would help in the identification of the levels and their semantics.

In the more practical area of the design of ecological interfaces, Terrier, Cellier and Carreras [TER 00] have pronounced doubts with regard to the use of the abstraction hierarchy as it is currently defined. Taking the example of a nuclear reactor, these authors note that the description of a domain depends on its state: the parameters to control change as a function of this state. They ask as a result, whether it is relevant and possible to give only one global representation of the process when the system is complex.

Also for these authors, the description of the process at high levels of abstraction must be motivated by the task that the operator has to realize, as well as by inherent limitations of the work domain. A minimal amount of knowledge of the strategies and procedures adopted would be necessary to avoid omitting any crucial limitations.

2.4.2. *Task analysis*

The analysis of the tasks is complementary to the analysis of the work domain. It consists of analyzing the state of a domain of work and then choosing an action to be carried out on this domain. These tasks are typically diagnosis (breakdown diagnosis, medical diagnosis, etc.) and decision-making.

The analysis of tasks aims to describe what needs to be done, independently of the manner in which it is done and of who does it. It is therefore not an analysis of activity, dealing with the real work of the operators.

The analysis consists of identifying the information necessary for the realization of a task and the information produced on output, as well as the limitations that must be taken into account during the selection of the actions that will allow it to be carried out. It is based on a descriptive tool developed by Rasmussen [RAS 86]: the double scale of decision.

2.4.2.1. *The decision ladder*

The scale of decision is not a model but, according to Vicente, it is a template offering a very simplified vision of the activities of information processing without prejudice to the way in which the information is processed.

It divides the different steps of information processing into two parts (Figure 2.11). The left part of the ladder (going up) contains the steps of the analysis of a situation (of the detection of abnormal conditions and the evaluations of the consequences on the state of the system). The right part of the ladder (going down) relates to the different steps of the planning of the action (definition of the task and then of a procedure) and finishes with the execution of the action itself. Each step of the processing of information (represented by a rectangle in Figure 2.11) leads to the acquisition of knowledge (represented by an oval).

To carry out a diagnosis and to make a decision a beginner will go through all of these steps. He starts a rational process, based on the knowledge that he has on the subject of the function and intentional properties of the work domain. But the decision ladder also brings up short cuts linking the two parts and bearing witness to the "heuristic" mode of decision that is observed in the expert operators. Taking "heuristic" style decisions corresponds to expertise and is based on inductions consisting of associating states of the environment with actions that have been proved effective in analogous situations. It depends on the empirical correlations between indices and actions observed in familiar scenarios.

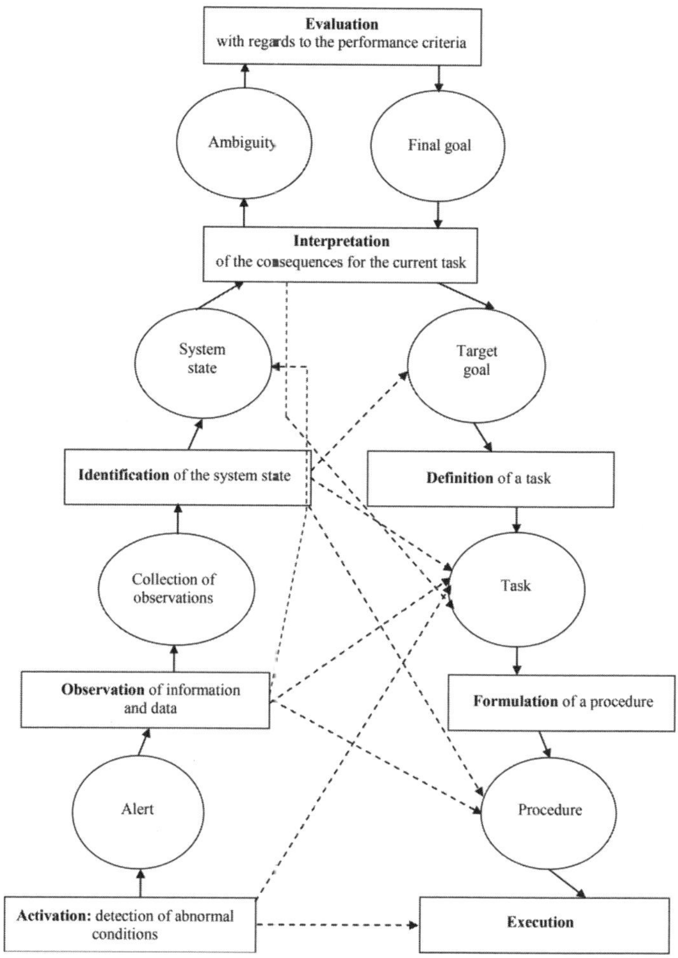

Figure 2.11. *Rasmussen's decision ladder (from Rasmussen [RAS 86])*

2.4.2.2. *Usefulness of the decision ladder*

Such a template has several advantages. It does not refer to any psychological mechanisms. It can therefore separate what needs to be done (which is in the model) from who needs to do it (a machine or the human agent).

Concretely, the decision ladder helps to identify expert behaviors and can lead, as a result, to the conception of a system that favors expertise. It highlights the sources of relevant information and is therefore also a primordial tool in the context of the design of information systems.

2.4.2.3. *Applications*

Jamieson *et al.* [JAM 07] show the complementarity of work domain analysis and of task analysis (task analysis relying on the use of the decision ladder but also hierarchical task analysis) in the design of ecological interfaces for the control of a reactor. The integration of three methods facilitates the specification of the information that is useful for the operator (as well as the probes necessary to acquire them), the contents of the interface, the structure and the mode of presentation of the information. Bisantz *et al.* [BIS 03] use these steps to provide recommendations to the designers of a war vessel; recommendations that concern the level of automation, the roles of operators and the interfaces of the weapons system.

Naikar [NAI 06a] also uses task analysis in conjunction with work domain analysis. She shows that these two methods can be used for applications other than the design of interfaces. She presents applications in the domain of training and the design of teams. Naikar, Moylan and Pearce [NAI 06b] also propose to complete the task analysis steps by introducing the notion of "context". They characterize the entirety of the tasks that need to be carried out in a system in terms of both work situations and work functions. Let us take the example of military aviation. They distinguish three situations (entrance into the combat zone, combat zone, exit from the combat zone); in the "combat zone" situation, they identify three functions: evaluating the tactical situation, firing on the target and avoiding threats. Each of the tasks identified is then represented using the decision ladder.

2.4.3. *Analysis of information-processing strategies*

This step deals with the different manners in which to accomplish the activities identified in the previous phase. All these activities consist of transforming input data into output data. Analysis of the task aims to specify this data, to identify the limitations associated with the transformations carried out, but it does not detail the manner in which the transformations take place. However, in real-work situations, the operators often have a working margin that allows them to choose different strategies to accomplish the same task, depending on the characteristics of the context but also on the individual characteristics. Sperandio showed in this way that the strategies used by air traffic controllers varied in function of their workload and, more exactly, in function of the number of airplanes to deal with [SPE 77].

In the context of the design of human–machine systems, analysis of the strategies is necessary as the systems will need to allow the users to use their normal strategies, but also to leave their usual strategies when they are not performant, to pass from one strategy to another, and will need to support them when they apply the strategies that are the most costly cognitively.

2.4.3.1. *Strategy analysis method*

Rasmussen, Pejtersen and Goodstein [RAS 94] state that the term "strategy" must be understood in the sense of a *category* of procedures to be able to understand the fact that the actions carried out, their order and occurrence are, in practice, difficult to predict. They explain that analysis must lead to the description of:

– all of the possible strategies;

– the criteria used by the operators to choose a strategy;

– the contextual indices at the origin of the choice of a strategy or of a change of strategy.

2.4.3.2. *Different types of strategy*

From studies conducted on fields as diverse as breakdown diagnosis or document research, Rasmussen, Pejtersen and Goodstein [RAS 94] identify two classes of strategy: "symptomatic" research strategies (researching the cause of the problem) and "topographical" research strategies (researching the faulty component).

Symptomatic research strategies rely on knowledge of the relations between symptoms and causes. They are divided into three sub-classes:

– recognition of configurations of data using an inductive reasoning, directed by data;

– research strategies that rely on decision tables to associate the representation of the problem with the different possible interpretations that can be made;

– hypo-thetico/deductive strategies that rely on knowledge of the functional structure of the work domain and consist of deducing symptoms from hypotheses related to the causes of the problem.

Topographical research strategies consist of locating the component at fault by evaluating the system outputs using knowledge available on its normal functioning. The investigatory field shrinks until the localization of the change is determined with enough precision to allow the choice of a corrective action. Research therefore relies on a map of the system that allows the localization of relevant sources of information.

These different strategies can be evaluated in terms of time, of taking or of processing information and of the knowledge necessary to apply them. Vicente underlines the fact that there is no intrinsically better strategy since each strategy has advantages and disadvantages with regard to each of the criteria and the characteristics of the context [VIC 99]. Rasmussen's approach can be the object of refinement. For example, Hoc and Carlier [HOC 00] evaluate the breakdown diagnosis strategies according to the characteristics of mental representation (e.g. functional or topographical) and the processes put into place (e.g. testing of one or more hypotheses at the same time); however, it does provide a useful basis for reflection in the design of human–machine systems.

2.4.3.3. *Implications for the design of human–machine systems*

Analysis of strategies provides a useful basis on which to design a support that will:

– allow users to adopt their usual strategies;

– manage the costly aspects of the strategy (by serving an external memory, for example, or by carrying out all or some of the information processing);

– as a result, make the adoption of strategies that might not otherwise have been spontaneously used by operators (like hypothetico/deductive strategies that are relevant to the resolution of non-familiar problems but which are very costly cognitively) easier;

– give operators the chance to change strategy in function of their preferences and of the characteristics of the context.

Jenkins *et al.* [JEN 10] use the first three phases of the outline of the CWA approach (work domain analysis, control task analysis and strategy analysis) to explore the flexibility of a system. They take the example of an iPod. Their analysis facilitates the modeling of this object in terms of the limitations that determine its use. Task analysis (coupling work situations and functions) highlights the situations in which the functions of the iPods are limited (e.g. charging the battery while moving or loading music while far from a computer). This analysis finally facilitates the suggestion of strategies to create distance from the limitations and to increase the flexibility of the object.

2.4.4. *Socio-organizational approach*

This step of the analysis consists of defining, on the one hand, the distribution of roles or functions and, on the other hand, communication and cooperation between the agents. It is based on the following postulation: since complex socio-technical systems are open, they require organization structures that are distributed and flexible to be able to deal with situations that were not planned during design.

2.4.4.1. *Method*

The distribution of roles relies on the previously conducted analyses: work domain analysis, task analysis and strategy analysis. It consists of distributing to different agents (human agent, groups of agents and automates) the responsibility of different sectors of the domain that appear in the abstraction hierarchy, the different activities that figure in the decision ladder and the different strategies identified.

The distribution of roles or of functions is conducted using six criteria identified by Rasmussen, Pejtersen and Goodstein [RAS 94]:

– the competence of agents;

– the access to information or to methods of action (it is astute to distribute the responsibilities in such a way so that decision-making comes back to the agents that have immediate access to the sources of information and to the commands);

– the demands of coordination (when coordination between agents is necessary, it is better that this take place within a same group of agents and not between different groups);

– the workload (the work is distributed when the demands of the task are too great for one agent to deal with);

– safety and reliability (the requirements that are related to these criteria in some domains lead to the establishment, in a deliberate manner, of a certain redundancy between the functions carried out by different agents);

– the existing regulations, finally, can limit the organizational structure.

With regard to the definition of the mode of communication between agents, Vicente [VIC 99] mentions three large categories of organization: autocratic (where an operator directs all the others), authoritative (where the information is transmitted along the length of a chain containing the hierarchical statuses of the agents) and hierarchical (where each agent must be able to communicate with all the others).

2.4.4.2. *Limits*

Rasmussen, Pejtersen and Goodstein [RAS 94] and Vicente [VIC 99] place the question of the distribution of roles in the stage of socio-organization analysis. However, since this is the design of a human–machine system, we must also deal with the distribution of the functions between human and machine (Hoc, in this book, [HOC 13]). Moreover, this must take place during task analysis. Indeed, it would seem a bit premature to go straight into the analysis of performant strategies without having dealt with this question.

To define human–machine cooperation, two approaches coexist [KAB 04]. A “static” approach that aims to optimize the distribution of the tasks between the human and the machine by allocating to the human and/or the machine tasks of the supervision of the process, of planning, or decision and execution and by measuring the impact of this distribution on performance, workload or even the mental representation that the operator has of the system. A "dynamic" approach aims to allocate different functions to the human or to the machine as a function of the requirements of the situation. The distribution of functions (see Hoc, in this book) leads to two successive questions.

First, there is some interrogation over the opportunity of a choice between static distribution and dynamic distribution of function between the human and the machine. Since the work by Fitts [FIT 51], the static distribution of the functions is understood to be a search for the functions carried out with the most performance by the human or by the machine. It is on this basis that the distribution is decided. All the same, this approach is a bit simple and causes a number of problems that need to be solved explicitly. Here we underline two of them.

The first problem concerns negative interference. In a given situation, the distribution of the functions on the indicated bases can lead to negative interference between the function carried out by the human and the one carried out by the machine. For example [DEA 00], it can be reasonable to trust the machine with the automatic extinguishing of a motor fire on board a plane. However, if the fire takes place during takeoff, the function carried out by the pilot (holding airplane in flight) can require, at certain stages, a lot of power. Indeed, even when on fire, a motor can still produce power. It would therefore be unreasonable to make the machine turn off the motor when the pilot needs it. The idea is then to let the pilot manage the extinguishing of the motor fire in function of the power requirements that only he knows.

The second problem is of assistance. The human operator can prove to be very bad in the execution of some tasks, because of a lack of appropriate assistance. For example, could we imagine nowadays an air traffic controller dealing with the density of European air traffic with the radars of World War II? However, automation does not seem to have been an acceptable solution.

Second, if we choose the dynamic distribution of functions, we are confronted with the question of methods to ensure good cooperation between human and machine (Hoc, in this book). Dynamic distribution aims to adapt the human–machine system to the variety of situations to which it is confronted. Depending on the situations, a distribution can be very good or very bad. Analysis in real time of the evolution of the situation (often in anticipation) can help to decide on adapted distributions of functions. The criteria proposed by Rasmussen, Pejtersen and Goodstein [RAS 94] – mental load, competence, access to information, etc. – could be applied to make this choice.

2.4.5. *Analysis of competences*

The competences expected of operators come from the limitations identified in the previous phases of the analysis but their definition must also take into account the capabilities and the limits of human cognition.

The objective of the fifth and final step of the analysis is, as a result, to integrate the requirement linked to the domain of application and the relevant sub-collections of the relevant knowledge that exist on human cognition. A model can be used to satisfy this objective: the taxonomy of "skills", "rules" and "knowledge" from Rasmussen [RAS 86] that we present now.

2.4.5.1. *The taxonomy of "skills", "rules" and "knowledge"*

Humans dispose of different modes of control of their interactions with their environment and with their cognitive resources. The taxonomy proposed by Rasmussen [RAS 86] distinguishes between three different levels of cognitive control.

The skill-based level is translated by the application, without any symbolic attention of cognitive automations (such as mental arithmetic) and of *patterns* of action that are automatic and highly integrated. In terms of actions, the existence of a dynamic model of the world facilitates synchronous coordination of physical movements with a dynamic environment. The flexibility of the performance, at this level, reveals the human capacity for adaption of a large repertoire of prototypical movements.

The rule-based level is defined by a correspondence of the "if-then" type, operated between the signs of an appropriate action (if "this sign", then "execution of this action"). A sign is a two-sided unit: expression (or form)/content. It is not the expressions that are immediately perceptible that trigger the rules, but the contents that they code [HOC 96]. No deep interpretation is necessary but the operators are conscious of their cognitive activities and can therefore verbalize them. At this level, the generation of movement *patterns* depends on access to stored rules and on past experience. Planning is not synchronous with the action: it intervenes beforehand.

The knowledge-based level is the level of problem solving. The knowledge applied is declarative in nature. This level is defined by a sequential, analytical reasoning that relies on an explicit representation of the goals for a mental model of the functional properties of the environment. Its application is costly, as it requires a focused symbolic attention.

The type of internal representations on which they rely can distinguish the three modes. However, they can also be distinguished by the manner in which the operators interpret the information that comes from the environment. At the skill-based levels, the information is received in the form of signals. The signals have a strong perceptive base since they are continuous and quantitative indicators of the spatio-temporal behavior of the environment. At the rule-based level, the information is interpreted in the form of signs. The signs are arbitrary but familiar indices of the environment. They are double sided "signifying" and "signified". They relate to the state of world by convention or because of past experience (e.g. the STOP sign). At the knowledge-based level, the information is interpreted in the form of concepts. The concepts are significant formal structures that represent the functional properties of the environment. They are accompanied by strong interpretive demands.

The relations between a particular task and the levels of cognitive control are not fixed; they depend on several variables: the level of expertise of the operators, the form in which the information is presented, the level of reflection of the operator on his own performance. The passing from one level to another takes place when an unusual situation is encountered. Furthermore, different modes can be active at the same time since activity is not always focused on the current activity.

2.4.5.2. *Use of the taxonomy of "skills", "rules", "knowledge" in the context of design*

Vicente [VIC 99] proposes three principles of design related to the three levels of the taxonomy.

For interaction at the skill-based level, the operator must be given the possibility to act directly on the interfaces (direct manipulation). Moreover, to provide a view of the dynamics of the work domain, the information must be presented in terms of spatio-temporal signals.

Favoring the application of "rules" presumes waiting until there exists some coherent correspondence between the limitations of the work domain and the signs provided by the interface.

Several reasons can lead the operator to applying the knowledge-based level. In these situations, he needs to be helped, as analytical reasoning is costly and subject to error. Seeing as this level involves reasoning based on a mental model, it would be useful for an interface to provide an external model, worthy of trust, in such a way that the operators need not encode, store or search for the information mentally. The abstraction hierarchy provides a useful base to reach this goal.

2.4.6. *Some general remarks on the integrated approach*

The CWA method is certainly the most complete in terms of analysis of a human–machine system. Indeed, it provides conceptual and methodological coherent tools that facilitate the description of the work domain, the tasks that need to be carried out in this domain, the strategies (or different way of accomplishing these tasks), the socio-organization aspects of the human–machine system that will act on the domain and the required competences.

However, it does prove to be lacking in some regards:

– The presented approach gives priority to the analysis of limitations of the domain, but it seems difficult, in practice, to analyze the constraints of a domain without envisaging the tasks that the operators will have to carry out and the agents that will be positioned in these tasks. The works already cited show that many applications do indeed rely on the two first steps of the method (analysis of the work domain and task analysis). However, only some authors have used it in its entirety and have taken into account the question of the distribution of the functions between agents; these applications concern the domain of "process-control" [JEN 08] and the domain of air traffic control [AHL 05].

– It leads to the establishment of the principle of general design and leaves the designer with the work of translating these principles into the form of concrete interfaces.

– It stresses the importance of the information necessary for the control of the work domain, and proposed recommendations concerning their organization, but deals less thoroughly with the distribution of the functions between the human and the machine. The effects of automation on the realization of tasks are, in particular, overshadowed; dynamic distribution of the functions is not envisaged.

– Finally, the formalisms proposed rely on notions whose semantics remain sometimes unclear; this is the case of the notion of "function" on which the abstraction hierarchy relies. It is also the case for notions of "knowledge" and "representation" that drive both task analysis and strategy analysis.

2.5. Conclusion

The two methods exposed – Long, Dowell and Timmer's on the one hand, and Rasmussen, Pejtersen, Goodstein and Vicente's on the other hand – take into account the human–machine system, without favoring one of its components (human or machine). The first of these methods relates to the performance of the system and defines a method that facilitates real-time evaluation of the efficiency of the system (with regard to the functional objectives of the work domain) and the costs incurred (in particular cognitive costs). It presents the advantage of stigmatizing drops in performance (performance being the ratio between efficiency and costs) and to point out, in the same way, the weaknesses of the system; it can thus be the starting point of a method of improvement, of correction of the human–machine system. The second method is usable in the context of design. It explicitly takes into consideration the notion of complexity of the system, since it is based on the postulation related to this dimension; a postulation according to which a complex system must be designed in a way that the human operator can face and deal with situations that the designers had not anticipated. It is very complete, seeing as it is composed of several steps covering both the work domain and the components of the human–machine system. Moreover, it proposes formalisms or concepts that enable us to organize data processing at each of the steps; the most important of these are abstraction hierarchy, decision ladder and the taxonomy of "skills", "rules" and "knowledge". It therefore makes up a framework and an aid for a design team.

These two methods therefore represent a definite advance and address the preoccupations of the authors who explain that tools of analysis and design of complex systems must integrate the notions of "system" and of "complexity". However, they still need to be improved in several aspects. They provide precise tools of analysis, but create questions related to 1) the processing of field data, 2) the translation of the results of the analysis into design principles and 3) the taking into account of risk management; this last dimension is a crucial element of complex situations.

The two methods involve the gathering of behavioral data and of verbal data. The analysis of cost, activities and cognitive strategies is led from this data. However, how its processed is not always exposed in a detailed manner. In particular, the criteria used to categorize verbal data and to deduce mental activities or strategies are not known.

The methods of design of human–machine systems should help the designer in answering two categories of questions: questions related to the information required by the agents of the system (what information must be taken from the environment? How must the information system be organized? What must the characteristics of the interfaces be?), and questions related to the sharing of responsibilities and the distribution of functions between the humans and the automated agents. The approach proposed by Long, Dowell and Timmer is not very generic in the manner in which the diagnosis can be the source of design decisions, but it is very precise inside a limited domain. The approach from Rasmussen, Pejtersen, Goodstein and Vicente is, on the contrary, more generic but undoubtedly less precise. It explains, in particular, that the abstraction hierarchy can be used in the design of an external model that would support diagnosis and the resolution of problems carried out by the operator in unknown situations, that the decision ladder highlights the information used as part of the different activities carried out and can constitute a support for the design of an information system and that the analysis of strategies and competencies will serve in the conception of tools that help in their application. Nonetheless, these principles are not very precise and, as remarked by Vicente himself, with regard to the designing of ecological interfaces [VIC 02], a lot of time and creativity is needed to translate them into a concrete form. Furthermore, it can be noted that they relate more to the information system than to the distribution of functions. This last point is dealt with late and quickly, without mentioning the effects of automation. This gap goes back to the last of the limitations mentioned.

Neither of the two methods directly deals with risk management. Despite being developed in contexts similar to dynamic situations, they still rely heavily on models that are quite determinist. However, in dynamic situations, there is a lot of emphasis on the necessity, for the human–machine system, to adequately manage the risks, notably through proper management of the available resources to deal with unplanned situations [AMA 96]. The results available on this theme of risk management should be integrated to deal with the distribution of functions between human operators and automations in this perspective. However, if it is already difficult to find the right indices for the evaluation of performance in certain domains, the question of the evaluation of the quality of risk management remains largely open. It is easier to evaluate what is going on than what would have happened if an unplanned event had occurred. In other terms, the performance can be excellent, but not very robust in the face of a disturbance. It is the question of robustness that is asked here.

2.6. Bibliography

[AHL 05] AHLSTROM U., "Work domain analysis for air traffic controller weather displays", *Journal of Safety Research*, vol. 36, pp. 159–169, 2005.

[AMA 96] AMALBERTI R., *La conduite de systèmes à risques*, Presses Universitaires de France, Paris, 1996.

[ANN 05] ANNETT A., "Hierarchical task analysis (HTA)", in STANTON N., HEDGE A., BROOKHUIS K., SALAS E., HENDRICK H. (dir.), *Handbook of Human Factors and Ergonomics Methods*, CRC Press, Boca Raton, pp. 33–1, 33–7, 2005.

[BAS 94] BASTIEN J.M.C., SCAPIN D.L., Evaluating a user interface with ergonomic criteria, Report No. 2326, Rocquencourt, INRIA, 1994.

[BAX 11] BAXTER G., SOMMERVILLE I., "Socio-technical systems: From design methods to systems engineering", *Interacting with Computers*, vol. 23, pp. 4–17, 2011.

[BIS 03] BISANTZ A.M., ROTH E., BRICKMAN B., *et al.*, "Integrating cognitive analysis in a large-scale system design process", *International Journal of Human-Computer Studies*, vol. 58, no. 2, pp. 177–206, 2003.

[BUR 04] BURNS C.M., HAJDUKIEWICZ J.R., *Ecological Interface Design*, CRC Press, Boca Raton, 2004.

[CAC 95] CACCIABUE P., COJAZZI G., "An integrated simulation approach for the analysis of pilot-aeroplane interaction", *Control engineering practice,* vol. 3, pp. 257–266, 1995.

[CHA 70] CHAPANIS A., "Human factors in systems engineering", in DE GREENE K.B. (ed.), *Systems Psychology*, McGraw-Hill, New York, pp. 51–78, 1970.

[CHI 00] CHIPMAN S.F., SCHRAAGEN J.M., SHALIN V.L., "Introduction to cognitive task analysis", in SCHRAAGEN J.M., CHIPMAN S.F., SHALIN V.L., (eds.), *Cognitive Task Analysis*, Lawrence Erlbaum Associates, Mahwah, pp. 3–23, 2000.

[DEA 00] DEARDEN A., HARRISON M., WRIGHT P., "Allocation of function: scenarios, context and the economics of effort", *International Journal of Human-Machine Studies*, vol. 52, pp. 289–318, 2000.

[DOW 98] DOWELL J., "Formulating the cognitive design problem of air traffic management", *International Journal of Human-Computer Studies*, vol. 49, pp. 743–766, 1998.

[DRI 09] DRIVALOU S., MARMARAS N., "Supporting skill-, rule-, and knowledge-based behaviour through an ecological interface: an industry-scale application", *International Journal of Industrial Ergonomics,* vol. 39, pp. 947–965, 2009.

[FIT 51] FITTS P.M., *Human Engineering for an Effective Air Navigation and Traffic Control System,* National Research Council, Washington, DC, 1951.

[GAI 94] GAILLARD I., "Analyse et modélisation de l'activité de contrôle du trafic aérien dans le cadre de la modernisation des systèmes de contrôle", in THEUREAU J., JEFFROY F., (eds.), *Ergonomie des situations informatisées,* Octarès, Toulouse, pp. 237–288, 1994.

[GRE 96] GREEN T.R.G., DAVIES S.P., GILMORE D.J., "Delivering cognitive psychology to HCI: the problems of common language and of knowledge transfer", *Interacting with Computers*, vol. 8, pp. 89–111, 1996.

[HAJ 01] HAJDUKIEWICZ J.R., VICENTE K.J., DOYLE D.J., *et al.*, "Modeling a medical environment: an ontology for integrated medical informatics design", *International Journal of Medical Informatics*, vol. 62, pp. 79–99, 2001.

[HAJ 04] HAJDUKIEWICZ J.R., VICENTE K.J., "A theoretical note on the relationship between work domain analysis and task analysis", *Theoretical Issues in Ergonomics Science*, vol. 5/6, pp. 527–538, 2004.

[HIL 95] HILL B., LONG J., SMITH W., *et al.*, "A model of medical reception: the planning and control of multiple task work", *Applied Cognitive Psychology,* vol. 9, pp. 81–114, 1995.

[HIL 98] HILL B., LONG J., "Diagnosing ineffective performance in the domain of emergency management: performance modelling and its decomposition", *Proceedings of ECCE9,* Limerick, Ireland, August 1998.

[HO 03] HO D., BURNS C., "Ecological interface design in aviation domains: work domain analysis of automated collision detection and avoidance", *Proceedings of the Human Factors and Ergonomics Society 47th Annual Meeting*, 2003.

[HOC 92] HOC J.M., *Psychologie cognitive de la planification*, Presses Universitaires de Grenoble, Grenoble, 1992.

[HOC 96] HOC J.M., *Supervision et contrôle de processus – La cognition en situation dynamique*, Presses Universitaires de Grenoble, Grenoble, 1996.

[HOC 00] HOC J.M., CARLIER X., "A method to describe human diagnostic strategies in relation to the design of human-machine cooperation", *International Journal of Cognitive Ergonomics,* vol. 4, no. 4, pp. 297–309, 2000.

[HOC 03] HOC J.M., "Coopération humaine et systèmes coopératifs", in BOY G. (ed.), *Ingénierie cognitive: IHM et cognition,* Hermès, Paris, pp. 139–187, 2003.

[HOC 04] HOC J.M., DARSES F. (eds.), *Psychologie Ergonomique: Tendances Actuelles,* Presses Universitaires de France, Paris, 2004.

[HOC 07] HOC J.M., AMALBERTI R., "Cognitive control dynamics for reaching a satisfying performance in complex dynamic situations", *Journal of Cognitive Engineering and Decision Making,* vol. 1, pp. 22–55, 2007.

[HOC 13] HOC J.M., "Human–Machine Cooperation", in LEE J.D., KIRLIK A., (eds.), *The Oxford Handbook of Cognitive Engineering*, Oxford University Press, Oxford, pp. 395–403, 2013.

[HOL 99] HOLLNAGEL E., WOODS D.D., "Cognitive systems engineering: new wine in new bottles", *International Journal of Human-Computer Studies*, vol. 51, pp. 339–356, 1999.

[HOL 05] HOLLNAGEL E., WOODS D.D., *Joint Cognitive Systems: Foundations of Cognitive Systems Engineering*, Taylor & Francis, Boca Raton, 2005.

[JAM 07] JAMIESON G.A., MILLER C.A., HO W.H., *et al.*, "Integrating task- and work domain-based work analyses in ecological interface design: a process control case study", *IEEE Transactions on Systems, Man, and Cybernetics – Part A: Systems and Humans,* vol. 37, no. 6, 2007.

[JEN 08] JENKINS D.P., STANTON N.A., WALKER G.H., *et al.*, "Applying cognitive work analysis to the design of rapidly reconfigurable interfaces in complex networks", *Theoretical Issues in Ergonomics Science,* vol. 9, no. 4, pp. 273–295, 2008.

[JEN 10] JENKINS D.P., STANTON N.A., WALKER G.H., *et al.*, "Using cognitive work analysis to explore system flexibility", *Theoretical Issues in Ergonomics Science,* vol. 11, no. 3, pp. 136–150, 2010.

[KAB 04] KABER D.B., ENDSLEY M.R., "The effects of level of automation and adaptive automation on human performance, situation awareness and workload in a dynamic control task", *Theoretical Issues in Ergonomics Science,* vol. 5, no. 2, pp. 113–153, 2004.

[LIN 99] LIND M., "Making sense of the abstraction hierarchy", *Proceedings of the 7th European Conference on Cognitive Science Approaches to Process Control*, Presses Universitaires de Valenciennes, Valenciennes, pp. 195–200, 1999.

[LON 96] LONG J., "Specifying relations between research and the design of human-computer interaction", *International Journal of Human-Computer Studies*, vol. 44, pp. 875–920, 1996.

[LON 01] LONG J., TIMMER P., "Design problems for cognitive ergonomics research: what can we learn from ATM-like micro-worlds", *Le Travail Humain,* vol. 64, pp. 197–221, 2001.

[MAR 04] MARTIN D., SOMMERVILLE I., "Patterns of cooperation interaction: linking ethnomethodology and design", *ACM Transactions on Computer-Human Interaction,* vol. 11, pp. 59–89, 2004.

[MEM 05] MEMISEVIC R., SANDERSON P., CHOUDHURY S., *et al.*, "Work domain analysis and ecological interface design for hydropower system monitoring and control", *Proceedings of the IEEE International Conference on Systems, Man, and Cybernetics*, Hawaii, HI, pp. 3580–3587, October 2005.

[MIL 97] MILLER T.E., WOODS D.D., "Key issues for naturalistic decision making researchers in system design", in ZSAMBOK C.E. (ed.), *Naturalistic Decision Making,* Lawrence Erlbaum Associates, Mahwah, NJ, pp. 141–149, 1997.

[NAI 06a] NAIKAR N., "Beyond interface design: further applications of cognitive work analysis", *International Journal of Industrial Ergonomics,* vol. 36, pp. 423–438, 2006.

[NAI 06b] NAIKAR N., MOYLAN A., PEARCE B., "Analysing activity in complex systems with cognitive work analysis: concepts, guidelines, and case study for control task analysis", *Theoretical Issues in Ergonomics Science,* vol. 7, pp. 371–394, 2006.

[RAS 86] RASMUSSEN J., *Information Processing and Human-Machine Interaction*, North Holland, Elsevier, 1986.

[RAS 94] RASMUSSEN J., PEJTERSEN A.M., GOODSTEIN L.P., *Cognitive Systems Engineering*, Wiley, New York, 1994.

[REA 88] REASON J., "Cognitive aids in process environments: prostheses or tools?" in HOLLNAGEL E., MANCINI G., WOODS D.D. (eds.), *Cognitive Engineering in Complex Dynamic Worlds*, Academic Press, London, pp. 7–14, 1988.

[SAL 10] SALMON P., JENKINS D., STANTON N., *et al.*, "Hierarchical task analysis vs. cognitive work analysis: comparison of theory, methodology and contribution to system design", *Theoretical Issues in Ergonomics Science,* vol. 11, no. 6, pp. 504–531, 2010.

[SUC 87] SUCHMAN L.A., *Plans and Situated Actions*, Cambridge University Press, Cambridge, 1987.

[SPE 77] SPERANDIO J.C., "La régulation des modes opératoires en fonction de la charge de travail chez les contrôleurs de trafic aérien", *Le Travail Humain,* vol. 40, no. 2, pp. 249–256, 1977.

[TER 00] TERRIER P., CELLIER J.M., CARRERAS O., "Task analysis does count in making the abstraction hierarchy concrete: evidence from a pressurized water reactor situation", *Proceedings of the IEA 2000/HFES 2000 Congress*, pp. 161–164, 2000.

[THE 94] THEUREAU J., JEFFROY F., *Ergonomie des situations informatisées*, Octarès, Toulouse, 1994.

[TIM 02] TIMMER P., LONG J., "Expressing the effectiveness of planning horizons", *Le Travail Humain,* vol. 65, pp. 103–126, 2002.

[VIC 99] VICENTE K.J., *Cognitive Work Analysis: Toward Safe, Productive, and Healthy Computer-Based Work*, Lawrence Erlbaum Associates, Mahwah, NJ, 1999.

[VIC 02] VICENTE K.J., "Ecological interface design: progress and challenges", *Human Factors*, vol. 44, pp. 62–78, 2002.

3

The Use of Accidents in Design: The Case of Road Accidents

3.1. Accidents, correction and prevention

The feedback experience proceeds from a process of common sense. It would seem not only logical but also inevitable to take into account experience and particularly dysfunction to improve existing systems, whether from the point of view of techniques, organization or the training of men. The corrective process and the design process come from the same logic and only differ by the degree of generalization that is given to the solutions that are brought, meaning that it is actually sometimes difficult to distinguish them. Indeed, design rarely takes place *ex nihilo*. It takes place in steps, by successive improvements, often after several cycles of tests and validation. This is particularly true in the domain of road transport. Vehicles evolve very slowly with regard to their basic characteristics. Security systems such as seat belts, airbags and others come from an accumulation of works in accidentology, biomechanics, modeling and crash testing. Secondary security enquiries, which combine types of lesion, colliding parts, kinetic energy and vehicle deformation, have played a large role in the design process. Their end product is the creation of certification standards and trial procedures, the comparative results of which are the object of regular publications (EuroNCAP tests, for example). The case of road infrastructure is even more exemplary: punctual structural improvements, statistical comparisons or even evaluation procedures facilitate the progressive generalization of the best performing solutions as much in terms

---

Chapter written by Gilles MALATERRE, Hélène FONTAINE and Marine MILLOT.

of traffic as in terms of security. We know for example that roundabouts are at least twice as safe as classical junctions in interurban areas, by limiting lateral conflicts and by decreasing the speed of approach of drivers [FLE 98]. It is still not perfect: these adaption processes are slow and are often questioned by evolutions in the context that can ruin the efforts undertaken. Increases in speed are a good example.

Each domain has its own specificities. Among them are the degree of gravity and the frequency of events that are considered as unacceptable for the user or for society. This is in no way objective data. Let us consider for example different causes of death (Table 3.1).

Type of accident	No. of deaths
Death from AIDS in 2005	392
Concorde accident in 2000	113
Deaths from the AZF catastrophe in September 2001	30
Total road deaths in 2005	5,318
Deaths in the 15–24 yr group on the road in 2005	1,482
Deaths due to the Creutzfeldt-Jakob variant since 1991	20

Table 3.1. *Accidents and causes of death in France*

Table 3.1 shows that the media coverage and public apprehension caused by these events and the possibility of them occurring again are not proportional to the number of deaths. Familiarity with the risk and the impression of being able to control them makes them more tolerable. This is the case with road accidents and it is more and more the case with AIDS. On the contrary, poorly understood risks, which are therefore potentially catastrophic, and against which there is a feeling of powerlessness, are much more worrying and warrant the implementation of precautionary methods without the possibility of using existing data. The principle of precaution, if fundamentally justified, gives rise to abuse, both in media and in politics. The type of event that is taken into account and its frequency are very different depending on the case. One single incident can form the base for analyses and have important consequences in terms of changes or political decisions if the potential threat is deemed intolerable, while road accidents pile up and are considered by some as the price to pay for mobility and pleasure.

Using accidents as instruments of diagnosis, or to predict future accidents, assumes that the hypothesis according to which the conjunction of circumstances and factors that caused them is not fortuitous is true. This hypothesis is all the easier to come up with as accidents are numerous, of the same type, or more generally, they have certain characteristics in common with each other.

3.2. Analysis of accidents specific to the road

In some domains, such as nuclear or aviation, accidents are rare and analysts use events that are more common and less dramatic, by assuming, sometimes wrongly, that there is a statistical continuity between the same types of accident and incident. They also use reliabilist methods. The problem is not usually presented in such terms for road accidents. Quite the contrary, the material here is overabundant, and most often the limiting factors are the sorting and aggregation methods that extract the most relevant information from what at first looks to be random distribution.

We will see that there are many different methods that exist. These, however, can be placed into two categories: quantitative methods, which helps in the determination of the stakes, risk factors, and their localization in time and space; qualitative methods, which allow local diagnoses or, more generally, allow us to move forward in the fine understanding of the mechanisms of accidents.

3.2.1. *Road accidents as a statistical unit*

Physical accidents in France, like in most countries, are recorded nearly systematically[1]. The police are called to the site of the accident, where they write up a report. To do this, they pick up evidence, note the positions of vehicles and put together the elements that characterize the infrastructure, signals and weather conditions. They note all of the administrative information with regard to the people involved and the vehicles, and describe the damage as well as any violations of the regulations in place that they notice. They take witness accounts, and list any injury or infraction. The

1 According to a comparison between the Rhône registers which list all the victims of road accidents and the BAAC (*Bulletin d'analyse des accidents corporels de la circulation*) of national file of corporal accidents, serious accidents as well as accidents involving motor vehicles are over-represented in the BAACs [UMR 96, AMO 08].

report is a legal document, more orientated toward establishing responsibility than understanding the finer details of the accident. The witnesses' reports are actually written up by the police officers based on the elements collected and signed several days after the actual incident by those involved. The report is sent to the courthouse after a bordereau (the BAAC[2]) is coded and then sent to SETRA[3] to be incorporated into the national file on physical accidents. Note that one in fifty reports is sent to INRETS[4] as a representative file of the national file [FON 01]. Since January 1st 2005, as part of a move to adhere with European definitions, France has moved to a new definition of the words killed and injured[5]:

– *killed*: anyone who dies on the spot or within 30 days after the accident (instead of six days before 2005);

– *injured*: more than 24 h in hospital, as opposed to *seriously injured*, which is six days.

Considering that the data we use are from before 2005, it is the six-day definition of killed and of injured that apply. Because it is the indicator that is the least subject to interpretation and also because it is the most symbolic, it is the number of deaths that is the most used. This number does not hold much significance on its own. It is most often used alongside a denominator that is the expression of a measurement of exposure to risk. Thus, the number of deaths depending on the population of a country is often used for international comparisons. It is mainly a criterion of public health, but it does not show disparities in terms of mobility. This is why it is preferable to take the number of deaths with the number of vehicles, or even the number of the distance traveled in vehicles of the country. These statistical indications serve to compare the risks of each country, network, vehicle, user, age range, gender, etc. They also serve to measure changes over time [FON 97].

---

2 BAAC: French acronym for *Bulletin of Physical Analysis of the Traffic.*

3 SETRA: French acronym of the former department of technical studies of roads and highways, now a branch of CEREMA (*Centre of studies of risks environment and amenagement http://www.cerema.fr/*).

4 INRETS: French acronym for Research Institute for Transport Safety, now a branch of IFSTTAR: *http://www.ifsttar.fr.*

5 *La sécurité routière en France, Bilan de l'année 2005*, Observatoire national interministériel de sécurité routière, La Documentation française, Paris, 2006.

Particular statistical tools allow us to identify the parts of the road over which risk is higher than the average for that route, and to detect congestion zones [ALL 01]. It is also possible to compare types of network or types of construction, or even to conduct before–after evaluations. The objective of this chapter is not to enter into the details of these methods. They are above all quantitative – nonetheless they do highlight certain risk factors, and, as such, they present a better understanding of security problems. In this way, they are useful in design, as much of the infrastructure as of vehicles. From the point of view of infrastructure, complete methods have been developed for an interurban context – the "*sécurité des usagers sur les routes existantes*" (security of users on existing roads, or SURE) protocol since 2005 for the national road network – and for urban areas – the "*Sécurité des déplacements en milieu urbain*" (security of urban travel) protocol since developed by CERTU[6].

There are also more clinical methods. Without going into a complete analysis of each individual case, these methods use mapping of different types of accidents in one location, one route or zone to highlight the similarities that are often linked to local activities or to principles of improvement [FLE 90, MIL 01].

3.2.2. *Accidents as diagnosis tools*

Reading and analyzing statements helps a lot in understanding the accident phenomenon, whether as part of punctual corrections, or as part of the gathering of more general knowledge and of methods. ONSER and INRETS (see footnote 4) have contributed to developing their use. "Commando" approaches consist of testing on the field hypotheses made on the basis of the details of accidents from reports and from observations of the behaviors of drivers. This testing is all the more fruitful as it is carried out by multi-disciplinary teams, a concept that is taken up later in the investigations project called "Réagir"[7]. These were not simple visits to the locations after reading the statements, but an analysis made after precise localization on a map of accidents of the same type and the listing of observations to be made so as to verify the hypotheses that were made [FER 79].

---

6 CERTU: another branch of CEREMA; http://www.certu.fr/.

7 This can be translated by "React".

Going further would require more detailed data than what is in typical reports. This can only be obtained through specific gathering of information, by qualified personnel. There is a distinction between studies carried out "in the field", which consist of going to the site of the accident afterward with all the available documents, and detailed studies that consist of going to the site of the accident at the same time as the emergency services and the police so as to gather all the elements that might otherwise be lost. A first study, called the "clinical study of accidents", was conducted by ONSER from 1965 to 1969 [ONS 70]. It gathered physicians, psychologists and engineers, all specialists in vehicles and infrastructure. Ten years later, the detailed study of accidents in Salon-de Provence [FER 85, FER 99] saw the light of day and is still ongoing, reinforced since 1993 by teams from LAB[8] working on the Amiens and Evreux sites using an identical method. The "Réagir" investigations carried out on the site, but at different times, also owe a lot to the methodologies developed by INRETS. The common factor for these methods is to go further than the local correcting specificities, and to have as an object the understanding of mechanisms of accidents described from typological analyses in the form of typical scenarios that have a global outreach [BRE 99, TIR 99]. A typology consists of gathering together accidents that have a certain number of properties in common. There exist as many typologies as research objectives. INTRETS has thus developed different groups of accident scenarios that can serve as a qualitative analysis: highway scenarios [MEG 99], urban scenarios with pedestrians [BRE 03], urban scenarios without pedestrians [CLA 10], etc. The "mechanism" is a more complex notion, with integrations of the idea of a temporal and causal occurrence, and the intervention of "factors" that are relevant to the infrastructure, the vehicle and the driver. Typologies based on mechanisms are founded, for example, on faulty functions or on the driver's errors, and on similarities in the organization of contributing factors [ELS 97].

Taking into account different sources of information and their confrontation enriches knowledge. Thus, the BAAC's analysis shows the correlation between the previously listed criteria. This correlation is the reflection of a "statistical proximity" and not the explanation of the cause of accidents. It is the detailed study of accidents that help us to go further in identifying the mechanisms, the accidents' factors and their consequences on a smaller number of cases. This identification then needs to be validated by

8 LAB: French acronym for accidentology and biomechanics laboratory from car manufacturers PSA and Renault.

statistical analyses carried out on samples that are more representative of the accidents by taking into account criteria that are shown through more clinical approaches. Such a back and forward process between qualitative and quantitative is vital for the validation of hypotheses.

3.3. Need-driven approach

3.3.1. *Definition of needs from the analysis of accidents*

The notion of *need* is hard to define, as can be seen in the number of dictionary entries for the word. Among these, there is "something necessary to existence". In the case considered here, it could therefore be said: "something necessary for security", as only this aspect of the problem is taken into account, excluding comfort, ease of driving and fluidity needs, which, as legitimate as they may be, are not relevant to an accident-based approach. It is therefore both as a deliberated choice and for technical reasons that the needs have been defined only from accidents. The advantage of this approach is the availability of objective and quantifiable data. The inconvenience lies in the lack of finesse in the data concerning the psychological processes caused by the drivers during the chain events that resulted in an accident. This is why, first of all, we use data from the detailed Salon-de-Provence study. The method of information gathering (on the sites, in real time) and the finesse of the analyses (cinematic reconstitutions, interviews of those involved) are in a sense an inventory of the mechanisms of accidents, i.e. of the ways of combining situations, actions and factors so that they result in an accident.

This notion of need assumes that the driver could have avoided the accident if he had the information or the assistance that was lacking. By definition, these needs concern only the active users in the accident process. Only the operators in the loop, meaning those whose actions or non-actions can have an effect on whether the accident takes place or not, are concerned[9]. It is to be articulated with the definition of error and of failure. The driver is the final regulator of the circulation system, and we can talk of failure as long as a function is not finished. It is generally possible to identify "errors" in the perception–decision–action chain that operates, this notion

9 Which excludes the driver stopped at a light if he is hit by someone else, as long as he did not stop in a sudden or unpredicted way

does not necessarily imply the idea of responsibility in the penal sense of the term. The idea is not to refer to an ideal driver who would never make any errors, but rather a driver who, once confronted with a number of difficulties, would be able to either avoid them or deal with them using an appropriate mechanism of recuperation of his own mistakes or a corrective response in the face of unpredicted behavior from someone else. In this sense, the needs of a driver concern what would have allowed him to maintain or to regain control of the situation, and therefore to avoid the particular accident analyzed in the accident file. This is the analyst's point of view, and not the driver's [ELS 08].

We must nonetheless remain aware of the fact that a driver partially regulates his own risk levels in function of the perception he has of his own capabilities. Any help that could be brought to him (satisfaction of the identified needs) could therefore cause new behaviors and new needs. If this method is therefore imperfectly predictive of the security level that could be reached by an approach aiming to satisfy all the identified needs, it would still be able to hierarchically order the needs depending on the encountered situations and on the groups of users.

We already used this notion of needs in two successive studies [MAL 91, MAL 99]. The definitions used are different in these two studies as the first was carried out on a sample representative of physical accidents that occurred in France in 1989, which were 3,179 police reports, while the second study was solely aimed at one type of accident in particular, which were rear-end collisions. For this, 329 accidents of this type that took place in France in 1995 and 1996 were analyzed. This gave rise to definitions of need that were more and more precise, particularly adapted to this type of accident. In Figures 3.1 and 3.2, the distribution of needs identified during these two studies are given.

Pedestrians essentially have needs of detection with regard to the coming of a motorized user. Motorcycles have proportionally more needs in predicting the behavior of another user. This is due to the fact that they are often inadequately taken into account by four-wheel vehicle users, who, for example, often cut them off when turning left, or pull over when they are overtaking.

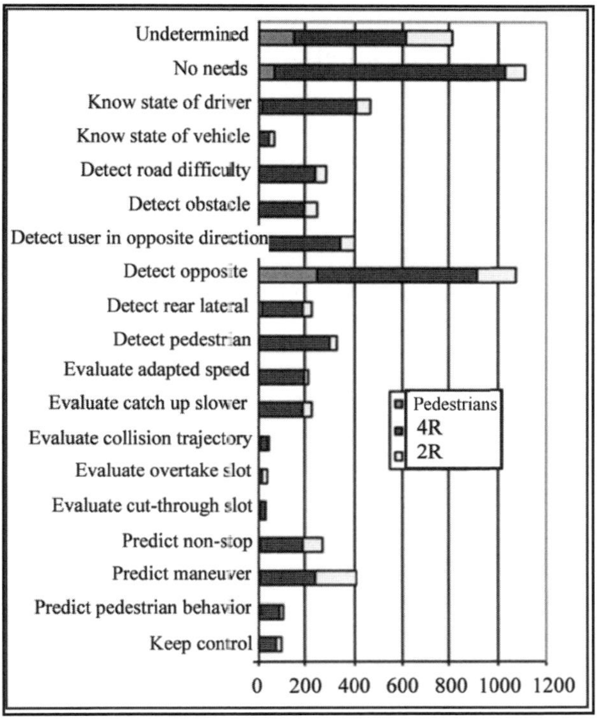

Figure 3.1. *Distribution of the needs of 6,049 users involved in the 3,179 accidents taking place in France in 1989 (all types of networks). The same user can have several needs, if satisfying any one of them would have prevented the accident*

For the rest, let us note the large number of users with no needs. They are generally passive during the accident, with no direct responsibility and with no possibility of action. Their own fault was to be there at that moment. Among the needs identified the largest involve detection, but the reports do not always make the distinction between pure detection problems and problems of not taking into account another user seen briefly and then wrongly assumed to not constitute a potential danger and "forgotten".

Toward what type of solution should we orient ourselves? The problem is knowing whether to prefer very targeted actions on well-identified particular situations (night accidents on bends, taking into account the corresponding needs), or whether it is better to come up with more generic assistance that can be efficient in a large amount of situations (improvements in vision, for example). We can thus solve the problem of a sharp angle turn with appropriate signs, or even improve vehicles' projectors (movable lights, etc.). The second solution would allow gains on all turns, whatever their angle and size, but would also be more likely to cause behavioral adaptions in drivers benefiting from this assistance. By vocation, action on infrastructure would tend to be more specific, and action on the vehicle more generic [MEI 08].

The second study was oriented specifically on rear-end collisions, on highways or on the national and departmental network. The goal was to get to know this category of accident better, which represents 14% of physical accidents and 7% of those killed outside built-up areas. The main idea was that these were often chain collisions, occurring on the highway in conditions of dense traffic. Some spectacular accidents that occurred in times of fog have certainly shocked public opinion.

While the behavior of drivers was indeed implicated each time (speed, intervals), the public powers could not remain indifferent inasmuch as it is known that traffic and short intervals are correlated [ARO 96] and in these conditions there are few ways of preventing chain collisions or over-accidents. Figure 3.2 shows a need for the anticipation of difficulties, whether this need is due to an absence of visibility linked to the outline of the road or to the presence of vehicles ahead. This is notably the case in situations of lane driving, or when the vehicle ahead is a heavy vehicle. Detection needs remain clearly in the majority. The smallest number of users with no needs is linked to the fact that we are only considering here active users.

The use of these results is only effective when a difference is made between networks and accident situations, which we will attempt to do later on.

3.3.2. *Particular case of urban areas*

The notion of need is more complex in urban areas. Indeed, while the accidents are less serious in urban areas than in interurban areas, there are more of them and they involve more users, in particular, users of non-motorized transportation [FLE 85, LEV 95]. Pedestrians and motorized two-wheel vehicles remain the most vulnerable, notably in terms of the seriousness of the injuries [OBS 11].

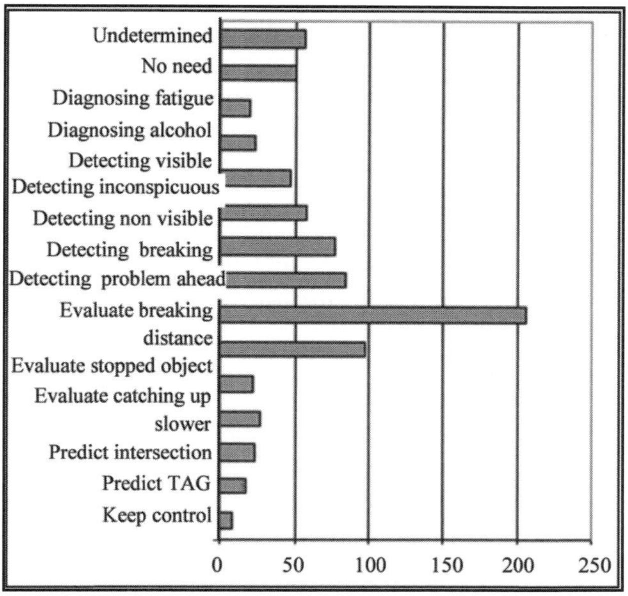

Figure 3.2. *Distribution of the needs of 1,098 users involved in 329 rear-end collision in France outside of built-up areas in 1995 and 1996*

Generally, conflicts are linked to differences in speed of different users. This has linked to a reduction of the speed limit in towns (down to 50 km/h in 1990), followed by the implementation of zones of slowed circulation at 30 km/h, or even 20 km/h in new traffic zones. Do users understand these spaces, and do they understand the speed limits? This question is rarely asked in the literature.

Conflict management between users has also led to the sharing of the road with installations dedicated to each mode: sidewalk for the pedestrian, cycle path for bicycles, dedicated areas for collective transport. Only motorized two-wheel vehicles have remained mixed in with general circulation. Some dedicated sites are open to other modes (taxis and/or bicycles in the bus route, for example). How visible and detectible are these different sites? Most importantly, how are their interactions with the rest of the road managed? Indeed, several recent studies have highlighted accidents at junctions between motorcycles using the bicycle lane due to the lack of proper developments, and drivers turning and being surprised by the arrival of these vehicles [MIL 11].

Conflicts can also occur during use. Indeed, the public space including the road receives the function of movement, but it can also support many other activities: parking, children's games, market, etc. The question of interaction between these different uses can cause security problems. For example, parking can cause direct conflicts because of the maneuvers it requires (U-turn, reverse, etc.). It can block visibility, notably between pedestrians and vehicles circulating on the road, according to the configuration of the spaces that are dedicated to parking. Parallel parking alongside bicycle lanes can lead to the opening of car doors as a bicycle is passing [PIA 02].

Movement in an urban area is therefore very complex. For vehicles, this involves avoiding potential conflicts with the different users and different uses of the public space (parking, local life, etc.), while respecting the rules that are imposed (speed limits, dedicated lanes, etc.). Pedestrians must be visible and detectable by other users, but also be able to detect these users. Indeed, pedestrians themselves rely a lot on the approach of circulating vehicles in making decisions on when to cross the road, including in the presence of lights, notably, to reduce waiting time in the case of a red light [ISH 08]. For two-wheel vehicles, on top of their visibility, their perceptibility is important. Also, the intrinsic dangers introduced by these vehicles, notably motorcycles, needs to be dealt with.

We now look at the problem of the *a priori* evaluation of the different types of assistance on which automobile manufacturers have been working for several years in collaboration with research institutes.

3.4. *A priori* analyses

A priori studies go from a seemingly simple principle. They calculate the influence that a technical device would have had on the occurrence of each accident examined individually, all other things being equal. This introduces three main problems: knowledge of the device in question (or of the measurement) is usually imprecise, as the projects are most often ones that have not yet been tested at actual scale; the condition "all things being equal" assumes that the situation in which the accident took place has not been modified by the device (which would not be the case for an inter-distance regulator that would be in all vehicles, for example); and the behavior of the driver must not be modified by the knowledge he has of the device. However, we know that such adaption mechanisms, which are commonly called "perverse effects" when they go the wrong way, can exist. The case of ABS[10] has been much discussed [EVA 96, FAR 97]. In practice, this method is only applicable for devices that do not fundamentally modify the conditions of circulation, and for which an individual reasoning can be applied (this device, for this vehicle, *hic et nunc*). We will see later that the estimation of effects of devices that are to equip several vehicles to function (cooperative systems) assumes more complex reasoning. A good example of *a priori* reasoning is the one that was made in 1973 on the effects of vehicle services by Ferrandez [ONS 73]. For each accident, taken in a representative sample, this author researched if a fault in one of the vehicles could have played a role in the accident, and then he determined whether this fault could have been detected by a service. He concluded that there was a certain rate of efficiency, to be balanced against the cost of the measurement.

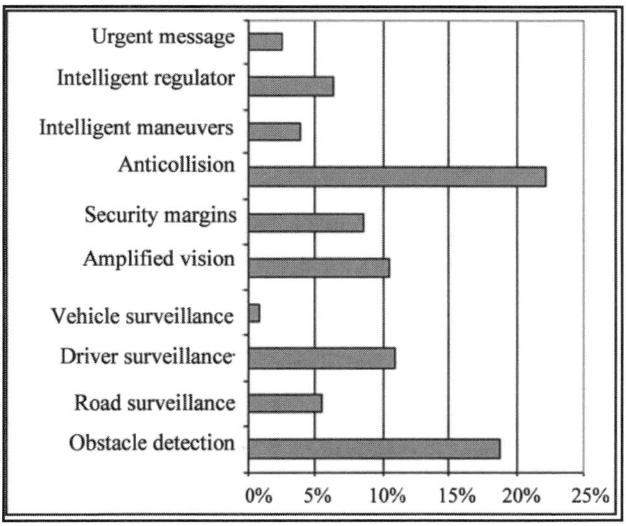

Figure 3.3. *A priori effectiveness of 10 functions for 3,179 cases of accidents involving 1,765 active users outside of built-up areas acting here as denominator*

The same reasoning has been carried out by the same author [FER 73] on the potential effects of ABS brakes. More recently, Malaterre and Fontaine and Van Elslande [MAL 91, ELS 92, MAL 99] estimated the potential effectiveness of assistance concepts for driving. It involves determining

10 ABS, Anti-blocking system: wheel anti-blocking system.

whether the needs identified in the previous section could have been satisfied by systems existing only on paper or as prototypes. These estimations are imprecise because of the very rough definition of the presumed function of these assistance tools, and also due to the weakness of the information that is sometimes available in police reports [IFS 11]. Nevertheless, this method, applied to a representative sample of accidents, facilitates the hierarchical ordering of stakes, which provides the advantages of one type of assistance compared to another.

It is the anti-collision function that seems to have the highest accident-reducing potential. Unfortunately, there is no device that fulfills this function to date. The focus is by default on systems that activate brakes when a collision is imminent and that set off the airbags and pretensioners in anticipation, so as to gain several fractions of a second before the crash. Next come the functions that help in the detection of obstacles, and those that monitor the state of the driver (such as lack of vigilance, and alcoholization), which can nevertheless give rise to changes in behavior in the drivers.

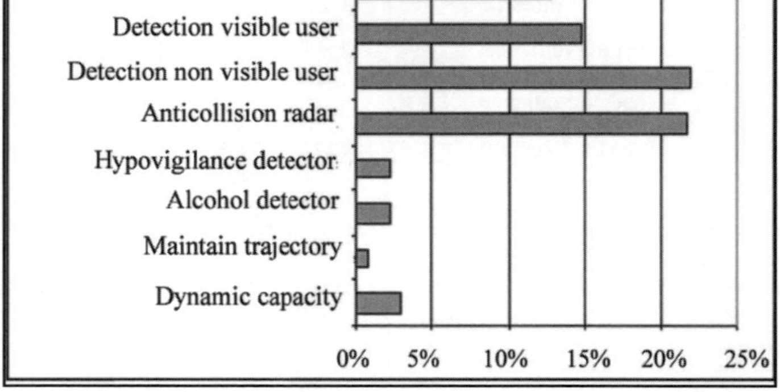

Figure 3.4. *A priori effectiveness of 12 functions for 329 rear-end collisions involving 1,002 users*

This is coherent with the previous study, with excellent potential effectiveness for the anti-collision radar, but we can also see that all the devices that would allow the anticipation of an incident that is not directly visible (corresponding to the need: detection of a problem upstream) could be very effective. In the next section, we present the situations and conditions in which these functions could be effective.

3.5. What assistance for which needs?

The global analysis of rear-end collisions outside of built-up areas shows that in 4 cases out of 10, there is at least one type of assistance that would have allowed an active user involved to avoid damage. This effectiveness is of 8 cases out of 10 if the effects of active assistance tools are cumulated (equipping one's own vehicle) with those in others' vehicles (usually the colliding vehicle). This is still without taking into consideration any possible perverse effects. On the highway, the anti-collision radar has the highest effectiveness score, followed closely by the presence detector of a non-visible user. On a national road (NR) or a departmental road (DR), it is the presence detector of a non-visible user that has the greatest potential for effectiveness.

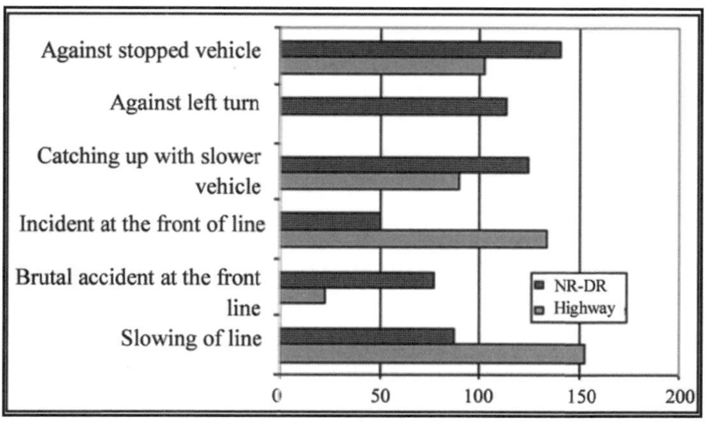

Figure 3.5. *Distribution of rear-end collisions into six categories, expressed as the number of drivers involved*

The global evaluations are very dependent on the hypotheses made. It is important to analyze the adequations between types of assistance tools and types of accident. In the case of rear-end collisions outside of built-up areas, a method of classification allows the grouping of accidents into the six most homogenous classes possible (Figure 3.5). Here, we present the main characteristics of these six categories of accidents, as well as the assistance systems whose effectiveness and relevance are significantly different from the entirety of the analyzed population (statistical over-representation). Stress is placed on researching common points in the accidents of the same class of the typologies, at the level of occurrences, circumstances involved types and mechanisms.

The first three categories concern accidents between vehicles colliding with each other, but that were not following each other, whereas the three following ones concern driving in a line. We will now examine each of these categories as well as the assistance tools that are best adapted to each of them.

3.5.1. *Collision with a stationary vehicle*

This involves a user hitting a vehicle that has stopped for several seconds or minutes, and which he was not following. The immobilization of the struck vehicle can be due to a queue, a primary accident, a breakdown or parking.

When the immobilization of the struck vehicle is due to a queue, the most common reason is that the vehicles hit had been stationary for a while, usually in all lanes, making avoidance impossible (the only avoidance option is to aim to pass between two lanes). In any case, the difficulty in predicting this type of situation is important. Other factors come into play, such as distraction or lack of attention in driver, and/or reduced visibility either inherent to the configuration of the location or due to weather conditions. Approach speeds are high, especially on a highway. It can also be remarked on such a network that these accidents mainly take place at night, in conditions of dense traffic – half of them involving a heavy goods vehicle (HGV).

When the immobilization of the struck vehicle is due to a primary accident, on whichever network, the recurring situation is the following: the visibility (night or fog) or road tire grip is bad. Following the first accident, some vehicles can perhaps stop in time, but others coming after hit them, either due to bad tire grip, late detection or even reduced visibility. This phenomenon can take place several times on all the lanes. As a result, the time between successive crashes can be considerable.

Finally, when the immobilization of the struck vehicle is due to a breakdown or to parking, the main difficulty is linked to the lack of visibility which delays the detection of stationary vehicles, most often on road, or partially on it. The vehicles are often not illuminated (at night) and not signaling (no hazard warning lights). Visibility problems proper to vehicles can combine with visibility problems linked to the location (corners).

The system of accident detection assistance presents the most specific effectiveness for this type of collision: one driver out of two could have benefited from this assistance. Other systems also intervene, but in a less specific manner than this category. These are the anti-collision radar, the detection of a visible or non-visible user, and the detections of incidents, notably on NR for active drivers.

3.5.2. *The struck vehicle is waiting to turn on an NR or a DR*

The main difficulty is linked to the detection of turning vehicles either because they are hidden by other vehicles, or because the change in direction is sudden or its late detection is linked to other factors (distraction, lack of obvious signs).

Most of these turning movements do not take place at junctions but rather toward other types of locations such a gas stations, service stations, property entrances, and rural paths. This can accentuate the unpredictable nature of the maneuver. In most cases, this turning movement takes place toward the left. No system appears to present a level of effectiveness significantly superior to the rear-end collisions on NR and DR in this type of accident. The detection of users hidden by another vehicle is effective for almost six out of ten drivers.

3.5.3. *Catching up with a slower vehicle*

In this group, the striking vehicles move at a speed that has been stable for some time. This is also the case for most of the struck vehicles, which are in no case carrying out a sudden maneuver. They are caught up with, not followed. Detection of the slower vehicle by the striking vehicle is non-existent or late. Sometimes the struck vehicle is well seen but the difference in speeds is badly estimated. For each type belonging in this category, the similarities are strongly linked to the types of vehicles caught up (visibility, vulnerability, aggressiveness, design speeds). Four types of accident can be identified:

– *Slower user hit by a user falling asleep.* This type of accident takes place mainly at night on the right/slow lane. Most of the time, the drivers of light vehicles (LVs) fall asleep and hit heavy vehicles.

– *Slower user hit by alcoholized user.* This type of accident takes place mainly at night and essentially involves LVs between themselves. The speed difference between struck and striking is important and especially on the highway where crashes rarely take place in the slow lane. On NRs and DRs, they can involve light two-wheeled vehicles. Little anticipatory and recuperative behaviors are observed because the time available is negligible and the drivers' reaction time is lengthened by alcohol. Over-accidents are rare but resemble each other in that, when they do take place, they affect all lanes of circulation.

– *Hitting a slower and poorly visible user.* Visibility problems dominate here in that these accidents take place mainly at night, with the struck vehicle often not being illuminated. On NR and DR, these are often light two-wheeled vehicles or agricultural machines that are either not detected or detected late. In any case, the speed difference between the struck and the striking is great.

– *Slower user hit by a distracted user, and other cases*. Here again the importance of the speed difference between the struck and the striking must be stressed. The striking vehicle's driver is often surprised by the slow speed of the vehicles ahead. These situations are sometime aggravated by visibility problems (around corners) or tire grip problems. The slow speed of the struck vehicle can be due to the type of vehicle or to a maneuver, when the vehicle is in a phase of re-acceleration or of slowing down, following or preceding a change in direction at a junction. The late detection of the slower vehicle can also come from a problem of attention of the hitting driver who focuses either on an internal element of the vehicle (radio, mirrors) or on an external element (vehicle going the wrong way, elements of context). Several accidents take place during badly calculated maneuvers in terms of either time or distance. This problem is often linked to the large speed difference or even, on the highway, to pressure from behind, forcing the driver to pull over behind a much slower vehicle.

The potential effectiveness of two assistance systems is over-represented in this type of accident: assistance tool detecting driver drowsiness or blocking the vehicle in case of a high alcohol level. The effectiveness of the anti-collision radar must also be noted, although it does not distinguish itself from the average of rear-end collisions; as well as for highways, the detection of visible users or those hidden by other vehicles.

3.5.4. *Dense lines: major incident at the front*

Two sub-categories can be distinguished here: the loss of control of a vehicle ahead and a sudden cutting off just in front (changing lane, cutting in front).

The problem of the loss of control of a vehicle in front is linked to the unpredictability of this event and its consequences, often causing the immobilization of the vehicle on the fast lane of the highway, or across the lanes on the NR or the DR. These situations are made worse on the highway by high speeds and on NR and DR by the frontal collisions that can take place there. Avoidance maneuvers are bound to fail due to the lack of space between vehicles, whether over one lane or two. Indeed, the drivers do not integrate into their driving strategy the possibility of such brutal incidents. The origin of the loss of control can be technical (burst tire), linked to poor road tire grip or even a problem of the configuration of the road.

Insertions ahead essentially highlight the lack of consideration of existing information before carrying out a maneuver. The cases on the highway are more numerous than on NR or on DR. This can be explained by the type of network that may be more likely to provoke this type of conflict (circulation over several lanes, access lanes). Sometimes these are forceful (voluntary). On the highway, (heavy) two-wheelers are often involved in the fast or middle lane following the movement of an LV. On NR or DR, it is often badly anticipated maneuvers, or a bad control of the blind spots.

On the highway, the effectiveness of the lane change manager or the overtaking manager is over-represented, notably for active drivers. Other assistance systems would generally be ineffective taking into account an insufficient time to react: assistance in detecting an accident, assistance in detecting inanimate objects and assistance in detecting an incident in the line.

3.5.5. *Dense line: violent accident happening just in front*

This type of rear-end collision is caused by a very violent shock (most often a frontal collision) and is obviously less frequent on the highway than on NR or DR. Given the violence of the original shock, this type of accident leaves practically no possibility of avoidance to the vehicles immediately following the first ones involved.

The effectiveness of three systems of assistance is over-represented on the highway: accident detection assistance, notably for active users; lane changing management or overtaking management, notably for passive users; and incident detection assistance. Let us also note the effectiveness of non-visible user detection system or of anti-collision radar, which still is not distinct from the effectiveness for all the rear-end collisions on the highway.

On NR or DR, the system that comes into play the most is the automatic control of the trajectory. The effectiveness of this device essentially concerns the vehicle at the origin of the frontal collision. The detection of a hidden user or of a visible user is most often not relevant. The ineffectiveness of some systems, such as obstacle detection, accident or incident detection, inter-distance regulation assistance, is essentially due to an available time that is too short.

3.5.6. *Dense line: sudden slowing*

This can be a traffic jam forming, or an accordion effect running through very dense traffic.

Accidents linked to the formation of traffic jams obviously take place in very dense traffic. The event that causes them would seem to be an effective way of differentiating them. Thus, the most common causes are deceleration, traffic jams and a primary accident on the highway. On NR and DR, changes in direction are also included. The most frequent aggravating element is bad tire grip on the road caused by wet weather conditions. Moreover, visibility is often reduced by the presence of fog or rain. All this quickly increases the number of those involved. Conversely, in dry weather conditions during the day, accidents caused by deceleration generally involve fewer users. This is less true for accidents caused by a traffic jam, which take place in even denser traffic conditions. Often, the initial shock takes place right at the end of the line, causing multiple project phenomena. Considering the frailness of the situation, it is sufficient for one user to have a slow reaction time to involve an entire line.

Whether it is on the highway or on NR or DR, it can be noted that there are few collisions from the category of accidents linked to the accordion effect circulation in very dense traffic. When the traffic is close to saturation, physical accidents become more rare. These conditions of circulation generate instability. The intervals are small and any delay in braking from one user causes the other users to brake more energetically.

On the highway, two devices appear to be more frequently effective in this type of accident, as much for active users as for passive ones: incident detection system and inter-distance regulation assistance. Two other systems are also effective, without distinguishing them from all the collisions on the highway: the detection of users by another vehicle and the anti-collision radar.

On NR or on DR, there is always some sort of effective assistance tool. This is most often the detection of accidents, as well as the detection of incidents, the anti-collision radar and the detection of visible or non-visible users, notably for passive drivers. The interval regulator also intervenes in this type of accident without being distinct from the entirety of the rear-end collisions on NR or DR.

3.6. Case of cooperative systems

The effectiveness of cooperative systems, meaning systems involving the transmission of information between several equipped vehicles, is a lot more complicated to estimate. If we hypothesize that all automobiles are equipped, the method is the same as the one exposed previously. However, it is not possible to make estimates of the effects resulting from partial equipment by simple cross-multiplication. Indeed, for each rate of equipment, we need to take into account the probability of having both an emitting vehicle and a receiving vehicle among those involved in the accident, or among the vehicles that are close if the device is, for example, a system of alarm transmission that is likely to have an effect on the chain reaction. Several starting hypotheses must therefore be considered: a rate of equipment on new vehicles; the number of years after which the estimation is to be made; the number of vehicles that are close but not equipped that could benefit by contagion. We can then come up with a probability p_1 that an active vehicle is equipped with the device and therefore likely to receive the alarm, and then the probability p_2 that a vehicle in front of it is also equipped and therefore susceptible to emit the message. For example, if we consider all new French vehicles to be equipped with the device, it will follow, considering the nature of the market, that after 10 years, 68% of French vehicles will be equipped. For an accident involving n vehicles, the number of combinations of numbers and positions in the line of equipped vehicles is given by the general formula:

$$C_n^q = \frac{A_n^q}{q!} = \frac{n!}{q!(n-q)!}$$

The formula that gives the probability of each combination is:

$$p_n^q = r^q.(1-r)^{n-q}$$

where n is the total number of vehicles, q is the number of vehicles equipped and r is the rate of equipment. Probability p_1 is equal to the rate of equipment. p_2, the probability of having an emitter *in front* of the receiver is a function of the number of vehicles involved. The probability of having both an emitter and a receiver equipped is equal to the product of p_1 and p_2. In this way, it is possible to calculate that a cooperative system of alarm transmission could avoid roughly 3% of the deaths annually, using the hypotheses of installation mentioned above, and on the condition that there is not adaption of behavior, often designated as a "perverse effect".

3.7. Using results in design

Here, we try to show which elements are likely to orient the actions to carry out in the domain of rear-end collisions discussed, and summarize the main problems identified.

The detection of a slower user is a difficult task, especially when the context leads to thinking that this user is simply following the average speed. This is the case on a fluid highway, particularly at night when the visual cues are less numerous and vigilance is lower, even more so if it is an LV for which a low speed is even less normal. Drivers with diminished capabilities (fatigue, alcohol) or beginners are particularly likely to encounter this kind of accident. The detection of a stopped vehicle is a similar problem. Let us note in these conditions that if the driver of the stopped vehicle has not kept his foot on the brake nor turned on his hazard warning lights, nothing is indicating that he is stopped, especially when visibility is poor (night, rain, fog). Over-accidents do not only concern highways. On NR and DR, the visibility distances are less, as well as the avoidance capabilities when there is only one lane per direction of circulation.

The problem of non-lit vehicles (breakdown, malfunction of the battery following the accident) has appeared several times, as well as the lateral visibility of a vehicle across the road or on the roof.

On NR and DR, the left turn is the most frequent maneuver at the origin of the rear-end collision, whether this takes place at a junction or toward a parking lot or toward a private road. This maneuver forces the total stopping of those that are following if there is only one lane per direction circulation. The problem is here of the late detection of the stopping of the vehicle, often aggravated by its unanticipated nature or by the lack of visibility caused by the profile.

On NR and DR, a lot of rear-end collisions follow a frontal collision that stops the involved vehicles on the spot. At present, preventing secondary rear-end collision immediately following a collision when circulation is dense is not feasible. Junction accidents can have similar consequences: a lateral collision immobilizes two vehicles, which are then struck secondarily.

On NR and DR, mopeds cause a particular problem of visibility at night and in bad weather conditions. This is due to their small surface, their slow speeds, and also lighting that is often insufficient (broken back light, or simply a weak bulb).

On the motorway, inter-vehicle times (IVTs) tend to decrease when traffic increases. In these conditions, users are often incapable of dealing with a major incident, especially on a wet road. The incident can be loss of control, sometimes following a wild lane change, requiring decelerations that are superior to the capacities of the vehicles that follow. No device seems truly capable of preventing this type of accident, as the times are too short.

Only a lengthening of the inter-distances, which limits capacity, seems to be able to improve the situation. However, a lot of over-accidents can be avoided. Following a disturbance (loss of control, primary accident), it often happens that a vehicle at the head of the line manages to avoid the obstacle by changing lane. The user of the vehicle behind then suddenly discovers the obstacle that he could not see before, and the surprise effect, coupled with the small distance has the result that he is no longer able to carry out a successful maneuver (something called dynamic masking).

From the inventory of these problems, we will attempt to formulate recommendations for action trying not to have any *a priori* in mind on possible technical solutions. These problems identified do not always cover all of the classes of accident considered.

3.7.1. *Detection of a slower user*

The problem comes from the association of a striking vehicle traveling at a stable speed, compatible with the infrastructure and the usual traffic conditions in the location and at that time, often not very vigilant and/or made fragile by various factors (inexperience, fatigue, alcohol) with a vehicle that is abnormally slow either due to its type or to a previous maneuver (tractor, two-wheeler, vehicle starting up again), and sometimes not very visible. It would therefore seem possible to act by reinforcing the visibility of certain categories of user (two-wheelers, tractors), or by detecting the following associations early enough in advance: circulation in the slow lane (when there are several lanes), speed difference greater than a threshold; speed of the caught up vehicle lower than a threshold (network function); speed of the faster vehicle above a threshold (network function); and stability of the speed of the faster vehicle. The rules of detections of situations that justify an action make up a research subject for themselves. Efforts must be made to have the least amount of untimely detections. The modes of action are to be studied (Alarm? Slowing of the vehicle? Different actions depending on the emergency detected?).

3.7.2. *Detection of several stopped vehicles blocking all the lanes*

The most critical problems come, as above, from the unpredictability of the event in function of the context of driving. The differences are that a complete stop is the only adapted maneuver; the detection of the disturbance seems a little easier considering the fact that there are crashes, emergency braking and complete obstruction of the traffic for a while. There are often aggravating factors (poor visibility, fog). The factors linked to the state of the driver are less frequent; they do not need to be diagnosed as finely as in the previous case. Seeing as the stop is the only maneuver possible, we can look at automatic stopping devices.

3.7.3. *Detection of a stopped vehicle completely or partially obstructing a road*

This variation is harder to deal with than the previous case. Here again the unpredictable character of the incident is determinant. The origin of the stopping is variable, we can think either of the automatic detection of incidents (ADI) by traffic surveillance or of a voluntary action from the stopped vehicle, which is a debatable solution as in a lot of cases the hazard warning lights are not used. Problems with the electrical supply of some vehicles need to be taken into account. An anomaly self-diagnosis could eventually activate an extra circuit and start processes of emergency signaling.

3.7.4. *Detection of a vehicle preparing to turn left*

The ambiguity of the turn signal is often cited, as it can mean overtaking or turning left. The problem is not limited to this as the turn signal is not always used, and when it is, it can be activated too late, or not be seen by other vehicles, in particular in the case of hiding from other vehicles. Once again, the unanticipated character is reinforced by elements of the context (left turn outside of a junction). The most accident-prone cocktail that needs to be detected seems to be this one: the deceleration or stopping of a vehicle outside of an intersection, on a road with only one lane per direction of circulation, with or without a turn signal; absence of consecutive slowing of the following traffic showing that the event has been taken into account; visibility problems linked to the profile of the road or the presence of groups of vehicles causing phenomena of hiding for those who are not at the head of the group. An automatic device seems hard to conceive, as avoidance is not necessarily stopping, but rather undertaking by the right, when space allows.

3.7.5. *Detection of light two-wheelers circulating on the right-hand side of the road*

These accidents often take place at night or in conditions of poor visibility. The problem has three distinct aspects: their small size makes them less visible; their small size makes it sometimes possible to pass them within their own lane, and as a result the following vehicles do not pick up on their presence; their lighting is often weak or inexistent (mountain bikes, etc.). This could lead to an improvement in lighting devices, or use of cooperative systems.

3.7.6. *Detection of a disturbance at the front of the line*

When traffic is dense, IVTs tend to decrease, eventually stabilizing around the minimum values that the drivers estimate corresponds to their reaction time. In this way, they are making the hypothesis that the deceleration of the vehicles in front of them will not be greater than their own deceleration (which can be wrong in the case of a crash) and that their reaction time will never be greater than this value, knowing that they rely on their capacity of anticipation to adopt low values. However, this anticipation can be scuppered by unexpected behavior from the driver in front, by a lack of visibility, or by inattention (looking at surrounding lanes, rear-view mirror). The driver in this situation can only attempt to make up for the lateness of his reaction by braking harder than the user in front, which makes it harder for the next person. These accidents take place in dense traffic, more often when the road is wet and in bad weather conditions. The use of greater IVTs would help avoid most of these accidents, but is this objective reasonable, as it is these very traffic conditions that cause drivers to choose these very small IVTs? We can think of three principal modes of action: get rid of the disturbances that destabilize the line (for example decreasing or regulating lane changes, which would remove some of the original incidents); give the drivers the capability of better dealing with incidents (by encouraging them to adopt longer IVTs, or by giving them better ways of anticipating the incidents, for example by warning them as soon as an anomaly takes place at the head of the group, beyond what they can see directly) or automate vehicle braking as a function of what the vehicle in front is doing. However, it must be considered that some of these modes of action could cause perverse effects, and lead to users adopting even shorter IVTs.

3.7.7. *Prevention of wild insertions*

Some rear-end collisions are the result of brutal incidents at the head of the line, and a line of action would involve treating the problem upstream by preventing these initial incidents. These are often insertions, lane changes or overtaking in too little space, or without taking into account the presence of other users in the lane. The victims are sometimes heavy two-wheelers, which are both less visible and going at different speeds from four-wheelers. These incidents result in emergency braking, a first collision, or even a loss of control leading to a poor attempt at avoidance. Three cases need to be distinguished, and all the three need to be dealt with differently: the original maneuver is due to the fact that the user changing lanes did not see the other (rear lateral vision problem of the blind spot); the original maneuver is due to the fact that the user changing lane wrongly evaluates speed differences; and the original maneuver is voluntary, whether a forced passing, or even aggressive behavior as "punishment" of a user deemed to be irritating or with whom there has been an earlier conflict. Cooperative systems can probably solve the two first cases. For the third, however, it is hard to see how it could be treated, if not through harsher punishment.

3.7.8. *Prevention of frontal collisions*

Drivers do not integrate the possibility of an instantaneous stop (which is very rare and unpredictable) when choosing a distance from the next vehicle. There is therefore no other solution here than the prevention of the original event, i.e. the frontal collision itself. The classical solutions are the improvement of markings, of tire grip, of signaling and, as a last resort, the installation of a central reservation.

3.8. Conclusion

In the mid-term, actual functional specifications of assistance tools adapted to the different classes of accidents would be needed, the few elements that we cover above being only a first step. This example clearly shows the complementarity of the detailed studies of accidents and of the studies from representative files. If the assistance tools whose effects are roughly evaluated can evolve in function of the technological progress, the needs evolve only slowly, in function of changes in traffic of infrastructure and of vehicles. These needs, even with the imprecision linked to the method of analysis, can serve as a basis for the design of assistance systems or systems of change to the driving environment in the sense of greater security, even if this means diverging from preconceived ideas or immediate wants of the user, who is often more interested in improvements in comfort or in ease of daily use than in improvements in security in the rare situations that they can remember.

This process is perfectible and requires several more trips between the files from detailed studies of accidents and representative police statement files. Even if the arrival of new technology can lead to completely unexpected solutions and create new needs, it is nevertheless necessary that the needs already identified through studies of accidents be correctly covered.

3.9. Bibliography

[ALL 01] ALLAIN E., BRENAC T., "Modèles linéaires généralisés appliqués à l'étude des nombres d'accidents sur des sites routiers", *Recherche Transport et Sécurité*, vol. 72, pp. 3–15, 2001.

[AMO 08] AMOROS E., MARTIN J.L., LAUMON B., "Estimation de la morbidité routière, France, 1996-2004", *Bulletin Epidémiologique Hebdomadaire*, vol. 19, pp. 157–160, 2008.

[ARO 96] ARON M., BIECHELER M.-B., PEYTAVIN J.-F., Interdistances, trafic et sécurité de la route, MAIA/DERA Report no. 9608, Convention DSCR/INRETS, no. 95-41019, December 1996.

[BRE 99] BRENAC T., FLEURY D., "Le concept de scénario type d'accident de la circulation et ses applications", *Recherche Transport et Sécurité*, vol. 63, pp. 63–76, 1999.

[BRE 03] BRENAC T., NATCHERGAELE C., REIGNER H., Scénarios-types d'accidents impliquant des piétons et éléments pour leur prévention, INRETS, Report no. 256, 2003.

[CLA 10] CLABAUX N., BRENAC T., "Scénarios-types d'accidents urbains n'impliquant pas de piétons et perspectives pour leur prévention", *Les Collections de l'INRETS*, vol. 274, 2010.

[ELS 92] ELSLANDE VAN P., NACHTERGAELE C., Aides à la conduite et fonctionnement du conducteur en situation. L'accident de la route, révélateur de limitations potentielles à la prise en compte des aides informatives, INRETS Report, no. 149, 1992.

[ELS 97] ELSLANDE VAN P., ALBERTON L., Scénarios-types de production de "l'erreur humaine" dans l'accident de la route. Problématique et analyse qualitative, INRETS, Report no. 218, June 1997.

[ELS 08] ELSLANDE VAN P., FOUQUET K., "Drivers'needs and safety systems", *European Conference on Cognitive Ergonomics,* Madeira, Portugal, September 2008.

[EVA 96] EVANS L., GERRISH P.H., "Antilock brakes and risk of front and rear impact in two-vehicles crashes", *Accident Analysis and Prevention*, vol. 28, no. 3, pp. 315–323, 1996.

[FAR 97] FARMER C.M., LUND A.K., TREMPEL R.E., *et al.*, "Fatal crashes of passenger vehicles before and after adding antilock braking systems", *Accident Analysis and Prevention*, vol. 29, no. 6, pp. 745–757, 1997.

[FER 73] FERRANDEZ F., Etude *a priori* des dispositifs anti-bloquants, ONSER Report, February 1973.

[FER 79] FERRANDEZ F., FLEURY D., MALATERRE G., "Dossier guide sur une procédure d'analyse des zones d'accumulation d'accidents en agglomération", *Cahier d'études ONSER,* vol. 48, October 1979.

[FER 85] FERRANDEZ F., FLEURY D., MALATERRE G., "L'étude détaillée d'accidents (EDA). Une nouvelle orientation de la recherche en sécurité routière", *Recherche Transport et Sécurité*, vol. 9–10, pp. 17–20, 1985.

[FER 99] FERRANDEZ F., "L'apport des études détaillées d'accidents aux retours d'expérience en sécurité routière", *Annales des Ponts et Chaussées*, vol. 91, pp. 36–42, 1999.

[FLE 85] FLEURY D., FONTAINE H., MALATERRE G., *et al.*, "La sécurité dans les petites agglomérations", *Recherche Transport et Sécurité*, vol. 5, pp. 11–16, 1985.

[FLE 90] FLEURY D., FLINE C., PEYTAVIN J.-F., Diagnostic de sécurité sur un département, application au cas de l'Eure-et-Loir, Rapport INRETS, no. 125, 1990.

[FLE 98] FLEURY D., *Sécurité et urbanisme. La prise en compte de la sécurité routière dans l'aménagement urbain,* Presses de l'Ecole nationale des ponts et chaussées, Paris, 1998.

[FON 01] FONTAINE H., GOURLET Y., Les bases de données issues des procès-verbaux d'accidents: évolution d'un outil de recherche en accidentologie, Outils et méthodes, INRETS Report, no. 9, 2001.

[FON 97] FONTAINE H., HUBERT J.P., "Evolution de l'usage de l'automobile et du risque d'accident", *Recherche Transports Sécurité*, vol. 57, pp. 7–20, 1997.

[IFS 11] "L'insécurité routière, facteurs et mesures: des enseignements pour la France", *Revue de littérature scientifique*, IFSTTAR, 2011.

[ISH 08] ISHAQUE M.M., NOLAND R.B., "Behavioural issues in pedestrian speed choice and street crossing behaviour: a review", *Transport Reviews*, vol. 28, no. 1, pp. 61–85, 2008.

[LEV 95] LEVINE N., KIM K.E., NITZ L.H., "Spatial analysis of Honolulu motor vehicle crashes: I. Spatial patterns", *Accident Analysis and Prevention*, vol. 27, no. 5, pp. 663–674, 1995.

[MAL 91] MALATERRE G., FONTAINE H., VAN ELSLANDE P., Analyse des besoins des conducteurs à partir de procès-verbaux d'accidents: évaluation *a priori* des fonctions Prometheus, INRETS Report, no. 139, 1991.

[MAL 99] MALATERRE G., FONTAINE H., MICHAUT H., *et al.*, Evaluation *a priori* à partir des PV d'accidents des actions dans le domaine de la route intelligente, Convention DSCR – INRETS, no. 9770013, study no. 4, Final report, 1999.

[MEG 99] MEGHERBI B., Scénarios-types d'accidents de la circulation sur autoroutes: élaboration, méthodes de reconnaissance et application pour le diagnostic et la prévention, Doctoral thesis, ENPC, 1999.

[MEI 08] MEILLAUD L., GALLENNE M.L., BLOSSEVILLE J.M., *et al., Quels avenirs pour les systèmes d'assistances technologiques à la conduite?*, La Documentation française, Paris, 2008.

[MIL 01] MILLOT M., BRENAC T., "Analyse qualitative et modélisation de l'influence des caractéristiques des voiries urbaines et de leur environnement sur les phénomènes d'accidents. Une première approche", *Bulletin des Laboratoires des Ponts et Chaussées*, vol. 232, pp. 5–18, 2001.

[MIL 11] MILLOT M., "Comment évaluer les impacts d'aménagement urbain sur la sécurité routière?", *XXIVe Congrès mondial de la route*, Mexico, 2011.

[OBS 11] OBSERVATOIRE NATIONAL INTERMINISTERIEL DE SECURITE ROUTIERE, *La sécurité routière en France – Bilan de l'année 2010*, La Documentation française, Paris, 2011.

[ONS 70] "Etude clinique d'accidents de la circulation routière", *Cahier d'études ONSER*, no. 25, April 1970.

[ONS 73] "Le contrôle technique des véhicules (Véhitest)", *Cahier d'études ONSER*, no. 31, November 1973.

[PIA 02] PIAT G., Stationnement et sécurité routière, Mémoire de travail de fin d'études de l'ENTPE, June 2002.

[UMR 96] "Registre médical des accidentés de la route, département du Rhône", *UMRETTE*, 1996.

[TIR 99] TIRA M., BRENAC T., MICHEL J.E., "Insécurité routière et aménagement de la ville. Etude et cartographie de scénarios d'accident", *TEC*, vol. 155, pp. 22–30, 1999.

PART 2

Evaluation Models of Human–Machine Systems

4

Models Based on the Analysis of Human Behavior: Example of the Detection of Hypo-vigilance in Automobile Driving

Humans only operate mentally thanks to models, which are only objective in researchers' ambitions, but never really are. Models will always need to be replaced by better models for the rest of eternity.

A. KAUFMANN, 1975

4.1. Introduction

Researching models of the human operator is a problem that the scientific community, in the domain of human–machine systems (HMS), has been focusing on for many years. This research is conducted to achieve two major categories of objectives. The first category mainly concerns human automation and cognitive psychology in that it aims to represent the structure of cognitive processes at a relatively high level of abstraction with the goal of better understanding the reasoning and processing mechanisms used by the human operator, for example the models by Baron (*Optimal Control Model*), Rasmussen, Hoc and others [MIL 03].

Chapter written by Jean-Christophe POPIEUL, Pierre LOSLEVER and Philippe SIMON.

The second category concerns automation specialists, or more generally the engineering sciences, where the objective is to have usable models with the goal of simulation or diagnosis.

The models of the first category have been mentioned in the previous chapters and in this chapter we discuss the models of the second category, more particularly with the goal of diagnosis of the operator's behaviors with the HMS.

4.2. The different models used in detection and diagnosis

The detection and the diagnosis of faults are two functions that are a part of the larger area of the supervision of systems. Their objective is to begin the activation of appropriate actions (reconfiguration, adjustments, maintenance, etc.) on a system in function of its state or mode of functioning [ISE 97].

The methods of fault detection can be structured into two main categories, in function of the models that they use: methods based on knowledge models, and methods based on behavior models.

4.2.1. *Methods based on knowledge models*

4.2.1.1. *Methods based on analytical models*

The existence of an analytical model that describes the system by a set of differential equations, difference equation, or any other format, provides the possibility of a comparison between the measurements carried out on the system and the information provided by the model. The analysis of a difference by the tools and theories of decision distinguishes normal hazards in the operation of the system of faults (Figure 4.1). Nevertheless, in any case, the quality of the model is of fundamental importance both for the detection and the isolation of faults, as well as for guaranteeing the absence of false alarms [FRA 00].

These methods can be placed in two families: analytical redundancy methods, which use state estimation techniques, and parameter estimation methods.

For the former group, the system is decomposed into sub-systems whose normal function is known and modeled. Variables measured for each real sub-system are constantly compared with those calculated by the model. A gap or residue indicates a failure in the sub-system [BER 96, ISE 97].

The methods based on the identification of parameters aim to estimate the coherence between the parameters of the model of correct functioning of the system and the parameters of the model of system identified in real time. They work in the structure space, as opposed to the previous ones that work in the state space, and are therefore more related to a fault linked to a structural change in the system (wear, breakage, etc.). In this way, they facilitate the diagnosis operation in the case where the parameters of the model present an immediate physical significance with regard to the process [BRU 90].

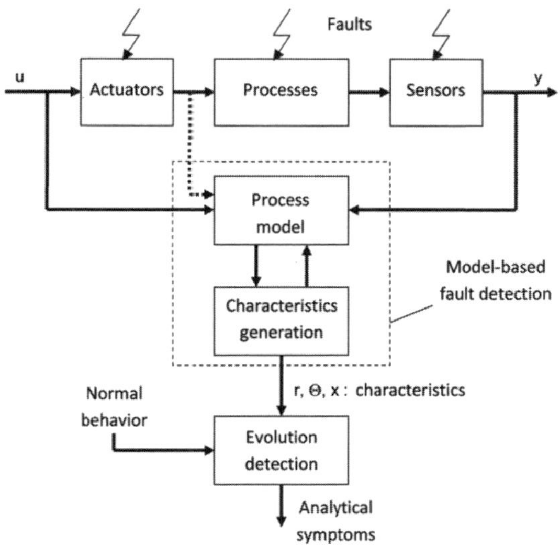

Figure 4.1. *Model-based fault detection (from [ISE 97])*

In both cases, a decision-making phase, which applies filtering techniques and a test of hypotheses, facilitates the detection and eventually the localization of a fault, based on gaps between parameters or residues.

Gaps or residues form a signature, and are an identifier of a fault. The process of diagnosis consists of comparing the signatures taken from the physical process with a library of theoretical signatures, representative of identified faults, so as to locate the origin of the fault to identify it precisely. It is therefore necessary to also have one or several default models.

4.2.1.2. *Symbolic model-based methods*

The application of diagnoses based on symbolic knowledge uses models that describe the relations of cause and effect within the studied system. Several approaches can be used in this goal. Two are presented, namely functional modeling of the system and representation in the form of production rules.

4.2.1.2.1. Functional modeling (functional analysis)

Methods of functional analysis were first developed to specify needs in the design phase of systems. They provide the representation of the system in the form of a hierarchy of functions, which interact between themselves by exchange of flows of "information". The successive progression of these functions results in the realization of the goal of the system [POP 94].

Generally, these methods are a considerable help in understanding and describing the functioning of the system [RIE 01]. For this reason, diagnostic approaches to industrial systems use it regularly. Among the methods used, we can cite:

– SADT [ROS 77, JAU 89];

– MFM [LIN 11a, LIN 11b];

– SA/RT [WAR 85, HAT 87];

– FAST [BYT 71, POP 94]; and

– GTST-MPLD [MOD 99, RIE 01].

These models are very useful in supporting a diagnostic approach carried out by a human operator; however, as they are, they are not very adapted to automatic diagnosis. The validation problems that they raise, of description completeness, of not taking explicitly into account temporal aspects, of subjectivity of the interpretation of diagrams, are at the origin of this.

4.2.1.2.2. Modeling by production rules

The representation of surface knowledge related to the bad functioning of the system in the form of production rules facilitates diagnostic operations to be carried out automatically, using inference reasoning.

The modeled knowledge is generally established empirically, by capitalization of experience on the functioning of the system or through a knowledge extraction process carried out by experts on the system [HOU 94]. The knowledge formalized in the form of production rules is originally described in the form of fault trees, fault graphs or FMCEA[1] tables. The translation of this knowledge into the form of a base of facts facilitates the use in a knowledge-based system through forward/backward/mixed chaining mechanisms.

A lot of work has been carried out in this domain, in medical diagnosis (MYCIN [SHO 74]) and in technical diagnosis (SACHEM [FRY 01]; SEDIAG [MAR 86]), initially with systems using binary logic, and recently with systems based on probability theories, or possibility theories, using Bayesian networks and conditional probabilities, or using the fuzzy sets theory [CAR 02, ARR 00, LEU 00].

However, these diagnosis systems generally remain in the state of laboratory prototypes, since even if we hypothesize that all the "good" fault indicators of the system are known and available, there remains the problem of representing all of the fault mechanisms and their causes. First of all, this assumes perfect knowledge of the system and a complete modeling of these fault mechanisms. If this is not the case, it must be accepted that the tool only knows how to deal with some of the faults in the system, and as a result, it presents an important risk of diagnosis error when faced with a fault that has not been modeled. If this risk is accepted, it must also be acknowledged that since the system is not able to learn on its own, its knowledge base must constantly be enriched by the capitalization of experience acquired throughout use.

4.2.2. *Classification methods: pattern recognition*

4.2.2.1. *Principles*

Rather than calling upon a reasoning based on the logic of predicates and sequentially exploring a multitude of hypotheses, to confirm or refute them and thus arrive at a valid diagnosis, it is possible to proceed, in the image of human reasoning in diagnosis, by making a parallel between the experience of faults already encountered, their effects and the element picked up in the

1 FMCEA: failures modes and critical effects analysis.

current situation [GRO 89]. For this strategy [REA 93], we can use pairing by similarity and selection by frequency.

The human expert puts forward a hypothesis of a certain pattern of fault by using his experience and a certain number of characters reckoned to be relevant that have been picked up on the system. In other words, he attempts to recognize the situation. The investigation is then pushed in this direction. The statistical techniques looking to imitate human function are techniques of "pattern recognition".

Pattern recognition can be seen as the "science of defining algorithms that allow the classification of objects whose aspect has varied from an object type" [DUB 90]. It is based on the search for similarities or dissimilarities between objects called patterns. Each pattern type, or prototype, defines a class. Any pattern belonging to a class has more similarities with the other patterns in the class than with patterns of other classes.

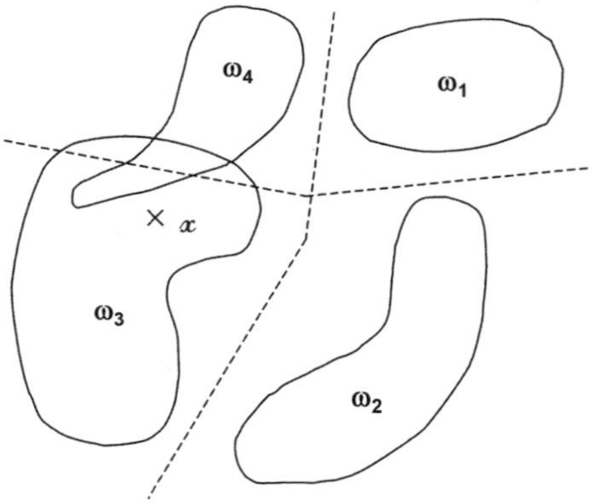

Figure 4.2. *Principle of statistical pattern recognition. With which class should observation x be associated among the four established (two of which overlap)?*

Considering a pattern x and M prototypes, a problem of pattern recognition can be written in the following manner: "associate x with a single class among the M possible classes". The problem being put in this way, its resolution passes by the definition of a decision space with a finite dimension $\Omega = \{\omega_i, i = 1, ..., M\}$, where ω_i designates class i.

Any process of pattern recognition therefore starts by a definition of patterns and of classes. In statistical pattern recognition, a pattern is represented by a pattern vector, which has d parameters, also called characters. Each parameter must have a signification for the characterization of the classes. The choice of parameters defines a space of dimension d called representation space, usually R^d, in which a pattern is represented by a point. A prototype is a pattern type, and is therefore also a point in this space. In order to represent the variability of the objects within a class, a geometrical zone of the representation space materializing the class is associated with each prototype. Any pattern x is therefore paired with the class associated with the geometrical zone in which it exists (Figure 4.2).

Therefore, the process of pattern recognition leads to the elaboration of a decision rule that establishes a relation between the representation space R^d and the decision space Ω. Its application requires:

– the choice of characters for the constitution of the vector type, i.e. the definition of a representation space; and

– the determination of boundaries between the classes, i.e. the elaboration of decision boundaries.

The definition of representation space conditions the separability of classes, i.e. the performance of the recognition process. The determination of decision boundaries creates a partition of R^d into M geometrical zones. These two steps are carried out from a set of initial data called a learning set, in which the M classes are represented. They involve three types of algorithmic methods [PEL]:

– methods of selection and extraction of characters: these intervene in the definition of the representation space;

– methods of coalescence, which deal with finding a structure in the learning set; and

– methods of classification, which result in the elaboration of a decision rule. They use the structure determined by coalescence to associate any new observation with one of the M classes. This phase is also called the discrimination phase.

A diagnosis problem can therefore be assimilated with a pattern recognition problem in which the classes represent functioning modes of the system. The pattern vector is composed of characters calculated from signals obtained by sensors. The detection of the functioning mode then consists of associating pattern x_t, evaluated at a moment t and representing the instantaneous state of the system, with a class ω_i. The definition of a pattern vector is done either when the number of observations is great, by researching correlation between the identified faults and the external signatures – this information coming from experience – or by extraction of the knowledge accumulated by a human expert who establishes correlation between faults and signatures. This step is crucial and largely conditions the effectiveness of the algorithm of pattern recognition [ZWI 95]. The problem then comes down to a problem of classification and of decision (Figure 4.3).

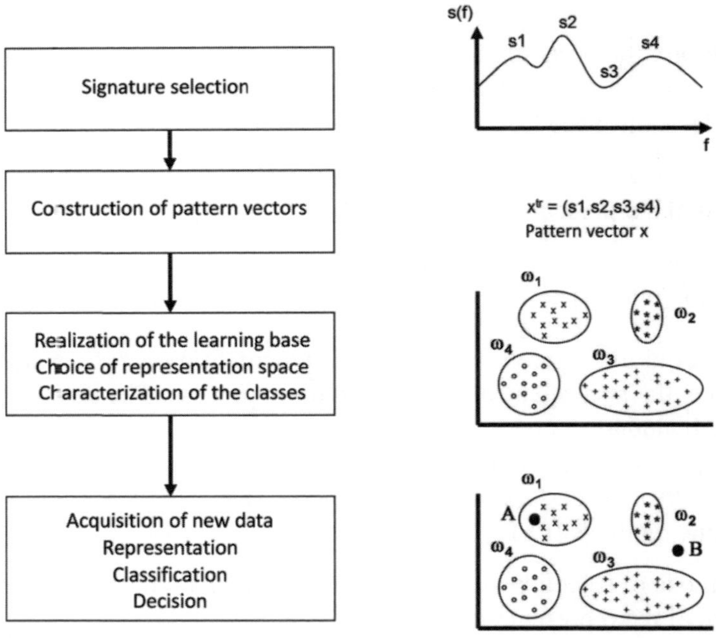

Figure 4.3. *Steps of diagnosis by pattern recognition [ZWI 95]*

It is rare to have available enough human expertise to determine the variables whose measurement relates to the fault or faults sought after and the thresholds to place on these variables to reach a conclusion. This is even more the case in the domain of HMSs where the human component still remains mostly misunderstood. For these reasons, the approach used is usually exploratory. Observation in real functioning conditions provides, using statistical methods, a definition of the relevant pattern. A step of classification then leads to the definition of a method of attributing a new observation to one of the classes identified during the first step.

4.2.2.2. *Methods of supervised classification*

4.2.2.2.1. Factor methods

The most used factor analyses in diagnosis are principal component analysis (PCA) and multiple correspondence analysis (MCA) [ESC 98].

PCA is applied to two-dimensional tables crossing individuals and quantitative variables. PCA is a descriptive method with a double objective [SAP 96]:

– on the one hand, establishing a list of resemblances between individuals (Who are those who resemble each other? Who are those who are different? Are there homogenous groups? etc.);

– on the other hand, establishing a list of links between variables (Which are those that correlate positively? Negatively? Does a typology exist? etc.).

To deal with the structure of the data in a large space, PCA looks for a projection of the cloud of points in a smaller space (generally, a plane). The adjustment criterion chosen is to maximize inertia in relation to the origin of the frame of the vector space of the projection (which corresponds to the center of gravity of the cloud). The metric used to do this is the Euclidian distance and the correlations calculated are linear.

MCA is a method that is also based on the geometrical representation that gives a global description of a set of observations taken from a system. It facilitates the study of a population of I individuals (the observations) described by J qualitative variables distributed into modalities. Each individual, characterized by N_j modalities of the variables, is associated with a point in space, RN_j. The whole of the observations corresponds to a cloud of points, noted N(I). In the same way, it is possible to define a cloud N(J), relative to the whole of the modalities of the variables, in the space RN_i defined by the N_i individuals. As in PCA, the objective of this method is to define a space of reduced dimensions, pointing out the relations between individuals, between the modalities of the variables and between individuals and modalities of the variables [ESC 98, SAY 91]. The first axis of the frame of projection is built in a way that maximizes the inertia of the projection of the cloud of points onto this axis. The second axis is orthogonal to the first and maximizes the inertia of the projection of the cloud of points onto the sub-space made by the two first axes (in this case, a plane). The process is repeated until the desired number of axes is obtained [ESC 98]. As a result, the projection of the cloud of points onto the first factorial plane (plane made by the two first axes) gives the broadest representation of the cloud of points. The metric used to calculate the inertias is the Khi2 distance.

Three types of relations are put forward by MCA:

– relations between individuals: two individuals are close if they are visually close on the factor plane (in the RN_j space). This visual proximity provides the definition of classes of individuals;

– relations between modalities of variables: just like for individuals, two modalities are close if they are visually close on the factor plane (this time in the RN_i space). In this case, the information provided by these modalities is redundant;

– relations between individuals and modalities of variables: MCA provides simultaneous representation of the projection of the cloud of individuals and projection of the cloud of modalities. If an individual is visually close to a modality, this means that the individual is characterized by this modality.

However, it cannot be forgotten that a factor plane provides only a reduced representation of a cloud of points located in a space with a big dimension (dimension N_i or N_j). For the updated relations to be valid, the cloud of points projected onto the factor plane needs to adequately represent the initial cloud of points. For this, Pau [PAU 97] proposed to fix a confidence threshold on the sum of the variances corresponding to the axes of the factorial plane below which a conclusion would be hazardous (for example 50% of the total variance).

The transformation matrix of the initial space toward the final sub-space is given by the eigenvectors [ESC 98, SAY 91]. The nth eigenvector elaborates the nth factorial axis from modalities of variables. This infers that the coefficients of the eigenvectors give information on the importance of the modalities of the variables with regard to the construction of the axes of the final sub-space. The modalities with a bigger coefficient contribute more to the elaboration of the considered axis. Most software of factorial analysis proposes a visualization of these contributions on the factorial plane. Thus, MCS gives the possibility of selecting the variables whose modalities contribute the most to the elaboration of the final sub-space.

For PCA as for MCA, the construction of classes is visual. It is done by grouping close points on the factor plane, taking into account the semantics of the data. It must be noted that to obtain a satisfactory result, several iterations are generally necessary, so as to discard the variables that are strongly correlated between themselves, or those that are only slightly involved in the construction of factor axes [POP 02].

The main advantage of MCA over PCA is that, on the one hand, it provides representation on the same plane of the individuals and the modalities (and therefore the variables), thus facilitating interpretation. On the other hand, following splitting into modalities, MCA can cause the appearance of nonlinear relations between the variables [ESC 98, LOS 03, SAY 91], giving it greater explanatory power.

4.2.2.2.2. Hierarchical classification

Purely graphical methods do not always show the splitting of the observation space into homogenous classes. The most commonly used technique in this case is hierarchical ascendant classification (HAC), or hierarchical clustering [ESC 98].

Hierarchical clustering is represented by a dendrogram or classification tree, obtained in the following manner: the two closest individuals are joined to form a summit, which is itself considered to be an individual in the following iteration of the algorithm [SAP 90].

The method of ascendant classification requires the definition of a measurement of distance between elements. Among the many measurements proposed in the literature, conducting a factorial analysis before an HAC results in the use of the Euclidian distance in the factorial plane considered [SAP 96].

Aggregation then requires the definition of a dissimilarity index between classes. Here again, several strategies exist: the minimal jump, diameter, etc. However, the Ward method for Euclidian distances, based on a criterion of minimization of the increase of intra-class inertia, is considered to be the best [SAP 90, ESC 98].

Cutting the hierarchical tree at a certain height therefore splits the observation space. It is done by taking into account the semantics of the variables and by ensuring that the split obtained presents both low intra-class variability and a high interclass variability.

4.2.2.3. *Methods of non-supervised classification*

According to this technique, the mechanism of classification establishes the different classes and simultaneously determines the relevant parameters. This technique is harder to put into practice than supervised learning in that neither the number of classes nor the characteristics of each of them are known in advance.

Generally, methods of classification based on this technique find the best possible groupings in terms of compactness [FAU 85]. They function in an iterative manner: going from the whole of the initial data, the different operations of processing lead to a classification. If this is not acceptable, the previous steps are reiterated in function of the new information made available by the previous iteration. There is no systematic convergence of the procedure.

In this category can be found algorithms such as:

– the K means [MAC 67];

– the ISODATA [BAL 67, FAU 85];

– the mobile centers [FOR 65].

Once one method or another has been able to obtain the most discriminating variables and patterns that need to be discerned, the problem of recognition of an observed pattern is a problem of classification.

4.2.2.3.1. Classification methods (non-parametric methods)

The classification used, which is because of the previous knowledge, is said to be guided by supervised learning. Following this technique, the *a priori* knowledge of the class of each measurement vector of the learning set facilitates construction of the classification mechanism that provides allocating a measurement vector whose class is unknown to one of the already defined classes [FAU 85].

The problem consists of establishing and optimizing the decision procedure, which is possible due to the learning set. It is based in general on the notion of proximity (characterized by a measurement of distance) between the candidate point and the perimeter of the representation space. This decision function can be linear or not: straight, hyper-planar, etc.

4.2.2.3.2. Visual exploratory methods

This approach, mainly used to help the work of a human operator making a diagnosis, generally relies on the use of multi-factorial methods such as PCA or MCA. Following the construction of factor planes from the reference observations (the learning set), the new observations to be classed are projection as illustrations into the plane, where their position in relation to the existing clouds enables them to be classed. It must be noted that the results obtained are strongly influenced by the user's subjectivity as soon as an ambiguity of classification appears.

4.2.2.3.3. Automatic methods

Here again, many methods exist. Among those best known, we can cite the following.

Multiple discriminant analysis: the goal of multiple discriminant analysis is to determine the discriminant linear functions that minimize the expectancy of the cost of classification of line 1 of an explicit table 1 of learning data (statistical correspondence, or random) among n number of E_i classes, described by vectors distributed in accordance with normal p-dimensional distributions [PAU 97]. The discriminant functions create a split of the p-dimensional observation space with hyper-planes. In practice, this linearity of the discriminant functions could prove to be a major inconvenience as the E_i state classes could not be linearly separable.

Ball discrimination: a given radius of r allows a ball B(x;r) to be traced in RN centered on observation x. The number of observations k_i of each of the classes C_j present in the ball are counted and lead to the attribution to x of the most represented class C_i in the B(x;r) [SAP 90]. The notion of rejection can be introduced by fixing a minimal threshold for k_i.

The k-nearest neighbors method: given a metric on RN, the idea is to attribute a class C_i, to which the majority of the k neighbors to x belong to a vector x. This takes place in two steps:

1) Choice of a minimal positive real number such that the ball B(x;r) of center x and of radius r contains exactly k samples.

2) Affectation to x of the class C(x) that is the most represented in B(x;r).

If we do not want to systematically operate the association with a class, a notion of rejection can be introduced. This takes the form of either an ambiguity rejection (a minimum of k' (< k) neighbors must be representants of x), or a distance rejection (the average distance to the nearest neighbors of x must be greater than a threshold of T).

Besides the delicate point of the determination of the list of the k-nearest neighbors of an observation in a space of dimension N, the choice of parameter k, has a critical incidence on the behavior of the classifier. A low value of k results in a classifier with a good resolution capable of defining complicated boundaries between classes, but very sensitive to noise on the samples and on the vector to be classed. A greater value of k would have an opposite behavior, smoothing the boundary but not being very sensitive to noise [MIL 93].

4.2.2.3.4. Connectionist methods

The shape of the networks used is often the same: multilayer perceptrons, which are discriminators that are able to approach the boundaries of nonlinear decision boundaries. The networks used are unidirectional and in general only contain one hidden layer for diagnosis [MAS 01]. Whatever the method used, the problem that remains is that of the evolution of the classification (creation of new classes).

The data of expertise or learning data used to split the workspace into classes are rarely representative of the whole of the patterns exhaustively characterizing all the functioning modes of the system. The tool used must therefore be capable of making the classification structuring the pattern space evolve over time so as to take into account the evolution of the observed system. Different works have proposed solutions unique to each of the methods exposed previously, taking into account the criteria of ambiguity or distance rejection during classification, rejections that lead to the creations of new classes when a new structured set appears in the shape space.

Despite these advanced techniques, the problem remains of the attribution of a semantic to these newly identified shapes. Does their origin lie in a new fault shape that is not identified in the learning base? If yes, which one? Or is this simply due to a change of the normal functioning point of the system because of the environment's evolution? On this point of semantic attribution, human intervention still remains indispensable.

4.3. The case of human–machine systems

The case of HMSs is a very particular case in the domain of fault detection and diagnosis. This particularity comes from the way in which the human's presence is considered in the system. It is generally justified by the fact that the task conferred to him is too costly, or impossible to automate, or otherwise for legal reasons, when the human is placed as the ultimate safeguard of the function of the system.

These characteristics lead to the observation that in the vast majority of cases, there is no real model of knowledge that can be used for the diagnosis of the task to be carried out. Indeed if the task is carried out by a human operator it is because it is not very (or badly) prescribed, or the model of the human operator himself as it is far too complex, at least globally (indeed, some models exist for certain simple situations). An additional factor that can further complicate the construction of models of knowledge comes from the existence of inter- and intra-individual differences that are very significant over a population of users.

As a result, approaches based on models of knowledge must be put aside, whether they are models of proper or wrong function. The only possible approach uses pattern recognition based on models of behavior. In a domain presenting some similarities in terms of looking for models (the pattern recognition done by humans), Milgram [MIL 93] compares cognitivist and behaviorist approaches. The first one looks for rules, no matter their complexity, that help in the determination of pattern recognition (a letter for example), whereas the second approach attempts to collect a maximum amount of prototypes and then try to reproduce the pattern recognition function, using a black box, and spread it to new examples. Cognitivists (AI researchers) look for rules, whereas behaviorists (statisticians) look for regularities.

Diagnosis in HMSs can only be based on this search for regularities, with the different regularities translating the different functioning modes of the system, normal or damaged. The problem that appears during the search for these regularities is that of the representation of pattern. If the pattern uses a "good" representation, it is not necessary to use an extremely high-performing learning technique, whereas in the opposite case, an inefficient representation requires the deployment of learning and discrimination efforts that are particularly big.

Therefore, the objective here is to work on the definition of the representation of pattern by using classical methods: multifactor exploratory methods. Preference is given to these methods, as opposed to connectionist methods, as they provide easy access to the variables discriminating the pattern and to their structure. They therefore make it easier to work on searching for a "good" representation.

The question that is then asked is: "how are the behaviors of a HMS characterized?" We propose the adoption of a classical approach in automatic control, consisting of observing the information exchanged between the components of the system. Each component is seen as a black box whose interactions with the other components of the system define its global function. In fact, we use observable input and output variables.

From a statistical point of view, this problem can be likened to a problem of characterization. The measurable signals within the system are numerous and diverse. Characterization has as a goal the definition: the signals to retain; and the indicators to consider for these signals, and these indicators are calculated for particular temporal windows.

It is obvious that the two sets of choices are not independent of each other. Characterization can then be envisaged from the engineer's point of view or from the statistician's point of view:

– engineer's point of view: by relying on the notion of state variable, "speed" and "position" signals can be considered. These signals can then be summarized in each temporal window by the temporal average or the effective value, for example. We can also search to characterize the amplitude or energy spectra;

– statistician's point of view: in considering a position signal, the arithmetic mean and the standard deviation must be used, for example, knowing that this second index can only be considered as an indicator of the gestural activity if the signal is relatively stationary in the temporal window.

We have carried out several comparisons of different methods of characterization. The reader can refer to [LOS 00, LOS 03] for details on these comparisons. The results that have been obtained on the considered examples show a definite interest in keeping a representation of the distribution of data in the characterization. If the mean and the standard deviation of the signals retained are satisfactory in the case of normal distributions, this is not the case for other forms of distribution.

At this point, it is important to add that the method of characterization must equally take into consideration another important aspect: the number of indicators used to characterize the data in a satisfactory manner. This manner, which can be called "level of data reduction", can be quantitatively translated by the ratio of:

$$r = \frac{\text{number of data output from the characterization}}{\text{number of data input into the characterization}}$$

In no way does this indicator translate the semantic contained in the data, but its minimization seems too often be used in practice. A use of the average is, for example, a lot more common than characterization by histogram or derivative, especially as the latter requires a calculation time that is clearly longer than the simple calculation for the average and/or the standard derivation. Nevertheless, it is important to remember that the histogram contains notions of central trends and of dispersion, the former not being translated by the average (for example, the histogram can make several modes where a strong asymmetry appears) and the latter not considered in relation to the average (as the standard derivation is). If it is undeniable that splitting into intervals constitutes a loss of information, it is possible to decrease this loss by using spatial windows like those shown in Figure 4.4 [LOS 03].

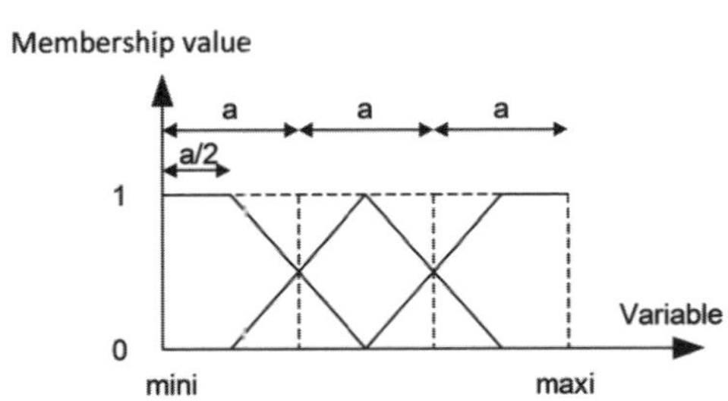

Figure 4.4. *Example of fuzzy spatial coding*

We have shown, in various application contexts, the advantage of using this technique, not only due to its performance in terms of characterization, but also in terms of results in pattern recognition.

4.4. Example of application: automobile driving

4.4.1. *Automobile driving*

Driving is the activity of the driver within the driver–vehicle–road environment system. This activity aims to deal with a main objective, which is to carry out a trip in a constantly evolving environment. This activity is orientated by sub-objectives that the driver gives himself (speed, security, ease, etc.) and is submitted to certain explicit and implicit rules (highway traffic act, common sense, etc.) [PER 97].

The activity of automobile driving is an activity that is considered as complex and not very structured. It places the driver in very diverse situations that demand continuous adaptation in an environment characterized by the management of multiple interactions under strong time constraints. Moreover, it is a unique activity in the context of HMS in the sense that the human operator involved is still a part of the control loop and is therefore not just a supervisor, acting only when function is impaired [RIE 93].

Automobile driving also represents a major stake in the eyes of society due to the high levels of mortality that it causes [TOD 99, POP 03]. Several studies have been conducted to better understand automobile driving with the goal of improving security. These studies have shown the limits of the effects of regulations and rules, and that it is necessary to develop onboard driving assistance systems, that are able to intervene at the right moment to parry inappropriate behavior or a form of failure on the driver's behalf [TOD 99, PAC 04] (see also Chapter 3 of this book).

This ability to intervene in an appropriate manner, at the right time, requires proper knowledge of the functioning of the driver–vehicle–environment system. Thus, a large part of the work has been dedicated to researching models of the driver's behavior so as to be able to describe and/or explain the processes that lead to the occurrence of road accidents, so as to propose ways to stop these. [BAD 91] and [TRI 05] conduct a detailed review of this. While these models do have different structures (models based on the management of physical values, risk management, cognitive function of the operator, etc.), a certain number of elements of consensus do result from them.

The first element is the hierarchical structure of the driving task. Generally taken from the Allen model [ALL 71], this hierarchical structure splits the task into three levels based on two criteria: the timescale necessary for carrying out sub-tasks and the level of consciousness required for carrying out sub-tasks by analogy with the cognitive model of the operator proposed by Rasmussen [RAS 83]:

– the strategic level, or navigation level, corresponding to the manner in which the driver plans his journey;

– the tactical level, or guiding level, limited to the management of the trajectory in the road environment (following vehicles, passing, crossing intersections, etc.);

– the operational level, or control level, consisting of regulating the lateral position and the speed of the vehicle.

The second element is the phenomenon of compensation. This phenomenon is characterized by the fact that the drivers change their behavior in the medium term so as to adapt it to their risk perception of a particular situation. Thus, the improvements brought to vehicles or infrastructure are partially counterbalanced by a change of behavior in the driver, which eliminates a lot of effectiveness.

The third element is the consideration of the driving task as a general process of information processing (perception–cognition–action). This approach highlights the essential role played by the perception of information in automobile driving [AUB 10].

Of these considerations, two forms of model appear: models of cognitive psychology and models of automation. The former are used to describe the global behaviors of drivers, by integrating complex mechanisms such as the ability to learn, to anticipate and to make errors that are inherent to human activity [POL 09, VAN 06]. The latter are mainly "mechanical" models that locally describe certain regulatory mechanisms such as longitudinal guiding or lateral guiding. They generally only cover the operational level and sometimes part of the tactical level in well-determined environmental conditions.

However, these different behavioral models are not adapted to the development of onboard help systems in the vehicle, either because they are too general to be able to be used to simulate the behavior of the driver (case of cognitive psychology models), or because they are too specific to certain phases of the driving task (case of the "engineer's models").

On the other hand, it is a well-adapted context for the application of a diagnosis process such as the one that we propose.

The different candidate variables that characterize the interactions between the components of the system and between the system and its environment are identified in Figure 4.5.

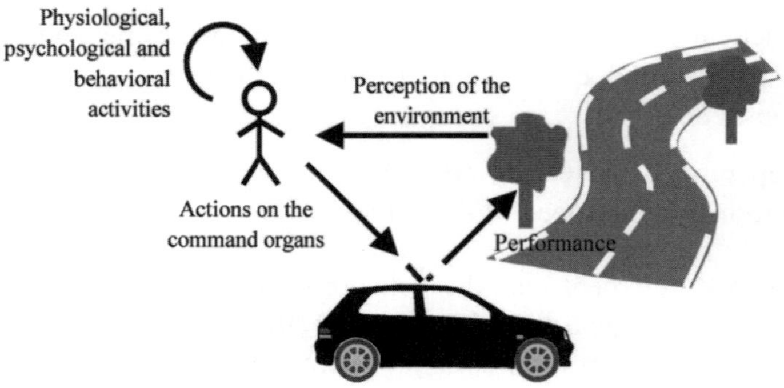

Figure 4.5. *Main categories of variables characterizing the interactions between different "components"*

The first category of variables characterizes the perception of the environment by the driver.

The second category represents the driver's physiological, psychological and behavioral activity. It is indicative of the level of awakeness of his nervous system and the level of attention and physical reactivity.

The third category mainly concerns the actions of the driver on the commands of the vehicle.

The fourth category concerns the indicators that are usually called indicators of the performance of the driver, and which apply to the kinematics of the vehicle in the road environment. It is actually the result of the actions of the driver on the commands of the vehicle within the road infrastructure.

In this vision of diagnosing the situations for which the driver requires assistance, the following section covers the problem of monotonous driving and phenomena of loss of vigilance and attention.

4.4.2. *Difficulties with diagnosing losses in vigilance*

Fatigue and losses in vigilance of the driver have always been considered to be important causes of accidents. A lot of research work has been carried out to evaluate the percentage of accidents that are due to problems of fatigue and reduced vigilance. The results from these studies vary greatly: 1.2–3.2% (the USA [KNI 95]), 2.5% (the UK [MAY 95]), 8.4% (France [THO 94]) when vigilance or fatigue are considered to be the main cause of the accident and 54% if they are considered to have contributed, among others (France [LEG 94]).

Various works have been carried out with the objective of improving public safety and reducing the cost for society of dealing with accidents. Two main routes have been explored. The first one aims to better know and understand the mechanisms of fatigue and vigilance in relation to numerous factors: the amplitude and organization of work periods, the time of day, the distance of the trip, its duration, etc., so as to propose recommendations aiming to reduce the risk taken by the driver [BRO 94, SAG 99]. The second route aims to develop systems of surveillance and/or alert of the driver so as to counter the effects of fatigue. These systems can be distributed between two categories. The first category is composed of systems designed to detect fatigue or reduced vigilance of the driver so as to trigger an alarm or even an intervention on the control of the vehicle. These systems are based on the real-time measurement either of representative indicators of the level of vigilance or of fatigue of the driver (for example the electroencephalogram (EEG) and gestural activity [ROG 01]), or of indicators of driving performance.

The second category is made of devices designed to sustain the driver's level of awakeness using reactivation mechanisms activated throughout the whole of the journey. This approach is based on the hypothesis that the execution of a secondary task requiring a certain amount of attention from the driver increases his awakeness and thus increases global performance. However, this complementary mental activity has to be induced at the right moments so as to not cancel, through distraction, the improvements gained by increased awakeness in the driver [VER 99].

These problems raised by the introduction of a secondary task have led the majority of works to concentrate on devices belonging to the first category. All the authors look to show the capacity of certain "behaviors" to provide a reliable diagnosis of high levels of fatigue or reduced vigilance in the driver. To this end, the "behaviors" of the driver have been characterized by various indicators. A first category concerns the psycho-physiological indicators like the EEG [ROG 01], variations in heart rate [MIL 97] or even variations in the frequency and duration of blinking [VER 00].

A second category concerns the variables related to driving performance (speed regulation, lateral position, heading management (angle of the steering wheel), time-to-collision, time-to-line-crossing, etc.) [VER 99, VER 00].

A third category concerns collateral activity (facial expression, yawning, posture adjustment, head movements, self-centered movements, etc.) [ROG 01].

However, all these indicators produced very variable results in function of the experimental conditions. This variability is caused by:

– the large inter- and intra-individual differences [VER 00]; and

– the influence of the environment on the task [MIL 77].

Considering the fact that in automobile driving, the majority of system measurements are sensitive to disturbances induced by the environment, the first approach to a solution consists of simultaneously using several different indicators whose changes caused by the environment are weakly correlated, while those induced by the temporal dimension of the drive, and therefore, fatigue, decreased vigilance and attention are more strongly correlated.

The following study details such an approach, looking to build a statistical model of the temporal evolution of indicators of the behavior of the driver which is able to provide a diagnosis of the reduction in vigilance during lengthy monotonous situations. Classical indicators from vigilance studies linked to driving performance have been used in conjunction with more original indicators, describing in real time the changes in posture of the driver, such as head movement. These indicators reflect the increase in postural adjustments described by many authors [GER 95, KIT 00].

4.4.3. *Approach applied*

To have enough observation to identify the variables representative of driving behavior and sensitive to the length of the drive, experiments must be conducted. To avoid any risk for the subjects to allow the easy gathering of large amounts of data, this experimental campaign was carried out in a driving simulator. A long highway trip in a simplified environment was used to favor the appearance of monotonous conditions.

During these experiments, several variables were recorded in real time and many indicators were calculated *a posteriori*, with the goal of identifying, using a multivariate exploratory method, those that changed significantly over time.

To do this, a certain number of statistical analyses were carried out and notably an exploration of the database was done in three steps using MCA.

This section presents these analyses in a synthetic manner, as well as the main results that they provided.

4.4.3.1. *Variables and indices retained*

This section succinctly presents the main indices that were retained in this study (see [TOD 99] for the details).

4.4.3.1.1. Variables and indices linked to perception of the environment

As vision is a sense that is particularly solicited in driving automobiles, measurements of eye movements have predictably imposed themselves as the main source of information on the perception of the environment by the driver. The conducting of this study therefore required the installing of an "eye tracker" as well as the development of tools to help in the extraction of data in a simulated environment.

The indices that were taken into consideration were oculomotor dynamics (frequency, length of the saccades, duration of the fixations), recording visual information (rate of consultation of a class of object, observation distances, etc.) and the physiology of the perception (blinking) at the same time.

4.4.3.1.2. Variables and indices linked to the driver's behavioral activity

The choice of variables relative to behavioral activity was limited to those whose measurement induces a minimal amount of disturbance on the driving task. Thus, certain relevant variables representing the decrease in vigilance in the driver were not examined, in particular physiological indices such as the EEG. Moreover, variables whose use in real driving conditions were systematically preferred.

According to these criteria, the driver's head movements and gestures were retained.

4.4.3.1.3. Variables and indices linked to the actions of the driver on the vehicle

Only actions on the steering wheel and the accelerator were retained as part of this study. Indeed, the driving context used required very little use of the brake and other commands (signals, etc.).

4.4.3.1.4. Variables and indices linked to the kinematics of the vehicle in the environment

Vehicle–environment interactions can be studied through driving performance, which is frequently used to detect deterioration in the driver's behavior. Performance has been characterized by the speed of the vehicle, the lateral position of the vehicle on the road and the "heading error" of the vehicle.

4.4.3.2. *Conducted experiments*

4.4.3.2.1. Experimental platform

The platform used is the SHERPA-Valenciennes driving simulator. It is a full-scale static simulator[2], built around an SGI Onyx "Infinite Reality 2" workstation and interfaced with a fully instrumented real vehicle "cockpit"

2 A dynamic version of simulator is described on the website: http://www.univ-valenciennes.fr/LAMIH/simulateur-de-conduite.

(Citroën Xantia): force feedback steering wheel, pedals, gearstick, dashboard, extra command, sounds, etc.

In terms of visuals, the scene of the road is projected on three flat screens (1,280 × 1,024 pixels) placed in a circle with a radius of 3.3 m and centered on the eye of the driver; the visual field provided is therefore of 180°. A single screen (640 × 480 pixels) behind the car is projected so as to simulate the rear-view (Figure 4.6).

Figure 4.6. *SHERPA-Valenciennes in static configuration (front field)*

The measurement of eye movements is done by an ASL 5000 HMO (Applied Science Laboratories) installed in the SHERPA cabin. This probe consists of a helmet for the measurement of the eye movements through the "cornea vector" technique and an electromagnetic ascension *Flock of Birds* system for the measurement of head movement. The recordings are provided in video format and as a digital data file.

To deal with the problem of prohibitive data extraction time, an automatic data extraction module (called SATMOS for "*système automatisé de traitement des mouvements oculaires en simulateur*" (automated system for the processing of eye movements in a simulator)) [SIM 99] was developed in the laboratory. Because of a particular design of the visual database and synchronization between the eye tracker and the generation of images by the simulator during the test phases, *a posteriori* playback facilitates automatic determination of the objects observed by the driver, as much inside the vehicle as in the scene observed outside.

4.4.3.2.2. Driver population

After two familiarization sessions, 41 out of 50 volunteer drivers were selected to participate in the experiment (13 students, 9 PhD students, 17 members of staff of the university and 2 external). The average age was 28 years (minimum: 21 years; maximum: 37 years). The population was male and characterized by regular highway driving, license held for over five years, non-corrected eyesight and a height of less than 185 cm (so as to facilitate the use of the eye tracker).

4.4.3.2.3. Running of the test

For the test itself, the subjects had to make a trip, exclusively on the highway, of 340 km in total. Considering the limitations linked to the use of the oculometer (optical system has a maximal use time of 45 minutes), the test was run as follows:

– 1st phase: 84 km with the oculometer;

– 2nd phase: 84 km without the oculometer (the helmet was still worn to conserve the head movements);

– 3rd phase: 84 km with the oculometer over a journey rigorously identical to the first phase; and

– 4th phase: 84 km without the oculometer (helmet worn for the measurement of the head movements).

4.4.3.2.4. Definition of the trip

The trip was defined so as to include a certain amount of constraint. Seeing as the analysis had to take into account the type of "section", it is interesting to use a limited number of types (straight line and left and right turns with only two bend radii), all of the same length. Thus, the highway journey specifically designed from the final experiment could be decomposed into a succession of 12 "standard modules" of 24 km, each module being itself composed of 12 sections of 2,000 m each.

The different types of section retained are:

– SL: straight line;

– LT5: left turn, bend radius of 5,000 m;

– RT5: right turn, bend radius of 5,000 m;

– radius of 2,000 m; and

– RT2: right turn, bend radius of 2,000 m.

The “standard module” is therefore made up of 12 sections of 2 km. It contains 6 straight sections, 4 “normal” turns and 2 “tight” turns. It is “arbitrarily” defined as follows:

RT5, SL, RT2, LT5, SL, SL, LT5, SL, RT5, SL, LT2, SL

Each of the four phases of the protocol contains three modules to which are added:

– at the start of the circuit, a gas station placed on a straight line of 2 km; it provides starting and an acceleration phase up to cruising speed; it is followed by a straight line of 5 km;

– at the end of the circuit, a straight line of 3 km (i.e. greater than the visibility distance of 2 km, as per the SHERPA simulator) followed by a gas station placed on a straight line of 2 km.

To increase realism, each phase starts and ends on a highway rest area. The total journey therefore contains four phases of 84 km, or a total of 336 km. The “decoration” (signs, houses, plants, gas stations, etc.) is distributed evenly over the four phases. However, to facilitate comparison in terms of eye activity, the first and the third trips are completely identical.

4.4.3.2.5. Used data

At the beginning of the journey, a period of training must be planned, during which the data are not analyzed.

The data gathered on the different equipment or *a posteriori* are (Figure 4.7):

– for the vehicle and the driver’s actions: 11 signals and their 11 derivatives in absolute value;

Meaning	Symbol	Unit (accuracy)
Speed (roughly **d**erived according to **X**) of the **V**ehicle	**dXV**[3]	m/s ()
Position along **Y** (lateral) of the **V**ehicle	**YV**	m ()
Linear position according to **X** of the **H**ead	**XH**	cm (0.3 mm)
Linear position according to **Y** of the **H**ead	**YH**	cm (0.3 mm)
Linear position according to **Z** of the **H**ead	**ZH**	cm (0.3 mm)
Angular position around X (**R**oll) of the **H**ead	**RH**	degree (0.1°)
Angular position around Y (**P**itch) of the **H**ead	**PH**	degree (0.1°)
Angular position around Z (**Y**aw) of the **H**ead	**YH**	degree (0.1°)
Relative **p**osition of the **A**ccelerator	**PA**	% (0.05%)
Steering Wheel **A**ngle	**WA**	degree (0.001°)
Steering Wheel **Torque**	**WC**	m.N (0.1 N.m)

– for visual perception: 27 variables (10 areas of interest are considered).

Meaning	**Symbol**	**Unit**
Length of **F**ixation	**LFY**	s
Membership value to the pointed **Z**one during **F**ixation	**ZFYi**	between 0 and 1
Sum of the angular gaps for **F**ixation	**SFY**	°
Duration of the **S**low movement	**DSY**	s
Membership value to the pointed **Z**one during **S**low movement	**ZSYi**	between 0 and 1
Amplitude of the **S**low movement (**s**um of the angular gaps)	**SSY**	°
Duration of the **S**accade	**DSY**	s
Sum of the angular gaps for the **S**accade	**SSY**	°
Duration of **B**linking	**DBY**	S

4.4.3.3. *Results obtained*

The scale model of the variables relative to the two sets is quantitative; nevertheless, the signals giving the ocular position make up a "semantic" linked to the image viewed, which makes them exploitable in rather the form of a qualitative scale. The combination of horizontal and vertical scales of

---

3 The letter "d" is used in an attempt to harmonize notation. The signal designated by dXV does not correspond to the tangential speed (derived from position X) but to an effective speed of the vehicle on the road. These two speeds being nearly identical, we shall use dXV.

the image is then replaced by a qualitative scale whose values (or modalities) are linked to the image. The presence of these scale models significantly complicates the statistical analysis.

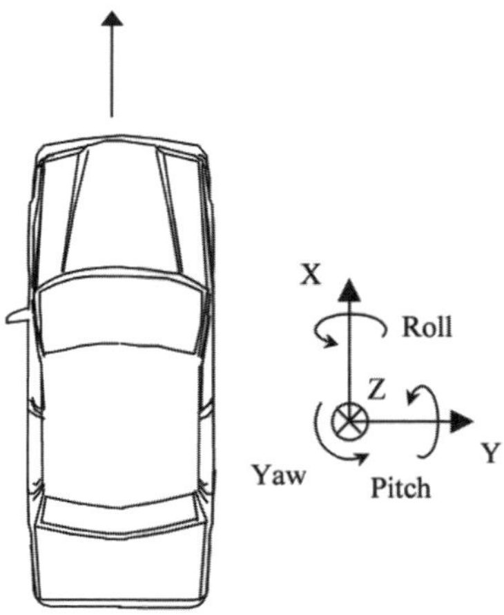

Figure 4.7. *Base variables concerning head movements and sign conventions*

As a result, the data are analyzed as two distinct sets:

– data relative to the kinematics of the vehicle and to the movements of the driver's head (quantitative scales);

– data relative to the ocular activity of the driver (qualitative scales).

4.4.3.3.1. Analysis of the data relative to the performance and the behavior of the driver

Characterization

The signals sampled at different frequencies were previously synchronized and filtered at 6 Hz (spectrum analysis shows that the energy of the signals is below 2 Hz [TOD 99]).

Spatial windows

The total number of variables analyzed being relatively high (about 40), only three spatial windows are retained; they correspond to the small, medium and large values. Let us remark that with classical windowing, i.e. with membership functions taking a value of 0 or 1, each window is characterized by the statistical units it contains; in the case of fuzzy windowing, this notion is replaced by that of average membership value (AMV).

The principle of spatial windowing relies on the shapes of the distributions of the filtered signals and on the method used for statistical analysis. The shapes belong essentially to two types: asymmetrical decreasing for signals corresponding to the absolute values of the derivatives and (near-)symmetrical of the Laplace–Gauss type for the others. Let us take note of the very rare presence of relatively high amplitudes. A windowing that only calls upon a spatial criterion is totally dependent on these extremes, which can generate problems as the latter can be abnormally high or low. Furthermore, with such a criterion, the spatial modalities have a different weight in the statistical analysis, the weight of the spatial window being the average of the membership values to this window. For these two reasons, a split giving near-identical weights to the three windows is preferred.

This choice being made, it is also important to take into account possible symmetry in the distribution. For example, if through construction the value zero corresponds to the absence of a phenomenon (like with a steering wheel angle equal to zero, which means that there is no rotation), it is important to center the intermediary modality on this value. In the other cases where symmetrical distribution is present, the intermediary modality is centered on the median.

Figure 4.8 shows the amplitude histograms and the respective windows for some variables, these being built from all the values considered in the analysis, in other words those relative to periods 1 and 3 (24564 values at 6 Hz).

In the presence of V = 22 variables and of E = 24564 temporal samples, the data from the spatial windowing are therefore placed in a table with E rows and 3*V columns containing membership values ranging from 0 to 1.

Temporal windows

The windowing can be solely based on a temporal criterion (for example with 10 minute windows) or call upon a spatial aspect (for example, with temporal windows whose boundaries are the changes of "section" or "turn"). From the perspective of a comparative analysis of the different drivers but also that of relating all of the variables – and therefore of the eye movements – the presence of a spatial aspect is vital. To be able to show possible resemblances for portions of the circuit that present identical geometrical characteristics, the temporal windows correspond to the sections rather than the turns. As a result, the membership values found are averaged within the same section, which is the same as characterizing each section with the 3-class histogram. In the presence of V temporal variables and TR sections, the previous table with E temporal samples and 3*V columns is therefore turned into a table with TR lines and 3*V columns.

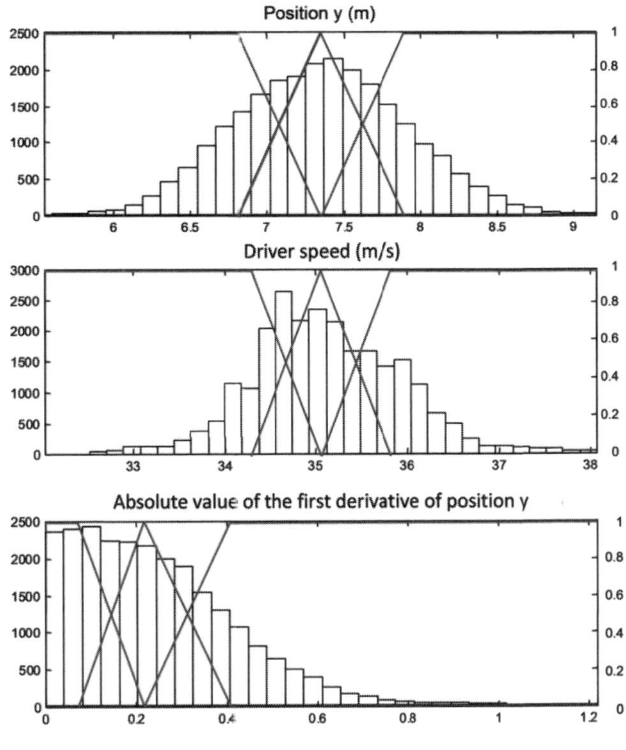

Figure 4.8. *Some examples of histograms of amplitude and the corresponding windows (the histogram is built from values relative to periods 1 and 3, or 24564 values at 6 Hz)*

MCA on all the temporal variables

Let us remind ourselves that the eye tracker was used during periods 1 and 3. As part of relating the data from this sensor to those relative to the vehicle and the head, only these periods 1 and 3 are considered. Each period contains three identical laps and each lap is itself composed of 12 sections, the total number of lines in the table is the nTR = 72. The number of columns corresponds to the number of spatial windows, which is $3 \times 22 = 66$.

The first factor plane shows the significance of the large and small modalities of variables relative to the position of the head along the Z axis and to the angular position of the head around the Y axis. Other modalities strongly contribute to the positioning of this axis; they are equally relative to the variables of the linear and angular positions of the head but present less obvious links. The observation of relative positions along axis 1 of points relative to the 72 sections shows that:

1) the influence of the time factor is clear but is more marked for the sequence of turns {1, 2, 3} than for sequence {7, 8, 9}. Taking into account the relative positions of the spatial windows that contribute the most to axis 1 and of the conventions taken for the variables, the head tends to drop and the head–chest angle opens with time; and

2) the influence of the section factor is not very clear.

Axis 2, which translates a two times smaller dispersion than the one translated by axis 1, is also essentially positioned by the variables relative to the positions of the head but also by the speed of the vehicle.

Because of the importance of the biomechanical aspect, two new analyses were conducted. For the first one, the variables relative to the linear and angular position of the head were removed from the analysis. There then remained V = 16 variables. This second MCA shows the influence of time on axis 1, which is essentially controlled by the speed of the vehicle and the position of the accelerator (increase in both cases). However, this influence is significantly less obvious and is a lot more disturbed by the influence of the section factor. The second analysis is focused on all of the variables that reflect the gestures of the driver.

MCA on only the head's mobility variables

The variables considered translate the mobility of the head (six variables: dXH, dYH, dZH, dRH, dPH and dYH), the activity at the level of the accelerator (dPA) and around the steering wheel (dWA and dWC). We must remember that these variables are obtained by considering the absolute value of the derivative of the base signals. The table considered crosses the 72 sections and the 3 × 10 = 30 spatial modalities. The percentage of inertia expressed by the first axis is clearly greater than the following values: 59% compared to 14%, 10% and 5% for axes 2–4.

Axis 1, which therefore represents on its own more than half of the dispersion of the initial cloud, is mainly positioned by the "large and small" modality of the variables relative to the gestural activity of the head (Figure 4.9(a)). This figure and the figure showing the points representing the sections (Figure 4.9(b)) show that with every turn, gestural activity in the head increases.

Axis 2 is practically controlled by the variable that represents the movement of the accelerator pedal, the small modality contributing four times less than the large modality. This axis translates the evolution during the second and third lap. Notably, the points that represent sections 7–12 present coordinates along axis 2 that are bigger than the previous turns. This figure shows that with every turn the dispersion of gestural activity changes and increases.

We can retain from these different analyses the importance of the variables that describe the position and the movement of the head. Notably, it would seem that with time, the driver tends to slump in the chair. Also, he seems to move more and more over time.

4.4.3.3.2. Analysis of the data linked to visual information

Characterization

The problems raised by characterization are very different from the previous ones in that the oculomotor system presents very specific movements. The data provided by the eye tracker is transformed after filtering into a succession of four base "entities": fixations (pause of the gaze on a static object), slow movements (following a moving object with the gaze), saccades (rapid jump from one fixation point to another) and blinking. Possible dysfunction of the sensor requires taking a fifth entity into account: other. Next is the problem of the characterization of each of these five entities in terms of indicators of temporal (in the sense of duration and chronology) and spatial aspects.

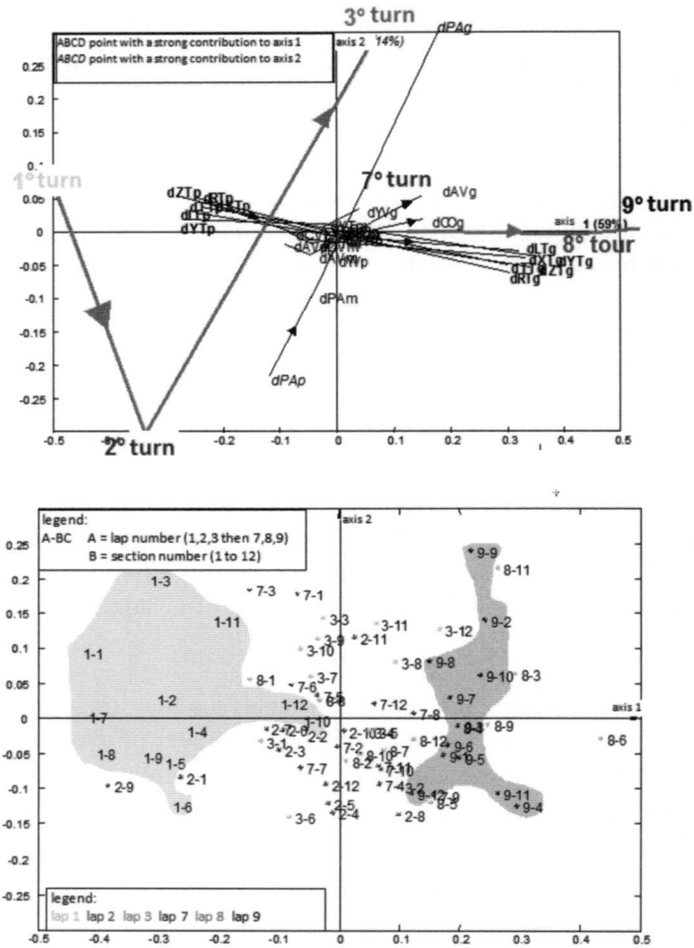

Figure 4.9. *First plane from the MCA of variables relative to head mobility: a) projection of the points corresponding to the 30 spatial windows (only variables with the modalities with strongest contributions are shown); b) projection of the points corresponding to the 76 sections*

Temporal aspects

Let us remember that in relating the ocular data with the previous data, the base temporal unit is a temporal window containing a section, or 72 windows in total (3 laps × 12 sections × 2 phases). The chronological aspect is therefore taken into account in the study of the evolution throughout the sections, themselves being grouped into laps.

When considering the aspect of duration, we can retain the duration itself (duration of fixations, slow movements, saccades and blinking) or introduce a notion of frequency or period. It is obvious that these three notions are not independent of each other and that they can be considered as absolute or as relative. Let us consider for example a fixation as an entity and the frequency as the indicator that we wish to retain. We can then calculate the number of fixations per temporal unit (the minute, for example) or relativize in a given temporal unit the number of fixations in relation to the total number of entities; in other words the rate of fixations, slow movements, saccades and blinks.

We can then note that some elements are not totally independent; for example, the curves of the evolution of the fixation elements are very close to those of saccades. Moreover, some elements evolve in a non-random manner with time, for example the number of blinks tends to increase with every lap.

More generally, let us assume that:

– fixations represent 41% of clearly identified entities;

– saccades 43%;

– slow movements 7%; and

– blinks 6%.

Having made these remarks, conscious of diversity in the manner of characterizing, it is preferable to retain the duration itself (of fixations, slow movements, saccades and blinks) as it seems better suited to show physiological aspects, a variable such as the number of occurrences being more statistical in nature. Just like with the previous data, the duration is split into three fuzzy classes so that the membership values are near identical. Let us note that the distributions are very far from Laplace–Gauss curves, which shows yet again the inaptitude of the average in characterizing a set of values of durations.

Spatial aspects

Concerning fixation and slow movement, characterization must above all end on the indication of the observed object. A first classification consists of distinguishing:

– little objects: signposts, emergency telephones, construction, tree, etc.;

– big objects, which make up the "back" of the image: the road, the scenery, the sky, the panorama (hills in the background), etc.

A deeper analysis of the objects observed and of ocular behavior shows that such a binary classification is not enough. Firstly, it is excessive to claim that the object that is "crossed by the gaze" is indeed the one observed, considering how the visual system works (cooperation between central and peripheral vision) and the uncertainness of the measurement (roughly). Secondly, there appears a restrained zone, looked at a lot but does not correspond to a defined object: this is the convergence zone of the two sides of the road ("expansion point" in the case of a straight light, the apex in the case of a turn).

These two remarks lead to adding a new zone to the previous classification and to retaining blurring limits between certain zones [LOS 07, LOS 09]. The principle of fuzzy spatial windowing is illustrated in Figure 4.10 for the case of the convergence zone. Intervals of hesitation ix and iy are chosen in function of the incertitude of the measurement (1 degree at an observation distance corresponds to an error of 6 cm) and the dimensions of the zone where the membership value 1 takes into account fluctuations in ocular positioning (lx = 48 cm and ly = 24 cm).

Let us note, in the perspective of remaining in a statistical context, that the sum of membership values must be equal to 1. For example, for a fixation f_i with coordinates $(x(f_i), y(f_i))$ and membership functions on the horizontal and vertical axes of X and Y respectively, the condition is obtained in the following manner:

$$\sum_{x=1}^{X} \mu_x(X(f_i)) = 1 \text{ and } \sum_{y=1}^{Y} \mu_y(Y(f_i)) = 1 \quad [4.1]$$

Both forms of [4.1] are of X + Y mono-dimensional membership functions:

$$\mu_{z=x \cap y}(\mathrm{x}(\mathrm{f_i}), y(\mathrm{f_i})) = \mu_x(\mathrm{x}(\mathrm{f_i})) * \mu_y(\mathrm{y}(\mathrm{f_i})) \quad [4.2]$$

Form [4.2] is a property of each of the Z = Y*X dimensional membership function, and therefore:

$$\sum_{z=1}^{Z} \mu_z (x(f_i), y(f_i)) = 1 \qquad [4.3]$$

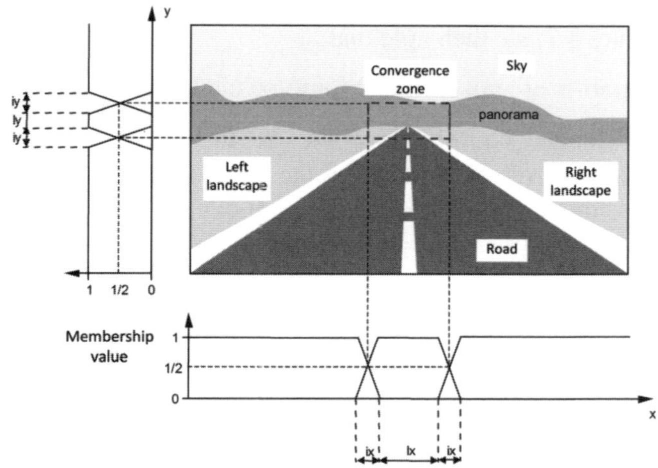

Figure 4.10. *Principal of fuzzing windowing defining the convergence zone*

If the zones are not included in a rectangular surface whose sides are parallel to the axes, it is relatively difficult to respect these conditions [4.3]. As a result, the calculation process for the membership value to a zone z of any shape is the following:

1) the zone is considered to be the juxtaposition of sub-zones, which requires windowing of the horizontal and vertical axes by using very small widths 1x and 1y and by respecting equation [4.1];

2) then the adding of the membership values of the sub-zones is done to calculate the membership value of the considered zone z.

The second spatial aspect taken into account is relative to the dispersion of angular gaps inside a fixation, a slow movement and a saccade. For each entity, the sum of the angular gaps must be calculated between two successive samples. In the case of an ocular saccade, this indicator plays the role of a notion of amplitude.

The characterization procedure is much more complex than the one used for signals relative to the head and the car. Here is a summary:

1) for periods 1 and 3 during which the eye tracker is used, determination of the different ocular entities: fixation, slow movements, saccades, blinking and others;

2) for each entity found, calculation of the 9 temporal variables from table 2 of Figure 4.7; we then note that:

– the variables are not identical for the 5 ocular entities,

– the variables do not respond to the same models of scale measurement: the 2 variables showing the zone observed (during a fixation or a slow movement) respond to the nominal qualitative mode, the others to the quantitative model;

3) for each ocular entity and each quantitative variable, calculation of the membership values to the three spatial (the qualitative variables having already been coded along 10 modalities) windows (or modalities);

4) for ocular entities belonging to a same section, calculation of the AMV at the different spatial modalities considered;

5) ordering of the AMVs in a 72 lines (each period containing three identical laps, each lap being itself composed of 12 sections) and 41 column ($7 \times 3 = 21$ window for initially quantitative variables and $2 \times 10 = 20$ windows for the zone-indicating variables) table.

Exploratory multidimensional analysis

Remember that the metric used in MCA is that of chi-squared, calling upon the inverse of the frequencies (here the inverse of AMV) in comparison to lines and columns. This justifies in part the act of favoring windowing into equal parts. However, the observation of histograms during the characterization step shows that certain zones are rarely looked at, either because they are very small, or because they contain elements that appear only rarely. In fact, the first MCA on all the temporal variables showed this phenomenon for three zones: individual, car in a parking space and emergency telephone. The results found related to these two modalities of very small parts were coherent. Then these 2 modalities were removed from the analysis because they hid other relational phenomena linked to modalities that were probably more frequent.

MCA covering all the modalities of conserved temporal variables

The table considered crosses 72 sections and 41 – 2 = 39 spatial modalities. As in the previous analysis, axis 1 is significantly more important than the next ones (18% of relative inertia compared to 13%, 11%, 10%, etc.). Most of the modalities that contributed to the positioning of axis 1, the large and small modalities of the durations of the blinks, are found again here as the main modalities; next is the large modality of amplitudes and duration of saccades; and finally the modality (landscape, sky, panorama) of the variable area-of-interest for fixations.

The respective positions of these modalities (see Figure 4.11(a)) actually translate two simultaneous temporal aspects:

1) temporal aspect of chronology, in that the turns are organized in a more or less coherent manner along axis 1 (see the considered turn variable illustrated, Figure 4.11(a), and the projections of the sections into two subsets with two different gray levels each, corresponding to the turns, Figure 4.11(b));

2) temporal aspect of duration, in that saccades of large amplitudes last longer than saccades of small amplitudes. This is coherent with the results found in the literature.

Axis 2 is essentially positioned by the modalities indicating area-of-interest (variables ZLY and ZFY). In the decreasing order of relative contributions can be found the signs, landscape, plants, construction modalities for slow movements and signs modality for fixations. Remember that, according to the barycentric principle, a modality is, through construction, closer to the center of gravity of the modality space – space of 72 dimensions – than is frequent. Although axes 3, 4, etc. play non-negligible roles, let us still note that this aspect of proximity is already well respected in the space built from the two first axes. Notably, the central modality is very close to the point (0,0). On the contrary, the sign modality is very far away. Let us also note that often identical modalities for slow movements and fixations are relatively close (see signaling, landscape or construction), and notably factors of the variation of an image of signaling or of construction.

This remark having been made, let us summarize axis 2 by saying that it brings out modern signs, plants and landscape (at the bottom of the landscape) once again shown by the absence of variables. The building of axis 2 can be explained once you have seen the deep way in which the 72 sections are placed and notably by studying certain factors of image variation, such as the number of signs or constructions. Indeed, the six number 8 sections and the six number 11 sections can all be found at the bottom of the plane (Figure 4.11(b)). However, these sections are those with the most signs. A check on the video recordings confirms that the driver is going to watch them. A same observation can be made for the higher (or top) part of axis 2 that is limited by constructions (sections 2 and 7).

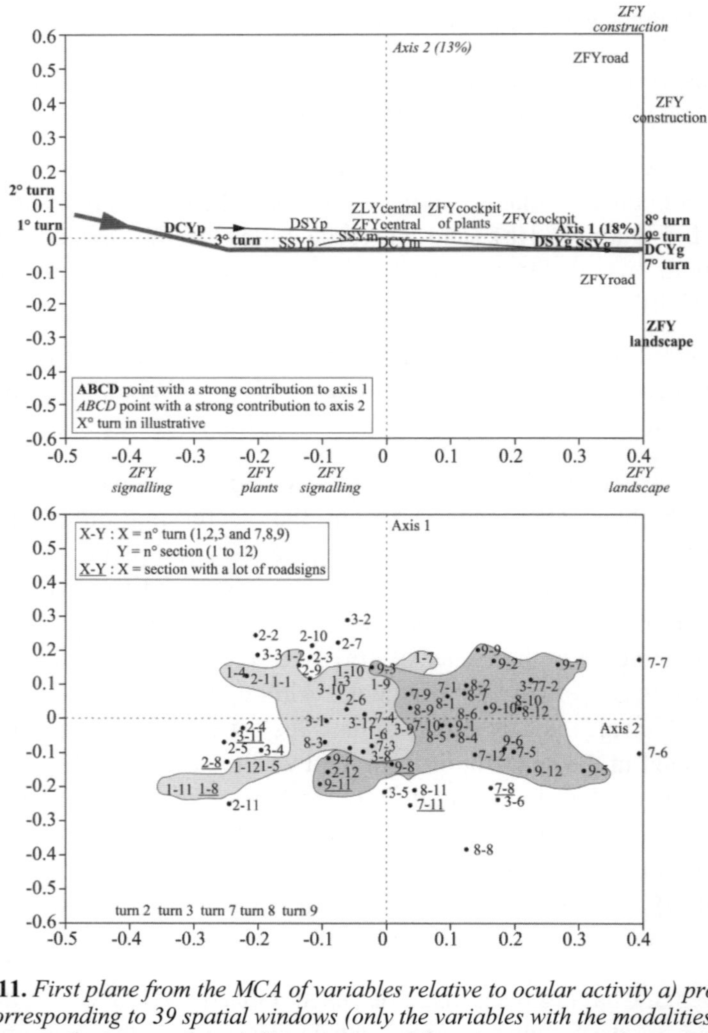

Figure 4.11. *First plane from the MCA of variables relative to ocular activity a) projection of points corresponding to 39 spatial windows (only the variables with the modalities with the strongest contribution are shown); b) projection of the points corresponding to 72 sections*

4.4.3.3.3. Joined analysis of the data relative to behavior and visual information

Seeing as the variables are numerous, it is necessary to only retain from each of these previous analyses a sub-group of variables appearing as preponderant and not very linked, so that intra-group relations does not mask inter-group relations (let us remember that certain variables appear as relatively linked together, over time, others not).

The search for two respective sub-groups relies on an exploratory approach that combines the visualization of dispersion diagrams and factorial planes. This research has led us to keep the following variables:

– for the data relative to the head, movements of rotation around the vertical axis Z, i.e. the angular position (yaw movement, written YH) and the absolute value of the derivative (DLT);

– for data relative to the car, no variable;

– for the data relative to the eyes, the duration of 4 ocular entities and the AMVs at the area-of-interest.

Generally, from this exploratory analysis in several steps we find that ocular variables are not very closely linked to other variables. Figure 4.12 presents the projections of statistical units in the first factor plane from the MCA of the variables cited above (8 variables). Axis 1 is exclusively positioned by the two extreme modalities of angular position of the head (YH); axis 2 is essentially controlled by the two extreme modalities of the duration of blinking (DBY) and, to a lesser extent, by the intermediate modality of YH. The results concerning the influence of the time factor are in accordance with those found in the separate analyses: at the beginning of the oculometric recording, the duration of the blinks is smaller than at the end, with an important jump between period 1 and the beginning of period 3 (see axis 1 in Figure 4.11); the angular position YH is clearly influenced by the time (see axis 2 in Figure 4.9).

We can observe once again on this plane the distribution of points corresponding to the start and the end of the test into two distinct classes. The proposed methods show its ability to choose from a large number of variables those that are the most discriminating with regard to the phenomena studied, here decreased vigilance, and therefore even to build a relevant representation space. The subsequent application of methods of classification in the representation space facilitates the attribution of observations to classes with minimal error.

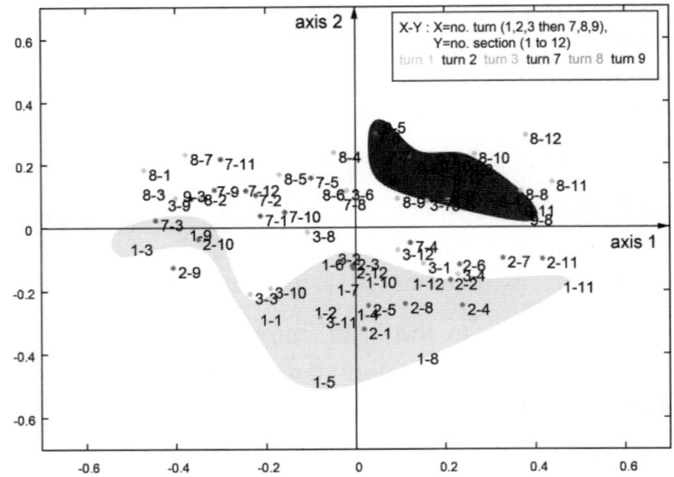

Figure 4.12. *Plane (1,2) from the MCA of variables relative to the eyes and the head. Projection of the points corresponding to the 72 sections*

4.5. Conclusion

Diagnosis in HMS is a complex issue for two main reasons: HMS are very often complex themselves, the presence of the human is usually justified by the difficulty of completely automating the system; next, human activity presents a strong inter- and intra-individual variability, making the search for behavioral regularities difficult.

There is no model of knowledge that provides “simulation” of the behavior of a human operator facing different situations in different contexts. Any diagnostic process in an HMS must call upon a model of behavior based on *in situ* observation.

This chapter has presented the use of a method based on statistical pattern recognition techniques and has shown that it is capable of providing a beginning of an answer to the problem of diagnosis in HMS. The success of the use of such methods strongly depends on the relevance of the indicators retained to build the pattern vector and to define the representation space of the state of the system. This study proposes, to do this, an exploratory statistical approach based on the analysis of information observable within the system after a step of characterization by AMVs to fuzzy modalities. This approach is illustrated with an example of HMS: the driver–vehicle–environment system.

More precisely, the example deals with the problem of vigilance and of fatigue in automobile driving through the analysis of a highway journey of 288 km, which equates to a drive time of roughly 3 hours. The existence of strong inter-individual differences led to the realization of a subject-by-subject analysis. The results provided come from one single driver, but the principle of analysis is generalizable to the whole tested population (the quantitative indications are given only for information's sake and have no statistical value on the whole population).

Extraction of the relevant information present in the data, both numerous and heterogeneous, gathered during the experiment, essentially consists of three steps:

1) analysis of data relative to performance and head movement;

2) analysis of the input of visual information;

3) the comparison of two groups of data.

A first analysis carried out on all of the variables relative to performance and to the driver's gestures showed the importance of changes in variables relative to the position of the head in relation to those of the variables describing the kinematics of the vehicle. On these "biomechanical" variables, analysis showed that time manifests itself through "crumpling" of the subject in the chair, but also by an increase in head instability, measured by indicators calculated on the derivative of base signals. To give some orders of magnitude of these changes, we can point out that during the first 72 km (first part) the vertical position of the head drops by roughly 2 cm. The absolute value of the derivative of this position increases by 100%. A new analysis carried out without the "biomechanical" variables showed a slight increase in speed of the vehicle during the journey (about 2 km/h between the first and the third parts). It is important to point out that these different evolutions are not due to the type of section, but to the time factor.

This second analysis carried out on the variables describing the input of visual information shows that time is manifested very clearly by an increase in the duration of blinks: 320 ms on average for the first part of the circuit against 410 ms for the third part. This is a well-known "physiological" effect, already shown in many works. Another influence of time, although only slight, concerns the amplitude of saccades, which also tend to increase over time: 5.4° angle on average for the first part against 6.8° for the third part. Simultaneously, we can observe an increase in the rate of consultation of not very informative zones: landscape, panorama and the sky.

The third analysis carried out aims to compare respective groups of variables chosen from the starting variables because of the preponderant character of their evolutions, revealed by the previous analyses, and which are not very linked between themselves (weak intra-group relations do not make intergroup relations). This analysis was carried out on "biomechanical" (head kinematics) variables and on data relative to the input of visual information. From this last analysis comes the observation that variables from the two previous analyses, appearing to change significantly over time, are poorly linked with each other.

Confirming analyses in the form of non-parametric tests (Wilcoxon tests) were carried out on this data to validate the evolution over the entire population of drivers. It results that all the variables identified as "preponderant" by factorial analyses present significant differences of around 5%.

The preponderant variables in the construction of axes provide the definition of the pattern vector, facilitating the best possible representation of the state of the system in the representation space.

In all the analyses, the influence of the time factor appeared clearly, facilitating the distinction in the form of clearly separated groups of the data relative to the start and the end of the experiment and, to a lesser extent, the evolution between these groups. The combination of fuzzy spatial modalities corresponding to these data groups facilitates the characterization of the different classes of behavior relative to driving at the beginning and at the end of the test [YOU 11].

On this basis, two methods of classing have been applied on the factorial coordinates coming from the analyses: the k-nearest neighbors and discriminant analysis give correct class results in more than 85% of cases.

While certainly needing a lot of improvement, concerning the necessity of working individual-by-individual or even the automatic evolution of classification, this study has shown the feasibility of the approach proposed for the diagnosis of HMSs. It has notably shown the essential character of the method used to carry out the characterization of temporal data with regard to the ability to discriminate between the different classes of behavior.

4.6. Bibliography

[ALL 71] ALLEN T.M., LUNENFELD H., ALEXANDER G.J., "Driver information needs", *Highway Research Board*, vol. 36, pp. 102–115, 1971.

[ARR 00] ARROYO-FIGUEROA G., ALVAREZA Y., SUCARB L.E., "SEDRET – an intelligent system for the diagnosis and prediction of events in power plants", *Expert Systems with Applications*, vol. 18, no. 2, pp. 75–86, 2000.

[AUB 10] AUBERLET J., PACAUX-LEMOINE M., ANCEAUX F., *et al.*, "The impact of perceptual treatments on lateral control: a study using fixed-base and motion-base driving simulators", *Accident Analysis and Prevention*, vol. 42, pp. 166–173, 2010.

[BAD 91] BADR Y., Influence de l'environnement routier sur le comportement des conducteurs, une contribution à la compréhension des déterminants du comportement des conducteurs: étude des facteurs influençant les vitesses pratiquées dans le cas de traversée de petites agglomérations, PhD Thesis, Ecole nationale des ponts et chaussées, 1991.

[BAL 67] BALL G.H., HALL D.J., "A clustering technique for summarizing multivariate data", *Behavioural Science*, vol. 12, pp. 153–155, 1967.

[BER 96] BERNAUER E., Les réseaux de neurones et l'aide au diagnostic: un modèle de neurones bouclés pour l'apprentissage de séquences temporelles, PhD Thesis, Paul Sabatier University, Toulouse, October 1996.

[BRO 94] BROWN I.D., "Driver fatigue", *Human Factors*, vol. 36, no. 2, pp. 298–314, 1994.

[BRU 90] BRUNET J., JAUME D., LABARRERE M., *et al.*, *Détection et diagnostic de pannes, approche par modélisation*, Hermès, Paris, 1990.

[BYT 71] BYTHEWAY C.W., "FAST: diagrams for creative function analysis", *Journal of Value Engineering*, vol. 71–3, pp. 6–10, 1971.

[CAR 02] CARRASCO E.F., RODRIGUEZ J., PUÑAL A., *et al.*, "Rule-based diagnosis and supervision of a pilot-scale wastewater treatment plant using fuzzy logic techniques", *Expert Systems with Applications*, vol. 22, no. 1, pp. 11–20, 2002.

[CHI 00] CHIANG L.H., RUSSELL E.L., BRAATZ R.D., "Fault diagnosis in chemical processes using Fisher discriminant analysis, discriminant partial least squares, and principal components analysis", *Chemometrics and Intelligent Laboratory Systems*, vol. 50, pp. 243–252, 2000.

[DUB 90] DUBUISSON B., *Diagnostic et reconnaissance des formes*, Hermès, Paris, 1990.

[ESC 98] ESCOFIER B., PAGES J., *Analyses factorielles simples et multiples*, Dunod, Paris, 1998.

[FAU 85] FAURE A., *Perception et reconnaissance des formes*, Editests, Paris, 1985.

[FOR 65] FORGY E.W., "Cluster analysis of multivariate data: efficiency vs interpretability of classifications", *Biometrics*, vol. 21, pp. 768–769, 1965.

[FRA 00] FRANK P.M., ALCORTA GARCIA E., KÖPPEN-SELIGER B., "Modelling for fault detection and isolation versus modelling for control", *Mathematics and Computer Simulation*, vol. 53, pp. 259–271, 2000.

[FRY 01] FRYDMAN C., LEGOC M., TORRES L., *et al.*, "The diagnosis approach used in SACHEM", *12th International Workshop on Principles of the Diagnosis*, Sansicario, Italy, 2001.

[GER 95] GERMAIN C., BLANCHET V., "La fatigue des routiers et ses conséquences en termes de sécurité", *Recherche, Transports et Sécurité*, vol. 49, pp. 63–72, 1995.

[GRO 89] GROEN G.J., PATEL V.L., "The relationship between comprehension and reasoning in medical expertise", in CHI M., GLASER R., FARR M., (ed.), *The Nature of Expertise*, Erlbaum, Hillsdale, NJ, 1989.

[HAT 87] HATLEY D.J., PIRBHAI I.A., *Strategies for Real-Time System Specification*, Dorset House, New York, 1987.

[HOU 94] HOURIEZ B., Acquisition de connaissances pour l'aide à la conduite et à la supervision des procédés industriels, Thesis, University of Valenciennes, 1994.

[ISE 97] ISERMANN R., "Supervision, fault detection and fault diagnosis methods: an introduction", *Control Engineering and Practice*, vol. 5, no. 5, pp. 639–652, 1997.

[JAU 89] JAULENT P., *IGL Technology, SADT: un langage pour communiquer*, Eyrolles, Paris, 1989.

[JOU 01] JOUGLET D., MILLOT P., "Performance improvement of technical diagnosis provided by human machine cooperation", *Proceedings of the IFAC/IFIP/IFORS/IEA Conference Analysis Design and Evaluation of Human Machine Systems*, Kassel, Germany, pp. 379–384, 2001.

[KAU 75] KAUFMANN A., *Introduction à la théorie des sous-ensembles flous: Applications à la classification et à la reconnaissance des formes, aux automates et aux systèmes, au choix des critères*, Masson, Paris, vol. 3, 1975.

[KIT 00] KITHIL P., "Driving simulator research projects using overhead mounted capacitive sensing array for detecting head motion indicators of drowsiness", *Proceedings of the Driving and Simulation Conference DSC'2000*, Paris, France, September 2000.

[KNI 95] KNIPLING R.R., WANG J.-S., "Revised estimates of the US drowsy driver crash problem size based on general estimates systems case reviews", *Proceedings of the 39th annual Meeting of the Association for Advancement of Automotive Medicine*, October 1995.

[LEG 94] LEGER D., "The cost of sleep-related accidents: a report for the National Commission on Sleep Disorders Research", *Sleep*, vol. 17, pp. 84–93, 1994.

[LEU 00] LEUNG D., ROMAGNOLI J., "Dynamic probabilistic model-based expert system for fault diagnosis", *Computers & Chemical Engineering*, vol. 24, no. 11, pp. 2473–2492, 2000.

[LIN 11a] LIND M., "An introduction to multilevel flow modeling", *International Journal of Nuclear Safety and Simulation*, vol. 2, no. 1, March 2011.

[LIN 11b] LIND M. "Control functions in MFM: basic principles", *International Journal of Nuclear Safety and Simulation,* vol. 2, no. 2, June 2011.

[LOS 00] LOSLEVER P., POPIEUL J.-C., SIMON P., *et al.*, "Les chemins de l'analyse statistique de données temporelles : exemple de l'analyse du système conducteur véhicule", *APII-JESA*, vol. 34, no. 8, pp. 979–1010, October 2000.

[LOS 03] LOSLEVER P., POPIEUL J.-C., SIMON P., "From classic statistical characterization to fuzzy windowing based characterization for the exploratory analysis of miscellaneous time variables: example in the field of car driving studies", *Fuzzy Sets and Systems*, vol. 137, pp. 271–296, 2003.

[LOS 07] LOSLEVER P., POPIEUL J.-C., SIMON P., "Using correspondence analysis when the system description yields several transition matrices: examples with simulated data and real driver-car environment system data", *Cybernetics and Systems*, vol. 38, no. 1, pp. 23–45, 2007.

[LOS 09] LOSLEVER P., POPIEUL J.-C., SIMON P., "Using correspondence analysis with a large set of Transition Matrices: example with eye movement data and fuzzy space windowing", *Intelligent Data Analysis*, vol. 13, pp. 917–942, 2009.

[MAC 67] MACQUEEN J.B., "Some methods for classification and analysis of multivariate observations", *Proceedings of the 5th Berkeley Symposium on Mathematical Statistics and Probability*, Berkeley, USA, 1967.

[MAR 86] MARRAKCHI M., Représentation des connaissances pour l'aide au diagnostic industriel: application au système expert SEDIAG, PhD Thesis, University of Valenciennes, 1986.

[MAS 01] MASSON M., GRANDVALET Y., "Réseaux de neurones pour le diagnostic", in DUBUISSON B., (ed.), *Diagnostic, Intelligence artificielle et reconnaissance des formes*, Hermès, Paris, 2001.

[MAY 95] MAYCOCK G., Driver sleepiness as a factor of in car and HGV accidents, Technical report 169, Crowthorne, UK Transport and Research Laboratory, 1995.

[MIL 93] MILGRAM M., *Reconnaissance des formes: Méthodes numériques et connexionnistes*, Armand Colin, Paris, 1993.

[MIL 03] MILLOT P., "Supervision et coopération homme-machine: approche système", in BOY G., (ed.), *Ingénierie Cognitive IHM et Cognition*, Hermès, Lavoisier, Paris, chap. 6, pp. 191–221, 2003.

[MIL 97] MILOSEVIC S., "Drivers' fatigue studies", *Ergonomics*, vol. 40, no. 3, pp. 381–389, 1997.

[MOD 99] MODARRES M., CHEON S.W., "Function-centered modeling of engineering systems using the goal tree-success tree techniques and functional primitives", *Reliability Engineering*, vol. 64, pp. 181–200, 1999.

[PAC 04] PACAUX M.P., ORDIONI J., POPIEUL J.C., *et al.*, "Design and evaluation of an advanced driving assistance tool", *IEEE International Conference on Vehicle Power and Propulsion*, Paris, France, October 2004.

[PAU 97] PAU L.F., *Diagnostic des pannes dans les systèmes: approche par la reconnaissance des formes*, Cépaduès Editions, Toulouse, 1997.

[PEL 93] PELTIER M.A., Un système adaptatif de diagnostic d'évolution basé sur la reconnaissance des formes floues – Application au diagnostic du comportement d'un conducteur automobile, PhD Thesis, University of Technology of Compiègne, December 7 1993.

[PER 97] PERRON T., Méthodologie d'analyse de sécurité primaire automobile pour la spécification fonctionnelle et l'évaluation prévisionnelle d'efficacité de systèmes d'évitement d'accidents, PhD Thesis, Ecole centrale de Paris, 1997.

[POL 09] POLET P., VANDERHAEGEN F., MILLOT P., "Human behaviour analysis of barrier deviations using a benefit-cost-deficit model", *Advances in Human-Computer Interaction*, pp. 10–19, 2009. DOI:10.1155/2009/642929.

[POP 94] POPIEUL J.-C., Contribution à l'étude et à l'implémentation d'une démarche de conception de machines et systèmes automatisés de production, PhD Thesis, University of Valenciennes (UVHC), February 1994.

[POP 02] POPIEUL J.C., SIMON P., LOSLEVER P., "Using failure detection and diagnosis methods to detect dangerous evolutions of the driver behaviour", *IFAC – Control Engineering Practice*, vol. 10, no. 5, pp. 577–583, May 2002.

[POP 03] POPIEUL J.C., SIMON P., LOSLEVER P., "Using driver's head movements evolution as a drowsiness indicator", *IEEE Intelligent Vehicles Symposium IV 2003*, Columbus, OH, 9–11 June 2003.

[RAS 83] RASMUSSEN J., "Skills, rules and knowledge, signals, signs and symbols and other distinctions in human performance models", *IEEE Transaction on Systems, Man and Cybernetics*, vol. 13, no. 3, pp. 257–266, 1983.

[REA 93] REASON J., *L'erreur humaine*, Presses Universitaires de France, Paris, 1993.

[RIE 93] RIERA B., Méthodologie d'évaluation des interfaces homme-véhicule automobile, PhD Thesis, University of Valenciennes, January 1993.

[RIE 01] RIERA B., Contribution à la conception d'outils de supervision centrés sur l'homme, Thesis, University of Valenciennes, December 2001.

[ROG 01] ROGE J., PEBAYLE T., MUZET A., "Variations of the level of vigilance and of behavioural activities during simulated automobile driving", *Accident Analysis and Prevention*, vol. 33, pp. 181–186, 2001.

[ROS 77] ROSS D.T., "Structured analysis (SA): a language for communicating ideas", *IEEE Transactions on Software Engineering*, vol. 3, no. 1, pp. 16–34, 1977.

[SAG 99] SAGBERG F., "Road accidents caused by drivers falling asleep", *Accident Analysis and Prevention*, vol. 31, pp. 639–649, 1999.

[SAP 90] SAPORTA G., *Probabilités, analyse des données et statistiques*, Editions Technip, Paris, 1990.

[SAP 96] SAPORTA G., LAVALLARD F., (eds.), *Analyse des données évolutives, méthodes et applications*, Editions Technip, Paris, 1996.

[SAY 91] SAYARH H., Contribution de l'analyse de données à l'étude locale et globale de signaux multidimensionnels: application à l'analyse de systèmes homme-machine, PhD thesis, University of Valenciennes, 1991.

[SHO 74] SHORTLIFFE E.H., MYCIN: a rule-based computer program for advising physicians regarding anti-microbial therapy selection, Memo AIM 251, Stanford AI Lab, USA, 1974.

[SIM 99] SIMON P., POPIEUL J.-C., VOILLEQUIN T., *et al.*, "Description d'un système automatisé de recueil de l'exploration visuelle du conducteur en simulateur", *Proceedings DSC'99 Driving Simulation Conference*, Paris, France, July 1999.

[THO 94] THOMAS C., ATTARD S., Drowsiness and fatal accidents, Technical report, Centre européen d'étude socioéconomiques et accidentologiques des risques (CEESAR), Nanterre, France, 1994.

[TOD 99] TODOSKOFF A., Etude des évolutions temporelles du comportement du conducteur sur autoroute : analyse multidimensionnelle de signaux relatifs au véhicule et aux mouvements de tête sur simulateur, PhD thesis, University of Valenciennes, 1999.

[TRI 05] TRICOT N., Conception et évaluation de systèmes coopératifs avancés : application à la régulation longitudinale en conduite automobile, PhD thesis, University of Valenciennes, décembre 2005.

[VAN 06] VANDERHAEGEN F., CHALME S., ANCEAUX F., *et al.*, "Principles of cooperation and competition: application to car driver behavior analysis", *Cognition, Technology, and Work*, vol. 8, pp. 183–192, 2006.

[VER 99] VERWEY W.B., ZAIDEL D.M., "Preventing drowsiness accidents by an alertness maintenance device", *Accident Analysis and Prevention*, vol. 31, pp. 199–211, 1999.

[VER 00] VERWEY W.B., ZAIDEL D.M., "Predicting drowsiness accidents from personal attributes, eye blinks and ongoing driving behaviour", *Personality and Individual Differences*, vol. 28, pp. 123–142, 2000.

[WAR 85] WARD P., MELLOR S., *Structured Development for Real-Time Systems*, vol. 1–3, Yourdon Press, Englewood Cliffs, 1985.

[YOU 11] YOUNSI K., LOSLEVER P., POPIEUL J.-C., *et al.*, "Fuzzy segmentation for the exploratory analysis of multidimensional signals: example from a study on driver overtaking behavior", *IEEE Systems Man and Cybernetics* (Part A), vol. 41, no. 5, pp. 892–904, 2011.

[ZWI 95] ZWINGELSTEIN G., *Diagnostic des défaillances,* Hermès, Paris, 1995.

5

Evaluation of Human Reliability in Systems Engineering

5.1. Introduction

Over the last 30 years, technology has gone through a lot of improvements, especially in terms of robustness, allowing machines to carry out a variety of operations without the support of human operators. As a result, the role of these operators has changed a lot and has transformed over the years into a function of supervision and control of the automatically executed procedures [SHE 92, SHE 99]. In such a configuration of human–machine systems, the design of automated systems and the control of interactions with human operators have become increasingly complex [VAN 97b, VAN 99]. The consequences of a human mistake or mutual misunderstanding between humans and automations can be irreversible or even catastrophic [NAG 88]. Two main factors justify such a lack of safety: the significant improvement in the reliability of technical systems and the intensive use of automation. Technological advances have resulted in a massive reduction of mechanical failures and have enabled the automated running of factories despite the presence of residual risks due to the existence of potential critical failures. In this way, the contribution of human factors to the maintenance or the safety of the system has been reinforced and human error has become the main cause of most accidents in all technically advanced fields. In our modern societies, following the occurrence of an accident, it is not rare to systematically voice an initial hypothesis of human error. Unfortunately, this is often only intuitive and in

---

Chapter written by Frédéric VANDERHAEGEN, Peter WIERINGA and Pietro Carlo CACCIABUE.

most cases inappropriate, the real explanation of the causes of an accident being often used rather for identifying legal responsibilities instead of determining future preventive measures. In some particular cases, mistakes made by humans in the field are wrongly designated as the primary causes by the highest level of the hierarchy so to protect themselves by masking these errors and minimizing their responsibility. In such a scenario, with the goal of offering reliable methods of evaluating safety and the control of human–machine systems, the role of decision-makers and the interaction between humans and machines are vital and require particular attention during system design.

The principles of human-centered design are useful because their application aims to give human operators a continued role in the supervision and control loop, and to notify these operators with regard to the performance of automated systems in the execution of all functions and procedures [BIL 97]. However, designers do not always follow this essential requirement. They define the rules and procedures that control the behavior of automated systems without dealing with the impact they have on human operators. Moreover, in conditions susceptible to accidents, the dynamic characteristics of a series of events complicate the work situation as well as the resolution of the associated problems. If a response is not predefined, a conflict can arise between the interpretation of the human operator on the dynamic evolution of the controlled process and the automated system's evolution. As a result, real work environments require the development of cognitive systems much more than the development of simple sensorimotor systems [RAN 98, HOL 93, CAC 10]. This discussion reflects two important points. When automation, on the one hand, is necessary to assist or replace the human operators in the execution of their tasks, the consequences of inappropriate decisions or conflicting interpretations should be taken into account. Furthermore, since the occurrence of a human mistake is one of the intrinsic characteristics of the management of any system, it is impossible to conceive of such a system devoid of error. As a result, it is obvious that the design and evaluation of the safe and high-performing systems not only belong to engineers, but also to other specialists, particularly those in the human sciences.

With regard to the evaluation of risks and human reliability, all of these remarks have been progressively taken into account in different methods that enable the association of a measurement of reliability, in this case the frequency of occurrence of inappropriate performance with the contribution of human factors in this measurement. In this chapter are developed these principles of the evaluation of human error. Some debate on methodological advances, but also on the future paths of research necessary to increase the integration of human factors in the analysis of safety of human–machine system is raised. Among the methodological advances, taking into account the dynamic characteristics of a given process is presented from research carried out by the authors themselves. Finally, suggestions for future development concern the extension of the concept of error so as to increase the scope of knowledge on deviations of human behavior and their impact on the process.

5.2. Principles of evaluating human reliability

After a definition of human reliability and human error, the general approach to analyzing them and a classification of the existing methods are presented. A discussion will introduce the key points of the authors' contribution in the domain of evaluating human reliability in human–machine systems.

5.2.1. *Human reliability versus human error*

The reliability of a technical component is measured by the probability of absence of a fault occurring in this component over a given timescale; a fault being defined as an alteration or cessation of the component's ability to perform a certain function. Human reliability can be assimilated to this probability. Nevertheless, it often concerns the task to be carried out, rather than the function, and can be defined as the probability of a task being carried out successfully over a given period of time. The task and the function have a common notion: the goal to be reached. The function is linked to the objective of the piloted process, or the service provided by this process, and the task is linked to the objective of the technical or human means carrying out this function [LAP 95, SHA 97]. To execute a given function, human behavior is subordinate to a prescription called a task, and what is put into practice to carry it out is the activity [LEP 85]. An error therefore involves a drift between the effective task, which is the model from the analysis of activity, and the prescribed task, which is a model of what is to be carried out.

All the tasks of the operator group include the tasks of control and supervision of the performance of a given process, the tasks linked to the safety of the human–machine system, the tasks of prevention and recovering of human or technical error. Indeed, an operator must not only control and supervise the process he[1] controls, but also recover deviations in the functioning of the process he controls, avoid the occurrence of any catastrophic events caused by these drifts and regulate his own activity so as to be ready to react or maintain his own knowledge. As a result, a measurement of human reliability is the probability of the absence of human error, but it extends to the probability of a human operator (1) correctly executing the prescribed tasks under the required conditions and (2) not executing any additional tasks that could be detrimental to the proper functioning of the human–machine system [SWA 83, VAN 11a].

There are two sets of human behavior: erroneous and reliable [MIL 87]. The measurement of human error is the probability of an error occurring in the execution of the tasks of a human operator. More precisely, it is the probability of not properly performing the prescribed tasks in the given conditions and executing additional tasks that could be detrimental to the functioning of the human–machine system [VAN 11a]. Taking into account the probability of the execution of unplanned tasks is very difficult, maybe even impossible; therefore, the calculation of this probability is often limited to the rate of success in performing predefined tasks.

5.2.2. *General approach for the analysis of human reliability*

The analysis of risks associated with the use of a human–machine system concerns the functioning safety with which a system must be able to satisfy one or required functions in a set of given conditions [VIL 88]. The analyzed risk is generally a combination of the gravity of the consequences of an undesirable event and a degree of certainty concerning the occurrence of the event.

The general approach adopted for most methods of analysis of the risk of human error in the design of safe human–machine systems is illustrated in Figure 5.1.

1 As previously mentioned in the introduction of this book, the masculine form *he* is used throughout the text to avoid weighing down the syntax of the text with the form *he/she*.

The structural and functional decomposition of the system being designed first depends on the objectives and limitations that it must satisfy. The analysis of tasks allows prescription of the contribution of human operators in the execution of previously defined functions, as well as the normal execution conditions of their respective tasks. This allows the determination of potential errors as well as their modes of control to return to a normal situation. Next, a quantitative analysis allows the calculation of the probability of occurrence of these errors. The consequences can be analyzed to evaluate the risk of human error.

As a function of the level of acceptability set out by the starting objectives and constraints, the risk of error or its probability of occurrence can be accepted, leading to the final specification of the human–machine system. In the opposite case, new recommendations must be issued. If tasks are added, modified or removed, their analysis must be renewed. The process of identifying and evaluating errors is done until the residual risk, or the residual probability of occurrence, is accepted.

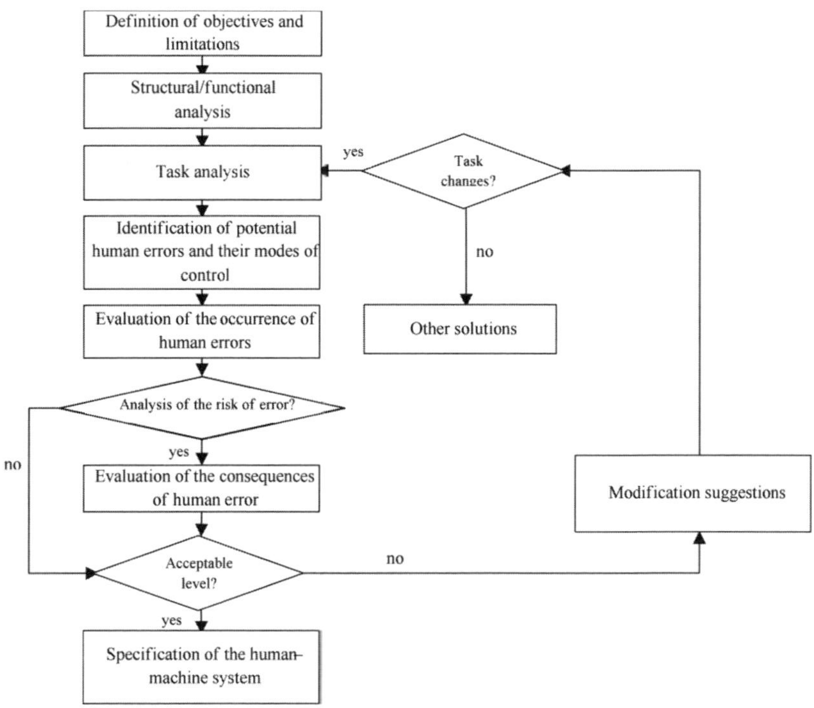

Figure 5.1. *Principles of human reliability analysis [GER 97]*

5.2.3. *Synthetic review of methods*

Methods of evaluating human reliability can be grouped into five distinct classes as presented in Table 5.1: methods based on the calculation of ratios, the analysis of faults, the probabilistic calculation of error, models of taxonomy and models of dysfunction.

The approaches of analyses based on ratios aim to determine, for example, a rate of error, failure, death or even of injury. They allow the calculation of the number of occurrences of undesirable events over a given period of time. The error rate can also be defined as the ratio between the number of occurrences of errors over the number of opportunities to make such errors [MIL 87]. It is often associated with a probability of human error. When the criterion of opportunity concerns failure in carrying out the given task, it becomes the number of failures over the number of times the task occurred. The rate of reliability is then the number of successes over the number of solicitations.

Example	Principle	Objective
Rate of error, failure, death, injury	Ratios	Statistical analysis
FMCEA, fault tree	Evaluation of dreaded events	Failure analysis
HCR, THERP, HEART	Probability of error occurrence	Quantitative error analysis
Pair comparisons, GEMS, SHERPA, CREAM	Taxonomy model of the error	Cognitive error analysis
BASIS, MAFERGO, PRISMA, RECUPERARE	Dysfunction model	Incident analysis

Table 5.1. *Example of approaches to the evaluation of human reliability*

The analysis of failures can be based on technical methods [VIL 88], such as the failure mode cause and effect analysis (FMCEA) or the fault tree described as follows:

– The FMCEA method determines the potential modes of failure, their possible causes and their effects on the human–machine system by evaluating for each of them the criticity from the gravity, the frequency of appearance and eventually the probability of non-detection.

– The fault tree method allows identification of combinations of elementary events that could lead to dreaded events and calculation of the probability of their occurrence.

Methods based on the probabilistic calculation of error are human cognitive reliability (HCR), the technique for human error rate prediction (THERP) or human error assessment and reduction technique (HEART) described as follows:

– The HCR method evaluates the probability of a lack of response, taking into account the relation between the time available to carry out a task and the time required to carry it out, weighting it with the competence of the operator, the level of stress and the quality of the human–machine interface [HAN 84].

– In the THERP method [SWA 83], the probability of error is evaluated from three factors: a base probability relative to the characteristics of the task being carried out, a corrective coefficient for taking into account the stress of the human operator and the probability of not recovering the error.

– The HEART method assesses the probability of error in the execution of a given tasks by integrating the impact of factors from an exhaustive list [HUM 88].

Methods based on models of error taxonomy include pair comparison, generic error-modeling system (GEMS), systematic human error reduction and prediction (SHERPA) or cognitive reliability and error analysis method (CREAM) described as follows:

– The pair comparison method enables the classification of a list of tasks from the least risky to the riskiest, by comparing them two-by-two [HUM 88].

– The GEMS model groups together the errors as slips (due to lack of attention), lapses (linked to memory) or mistakes (linked to reasoning based on the rules or on knowledge) [REA 90].

– The SHERPA model is based on Rasmussen's taxonomy [RAS 83] comprising three types of behavior (i.e. behaviors based on skill, rules and knowledge), by making the difference between the reasoning and the actions based on rules [EMB 86].

– In the CREAM method [HOL 98], errors are divided into two categories: genotypes, i.e. the causes of errors, and phenotypes, i.e. their manifestations.

Approaches based on models of dysfunction require the use of experience from the field and are described in the following:

– BASIS is a method for incident reporting so as to identify the factors that favor the occurrence or avoidance of incidents [O'LE 99];

– methodology of analysis of operational reliability and ergonomics (MAFERGO) combines work analysis and the analysis of the reliability of a human–machine system studied so as to evaluate its operational dysfunctions [FAD 97];

– PRISMA is an incident analysis method for Prevention and recovery of errors based on three functional levels of modeling: the technical system, human operators and organization [VAN 97a];

– RECUPERARE aims to study the possibilities for the recovery of incidents modeled in a predefined database [BAU 99].

5.2.4. *Discussion*

Methods such as the fault tree method evaluate the technical system and its behavior in binary terms. Thus, a pump is analyzed to determine whether it can be faulty and stop the achievement of a function such as "provide cooling water" with a certain probability. This probability is estimated from data from use or from tests. The human operator can be evaluated according to the same outline and his performance is then expressed in terms of fault, i.e. his capacity degradation to execute a given function in given time interval. For example, a human error can be a fault when the human operator is not able to provide cooling water before a given date.

Methods for the study of the reliability of technical systems can be adapted for the analysis of human reliability but remain insufficient. Indeed, they generally deal with independent events so as to simplify the calculation of the combinations of probabilities of occurrence of events. They also require detailed knowledge of human behavior, which cannot be the object of batteries of tests such as those done in the study of technical components.

Even if this knowledge can be built from expert judgment, human-centered methods seem best suited for the analysis of human reliability. Nevertheless, several comparative studies have shown the heterogeneity of the results obtained while using methods of quantitative analysis of human reliability by the same person or during the use of the same method by several people [HUM 88, SWA 90, REA 90, KIR 97]. Moreover, even if the other methods based on models seem more attractive, they are faced with the problem of the relevance of the model used and the difficulty of its use [DE 03].

From a prospective point of view, all of the problems of validity can be linked to the fact that methods of analysis of human reliability presented previously are not adapted to take into account the dynamic interaction between human operators and technical systems. However, this aspect is primordial as it is one of the important factors in the evolution of incidents, as much in terms of avoidance as in terms of the recovery of errors. Indeed, on the one hand, most of the conditions linked to the actions of the control system and human behaviors as well as the high-risk situations in which the operator plays an important role are not explicitly taken into account [SIU 94]. In this case, we have to wait for the occurrence of the incident or the accident to happen before beginning a retrospective analysis aiming to determine the interactions between the different elements of the system. Moreover, these methods allow the specifications of only one static system limiting themselves to improving the ergonomics of workstations or programs of training to prevent certain errors, but without taking into account the possibility of human operator to reproduce these errors on the field.

Thus, these gaps make it impossible to create a reliable model of the consequences of an error on a human–machine system and the dynamic evolutions that result. Moreover, these problems of validity are certainly linked to the limits of measurement of human error in most of the methods of analysis of human reliability. These are limited to evaluating the execution of the prescribed tasks without worrying about possible new tasks that the human operator could be called upon to carry out, or voluntarily altered tasks that can be detrimental to the safety of the global system. Reason [REA 90] distinguishes error from violation: there is an error when what is obtained or done is different from what is prescribed without any intention of divergence, and there is a violation when this divergence is voluntary. Altered tasks, which are violations, are often not considered in the analysis of human reliability. However, this type of behavior can be interpreted as a latent error in human resource management or in design or as an indicator of a judgment error from a human operator [DOU 95].

It is therefore necessary to adapt the current methods of the analysis of human reliability or to develop new approaches to help in the specification of the systems integrating dynamic characteristics of technical components and human operators who interact with these components, on the one hand, and tolerate erroneous tasks, as well as tasks coming from altered or added behavior, on the other hand. In the following, these two points are discussed.

5.3. Analysis of dynamic reliability

Techniques of evaluation of dynamic reliability allow the limits of the classical methods of probabilistic analysis to be recovered. They require a model of simulation and most of these techniques use the dynamic event tree method as an analysis tool. Contrary to a classical representation, which is limited to the determination of logical transition links between events or between the sequences of a procedure obtained from the process analysis, this method presents them chronologically. All the possible combinations of the states of the process can then appear at each branching point in time. From the simulation of the physical process, the dynamic event tree provides a support for the analysis of the impact of a given combination on another combination. The following two methods can be used: the dynamic logical analytical methodology (DYLAM) method [CAC 04] and the human interaction timeline (HITLINE) method [MAC 94].

5.3.1. *The DYLAM method*

The DYLAM method is based on a model of simulation that integrates the dynamic evolution of the variables of the piloted process and the behavior of the human–machine systems in normal and deteriorated conditions [CAC 98, CAC 04]. It was initially developed for the analysis of the dynamic behavior of production systems [AME 81]. It combines a physical simulation of the piloted process and a probabilistic evaluation without requiring complex analyses or important calculations of possible scenarios. It evaluates on a given timescale the dynamic evolutions of the human–machine system from predefined conditions of function and predefined modes of failure. By basing itself on a simulation of human–machine interactions, DYLAMS allows the combination of technical faults with erroneous human behaviors. DYLAM then generates a series of sequences of possible incidents that could take place over time due to faults or inappropriate behavior from technical or human components of the production system. The dates of occurrence of the faults are unpredictable at the beginning of the analysis because they are evaluated without taking into account the temporal dynamics using probabilistic algorithms or calculations of logical correlation between the events and their occurrences. The analyst defines the conditions of occurrence of a fault. Application of the DYLAM method is only possible when a simulation of human–machine interaction is available. It requires the following four iterative steps:

– identification of the main components of the system;

– analysis of the modes of failure and their effects;

– selection of the parameters and data of the simulation;

– choice of the type of analysis.

The logical and physical behavior of a technical component is generally represented by equations. However, it is possible to describe its states through heuristics or rules of production. The logical and physical behavior of a human operator can be associated not only with specific behavioral manifestations during the execution of procedures, but also with human errors that affect the result of cognitive functions such as perception, interpretation or planning. As for the technical components, human behavior can be described by algorithms or logical expressions. It can be identified as an acceptable or unacceptable state of performance, related to the cognitive or physical activity prescribed by the procedures. For example, a mal-adapted level of performance can be induced by behaviors such as "systematic omission of actions of regulation and control" or "confusion between two similar procedures". The causes of transition between the states of human behavior are linked to contextual and environmental conditions as well as unplanned events. Individual attitudes and internal factors that can affect them can generate inadequate levels of performance. The DYLAM method, used as a manager of dynamic simulation, allows the study of faults depending on time and the inappropriate behavior of human operators, and with respect to logical and probabilistic criteria. An example of temporal sequences generated by DYLAM is shown in Figure 5.2.

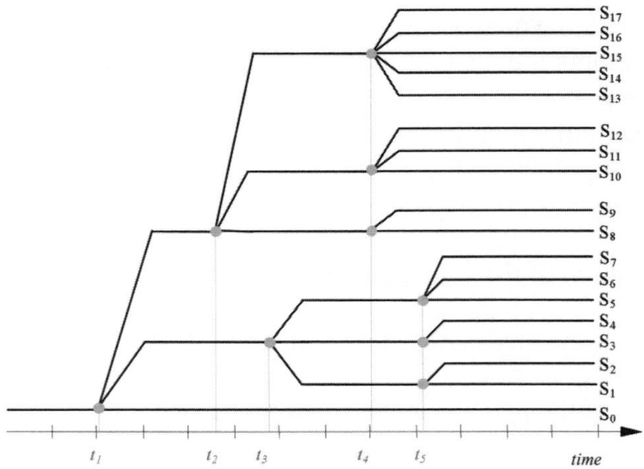

t_i Date of the occurrence of the fault

● Fault mode

Figure 5.2. *Sequence generation by DYLAM*

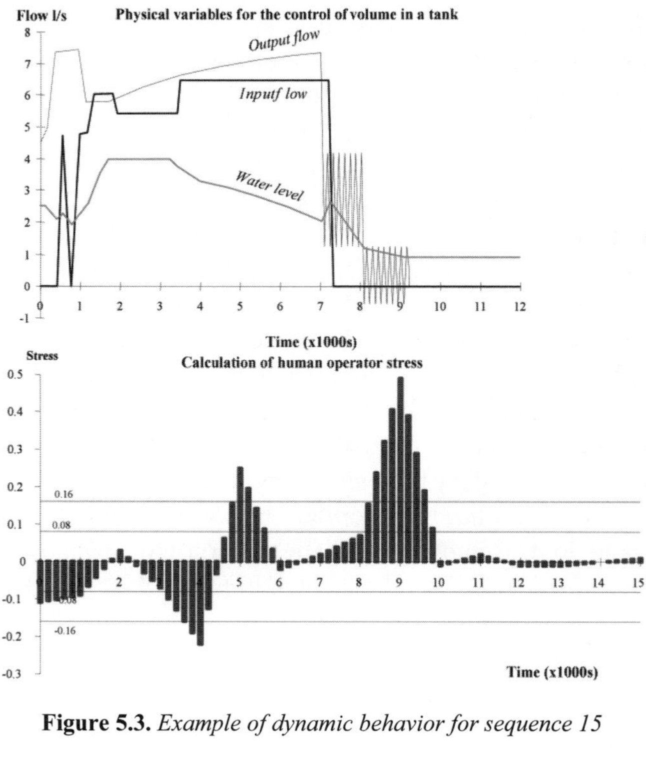

Figure 5.3. *Example of dynamic behavior for sequence 15*

The nominal sequence S_0 is due to a given initiating event and includes an adequate response of the physical components and human operators faced with this event. Each date t_i is then a point that can generate different error modes represented by a new sequence $S_{j\ (j>0)}$. For example, at the point t_4, the possible errors of human operators are illustrated by the series of blocks S_{13}–S_{17}, whereas the technical components are associated with S_{10}–S_{12} and S_8–S_9. Each sequence can be described over time and in relation to the evolution of the physical and human characteristics variables to represent the response of the human–machine system. Figure 5.3 shows the evolution of the system for the S_{15} sequence with regard to the variables to control water volume in a tank and the stress level of the operator. The stress level is obtained via an empirical function combining various factors such as the dependence of the stress in time, the state of human actions, alarm status and the state of the process.

5.3.2. *The HITLINE method*

The HITLINE method [MAC 94] is based on a model of simulation that integrates the factors that influence human performance and their effects on human behavior. It was developed to answer two fundamental needs in the analysis of human reliability: (1) evaluate the dynamic aspects of human–machine interaction during emergency situations; and (2) evaluate the consequences caused by the use of human–machine interfaces. As for the DYLAM method, it requires several models:

– A simulation model: it generates the information coming from the dynamic event trees by integrating modules such as the evolution of physical variables, the behavior of technical components and the characteristics of human operators, as well as the exchange of information between these modules. Moreover, it enables analysis of the sequences defined by the different transitions between the states of the components. It therefore takes into account the transitions associated with a given sequence and simulates the evolution of the physical process after each transition. As a result, it interacts with the other models.

– A model of the technical system, which consists of two parts: the first part simulates the probabilistic behavior of the technical components such as pumps or gates. The second part simulates the physical process from arithmetic equations.

– A model of the human operator in two main parts: in the first part, the human operators behave normally, i.e. by following a set of predefined procedures. A human error is then associated with deviations in the execution of these procedures. The second part concerns the cognitive models that simulate the information processing of the human operator.

– A model of human–machine interaction that allows the connection between the technical model and the human model. It is primordial, especially in order to point out a technical fault to the human operator.

These models are integrated all along the process of analysis of the method that contains the following four main steps:

– Preliminary screening of relevant scenarios for the study of errors. This step takes into account the fact that the method combines a classical evaluation of reliability, which is the input point for a dynamic simulation that allows the classification of scenarios.

– A detailed analysis of this screening by simulation so as to determine scenarios of accidents. Each scenario is then associated with a history of physical variables of the piloted process such as temperature or pressure, transitions between the states of the process and mental states of the human operator. It is represented using a dynamic event tree.

– A phase of reduction and grouping of the results enables the elaboration of a simplified but significant preview of the obtained dynamic event trees. Indeed, these can contain enormous amounts of information such as the total number and type of possible sequences and the history of each sequence. The simplification is done through previously defined criteria. Figure 5.4 gives an example of this type of simplification for an initial process E-0 and a recovering process E-1. The input points of E-1 come from transitions or shortcuts related to E-0. The terminal sequences linked to the end of procedure, premature termination of a procedure or even to fusions between branches must be anticipated.

– A qualitative and quantitative analysis of the results enables the evaluation of the degree of safety of the studied system on two levels: the level of the human operator (e.g. the impact of the procedures on human behavior, the impact of the training scenarios on security, the identification of new errors, etc.) and at the level of human–machine interaction (e.g. the ergonomic evaluation of the human–machine interface, the impact of the human–machine interface on the occurrence of errors, the impact of the human–machine interface on the control of errors, etc.).

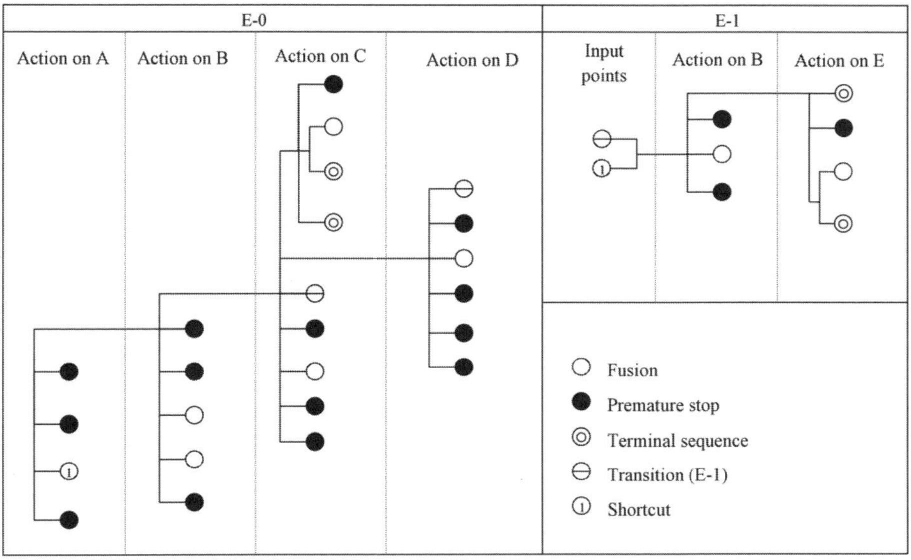

Figure 5.4. *Example of a simplified tree of dynamic events [MAC 92]*

The experimental validation of the HITLINE method was carried out on a simulation of the control of a nuclear power plant generator. This simulator, the details of which are given in [GOO 84] and [SAS 93], not only integrates models of primary, secondary and tertiary circuits, but also models of the reactor, the pressurizer, the vapor generator and several controllers such as the safety valves.

Different faults can be simulated and two scenarios have been used: the first is presented to the human operators without any procedure to be followed, and the second with a series of procedures to follow. These are defined beforehand as a function of particular events associated with behaviors of variables (Table 5.2).

The results of experiments with students at the University of Technology of Delft (NL) were automatically saved in terms of an events history. Using a replay of their experimentation, the human operators submitted to a questionnaire so as to explain their behaviors.

Procedures	
1.	Control power of the reactor (Qi)
2.	Control the control position of the reactor (XI)
3.	Control the temperature of the reactor (Ti)
4.	Control the electrical load of the turbine generator (WE6)
5.	Control the pressure of the pressurizer (P7)
6.	Control the level in the pressurizer (ALFA7)
7.	Control the pressurizer's heater (E7)
8.	Control the spray flow in the pressurizer (F15)
9.	Control the level in the steam generator (ALFA3)
10.	Control the pressure in the steam generator (P3)
11.	Control the pump in the volume control tank circuit (0M23)
12.	Control the feedwater pump (0M21)
13.	Control the flow in the reactor cooling system oil pump (F24)
14.	Alarm AC5 activated? Activate "SGTR" by pressing "d"
15.	Alarm F10 activated? Activate "SIG Open valve" by pressing "d"
16.	Alarm F15 activated? Activate "PRZ safety valve open" by pressing "d"
17.	Alarm ALFA8 activated? Activate "VCT tank empty" by pressing "d"
18.	Alarm AC10 activated? Activate "SIG valve open" by pressing "d"
19.	If all the faults from 14 to 18 have been solved, archive them by pressing "o", otherwise repeat steps 1-18.

Table 5.2. *Summary of the operational procedures for the control of the simulation*

The results coincide quite reliably with those anticipated from the HITLINE method by using the models of the human operator and human–machine interaction in terms of the occurrence of errors in the execution of procedures (i.e. omissions, shortcuts, diagnosis errors) and identification of factors influencing these errors (i.e. weight of the procedure, perception of the advantages of a procedure, confidence when faced with failures, temporal pressure).

In the DYLAM and HITLINE methods, the tasks coming from added or altered behaviors can be integrated in the dynamic event tree. Also, HITLINE can take into account a type of violation: shortcuts in the execution of procedures.

The following method does evaluate dynamic situations, but was designed to explicitly integrate not only the analysis of erroneous tasks, but also of additional tasks that could be detrimental to the reliability of the global system, i.e. altered or added tasks.

5.4. Analysis of altered or added tasks

The ACIH[2] is an approach for multicriteria and multireference evaluations of dynamic work situations regarding different indicators of comparison of these situations [VAN 03, VAN 10a]. This section summarizes the different steps of the method and develops these analysis indicators. An example illustrates the feasibility of its use.

5.4.1. *Principles of the ACIH method*

The ACIH method evaluates in a prospective or retrospective manner the consequences of human unreliability on the performance of human–machine systems, the frequency of occurrence not being taken into account.

The approach comprises several successive steps as follows:

– the step of functional analysis of the human–machine system concerns the analysis of the piloted process to identify its functions and the role of the human operators to carry out. Human functions are expressed in terms of procedures that are lists of tasks to be carried out depending on the context. The lowest level of functional decomposition therefore corresponds to all of the procedures to carry out a given function or subfunction;

– the procedural and contextual analysis step consists of identifying, for each procedure, the associated contexts and the tasks to carry out in these contexts. This step therefore determines the contexts and the tasks that result for each procedure. Any new event that could modify the list of initial tasks to be carried out in a normal work context determines a new context;

– analysis of the tasks and the activity. The analysis of tasks aims to determine the prescriptions and anticipate the real behaviors from possible deviations for a given task. The analysis of activity enables the identification of real behaviors;

– identification of modes of behavioral deviations. The comparison between effective tasks observed on the field or anticipated by predictive

2 ACIH: French acronym for analysis of the consequences of human unreliability.

analysis with the prescribed tasks allows the identification of three classes of modes of behavioral deviations: erroneous modes, altered modes and added modes;

– the final step is the analysis of consequences that consists of coupling modes of human unreliability and their impact on the system for a given procedure associated with its context, and equally for a combination of procedures associated with the same context. The analysis of the consequences is based on the concept of dynamic work situation with regard to its level of gravity, determined from predefined functions, so that the greater the value returned by these functions is, the greater the gravity of the situation. Thus, a given task is considered according to two points of view:

- in relation to what is done by the operator, this is his behavior;

- in relation to the result of this behavior, this is the work situation that results.

A situation at a given instant is defined as the collection of independent events or groups of events present at the same moment of which one or several tasks of the operator are required. As a result, the gravity of the situation is the sum of the gravities associated with these events or groups of events.

A situation can be evaluated with regard to several performance criteria relative to the process or to the operator, such as safety or production of the process and the safety or workload of the human operators.

From this simplified representation of a situation, the rules of acceptability of human behaviors are determined regarding different references, and require the definition of significant indicators to compare dependent situations (i.e. when a situation is the result of one action on another) or independent situations (i.e. when several actions plans are evaluated or the prescriptions are compared with real behaviors).

5.4.2. *Acceptability and evaluation of human behaviors*

The levels of acceptability of human behaviors in the use of a given human–machine system depend on points of view associated with distinct references [VAN 03] (see Figure 5.5).

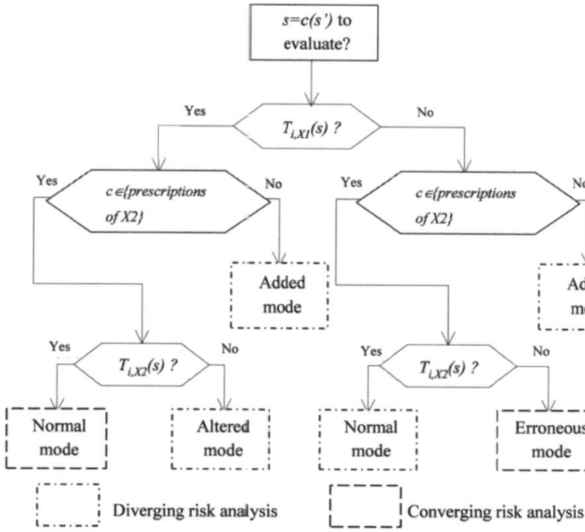

Figure 5.5. *Acceptability of human behaviors*

For example, considering references *X*1 and *X*2, faced with a common situation *s'*, a behavior *c* of *X*1, or the situation that results *s*, is prohibited or prescribed by *X*2 as a function of the consequences that it can induce on *s*. Thus, the acceptability of a given behavior *c* is associated with the gravity $g_{i,s}$ of the situation *s* that results at the instance *t*, and this for a given criterion *i*. The tolerance or acceptability of this situation, noted $T_{i,X}(s)$, is determined by the position of its gravity in relation to a given threshold of acceptability $S_{i,X}$ of the reference *X*. Below the threshold, which, to simplify things is considered to be constant and linear, the situation is tolerable for *X*, beyond which it becomes intolerable:

$$T_{i,X}(s) \leftrightarrow (g_{i,s}(t) < S_{i,X})$$

The erroneous mode is related to a prescription and is unacceptable for *X*1 and *X*2. The altered mode is a mode of violation for which the prescriptions of *X*2 are not followed by *X*1. These can be malevolent, sabotages or even violations with no harmful intentions. The added mode does not correspond to a mode prescribed by *X*2, and also does not correspond to an altered mode. It was previously analyzed by *X*1 and not by *X*2. It creates a new behavior, which consists of using a work support to reach an additional unplanned objective. An altered mode helps to reach a planned objective but by voluntarily changing the prescribed behavior.

We will not cover errors of design here. Nevertheless, they can be identified using divergent analyses. However, in no way does this exclude the possibility of the existence of this type of error, even when the analyses are convergent.

Several characteristics have been defined to compare dependent situations for the same criterion i, i.e. situations appearing at different moments but one is the result of an action on the other [POL 00, POL 02b]. They have been adapted so as to compare two independent situations, i.e. situations appearing at the same time and that could be related to different planes of action or to a comparison between the real situation and the one prescribed [VAN 01, VAN 04, VAN 11a]. Thus, a situation $s1$ can be compared to situation $s2$ from the associated gravity gaps denoted by $g_{i,s1}(t1)$ and $g_{i,s2}(t2)$ with $t1 = t2$ or $t1 \neq t2$, for the same criterion i. These gaps can be evaluated qualitatively or quantitatively and are interpreted in terms of benefits and costs:

– the benefit, written as $B_i\,(s1,s2)$. The benefit between two situations is defined by a decrease in gravity of $s2$ in comparison to $s1$;

– the cost, written as $C_i\,(s1,s2)$. The cost between two situations equates to an increase in the gravity of $s2$ in comparison to the gravity of $s1$.

It must be noted that for added modes, the prescribed mode does not exist, and the situation must be compared to a reference situation determined from the situation that should have occurred without applying the mode considered.

Two other characteristics have been defined so as to analyze the gravity of two situations in relation to the acceptability threshold $S_{i,X}$ of the tolerance function $T_{i,X}$:

– the deficit, written as $D_{i,X}\,(s1,s2)$. The deficit is a cost with crossing of the acceptability threshold $S_{i,X}$. Situation $s2$ is intolerable while situation $s1$ is not;

– the recovery, written as $R_{i,X}\,(s1,s2)$. The recovery is a benefit with crossing of the acceptability threshold $S_{i,X}$. Situation $s2$ is tolerable while situation $s1$ is not.

It is then possible to qualitatively or quantitatively compare different situations in relation to the same reference or in relation to different references.

Other numerical indicators associated with a series of situations have been defined: these are the control and loss of control dynamic situations (see Figure 5.6).

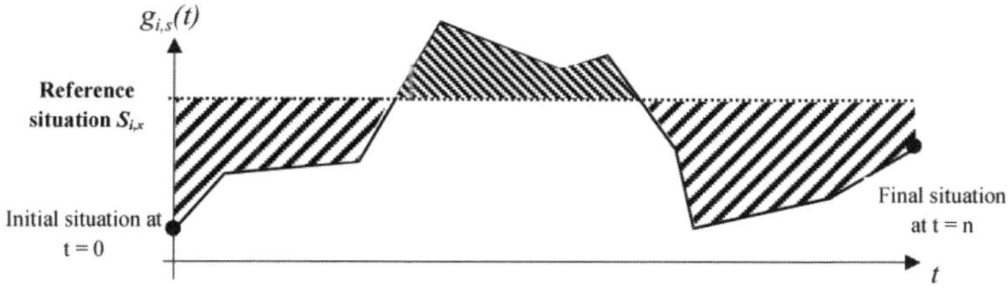

Control	$$\frac{\int_0^n \left|g_{i,s}(t)-S_{i,X}\right|.dt-\int_0^n \left(g_{i,s}(t)-S_{i,X}\right)dt}{2}$$
Loss of ontrol	$$\frac{\int_0^n \left|g_{i,s}(t)-S_{i,X}\right|.dt+\int_0^n \left(g_{i,s}(t)-S_{i,X}\right)dt}{2}$$

Figure 5.6. *Control and loss of control of situation*

For a list of consecutive situations over a time period *[0,n]*, the acceptable control surface, for a give criterion *i*, is the surface occupied between the points $(0, g_{i,s}(0))$ and $(n, g_{i,s}(n))$ and located below the acceptable limit given by a reference situation $S_{i,X}$. Similarly, the unacceptable control surface for a given criterion *i* is the surface occupied between the same points but located beyond the limit $S_{i,X}$. As a result, the acceptable surface is all the more important as the situation is under control.

5.4.3. *Example of application*

The ACIH method has been applied for the study of the use of large flow industrial rotaries, as part of the PROSPER program of the CNRS and in collaboration with INRS[3] [POL 03]. Observations were made on the field from two customers using this type of machine. Real behaviors were observed and compared with those prescribed, determined using user manuals, or rules of safety or of use provided by the designer of the rotaries. For the added modes, the reference behavior is the one that exists without

3 INRS: French acronym for National Research Institute for Safety: http://en.inrs.fr/.

applying the corresponding added mode. Altered and added modes were evaluated in terms of potential benefit and potential deficit. The analysis of these consequences relies on the following four criteria of performance:

– workload is defined as the number of operations on the machine;

– safety is analyzed in relation to the designer's instructions. These help avoid exposing the user to certain dangers: danger of falling, crashing, cuts, noise-related danger, use of toxic products, fire danger, electrocution, etc.;

– the quality of the copies is defined in relation to the printing and how it is followed. It takes into account the number of copies accepted;

– production takes into account the running time of the machine. It is related to the fact that stopping it costs a relatively large amount of money (around 1,550 euros per hour).

Considering that the same mode can affect several of these criteria, 20 altered modes and four added modes were identified [VAN 03]. The potential benefits and deficits that were predominant in all of the modes observed are listed in Table 5.3. Two contexts were retained: the individual context if it is an operator that is concerned and the collective context when it is a group of operators that is involved.

		Altered modes		Added modes	
		Individual context	Collective context	Individual context	Collective context
Immediate benefit	Workload	7	3	1	0
	Quality	2	0	1	0
	Production	4	4	2	0
	Security	0	0	0	0
Potential deficit	Workload	0	0	0	0
	Quality	2	2	1	0
	Production	4	2	2	1
	Security	9	5	0	1

Table 5.3. *Examples of altered and added modes*

Some of the altered modes that were observed resulted in an immediate benefit relative either to production or to the workload of the operators, and the potential deficit was to do with the safety of these operators. These modes are safety barrier removals [POL 02a], resulting in immediate benefit despite the existence of a potential deficit; they are associated with removal or disrespect of these barriers.

Each mode can also be evaluated in terms of maintaining or losing control. As an example, from a prescribed procedure of cleaning the printing cylinders, the following three deviations can be defined and studied [VAN 03], as presented in Table 5.4:

– in deviation P1, the operator is not wearing any individual protective clothing. The operator is exposing himself/herself to the dangers linked to the use of toxic products;

– in deviation P2, he/she is not following the regulations that state that no manual intervention should take place on a running working tool. The operator is exposing himself/herself to the dangers linked to intervening on a running working tool;

– deviation P3 is a combination of deviations P1 and P2.

Deviated procedure 1 (P1)	Deviated procedure 2 (P2)	Deviated procedure 3 (P3)
1. Press the PRINT button 2. Press the EMERGENCY STOP button 3. Wash the visible part with a sponge 4. Wipe the washed surface with a dry cloth 5. Deactivate the EMERGENCY STOP button 6. Press the HOLD FOR RELEASE button to obtain a visible non-treated part 7. Repeat steps 2 to 6 as much as necessary 8. Press the PRINT button	1. Wear individual protective clothing 2. Press the PRINT button 3. Press the HOLD FOR RELEASE button for continuous rotation 4. Wash the visible part with a sponge 5. Dry the washed surface with a dry cloth 6. Press the STOP button 7. Press the PRINT button	1. Press the PRINT button 2. Press the HOLD FOR RELEASE button for continuous rotation 3. Wash the visible part with a sponge 4. Dry the washed surface with a dry cloth 5. Press the STOP button 6. Press the PRINT button

Table 5.4. *Example of the comparison of altered modes*

Each procedure leads to the same result, i.e. cylinders are clean, but the result of their evaluation in terms of benefit, potential deficit, control and loss of control is different. The prescribed procedure, which is equivalent to procedure P1 but while wearing protection, takes more time, while the altered procedures reduced execution time. Moreover, while the level of safety is maximal for the prescribed procedure, it decreases with the altered procedures, which exposed the human operators to additional danger. The threshold from which having and losing control are evaluated is determined by the path the prescribed procedure follows.

5.5. Perspectives for the design of a safe system

Following risk analyses, but also through following rules and safety standards, through a logic of defenses in series, the designer equips the production tool with barriers. These are not only preventive methods, but also methods of protection against a given risk [HOL 99]. However, human operators can decide whether or not to use these barriers, which are related to the prescription of use that the designer defines for a given machine. Ultimately, the operators change the operatory modes of use of the machine and end up accepting them by considering them as normal modes.

Currently, there is no method of human reliability evaluation that is able to take into account such violations of the prescription, or additional tasks that were not planned during design. At this point, we can ask how a system can be considered safe when violations and added tasks are not evaluated.

This gap between the prescribed uses and the real uses of a given machine can appear for different reasons:

– the description of the procedures is lacking of precision. Their execution then depends on the human operators' interpretation;

– several shortcuts can exist in the execution of procedures. When these shortcuts are easily noticeable by the human operators, they test them and then apply them systematically as soon as they become convinced that they pose no problem;

– the human operator's knowledge about the safety of the system used can decrease over time, especially when the execution of modified procedures becomes a regular occurrence;

– work conditions such as a high production rate or a social atmosphere in a team can also encourage drifts.

Such deviations seem inevitable and must therefore be seriously considered in methods of analysis and reliability. Violations and added tasks taking place with no bad intentions must be considered as relevant indicators to increase the operational control of the human–machine system concerned. The designer should no longer prescribe procedures that indicate the different steps to obtain a given result, but should propose a protocol at a more abstract level, allowing control of behaviors faced with the results. A procedure is limited to a prescribed sequence of steps. Each step is either correct or incorrect, and is linked to other procedures related to the operational context. A protocol covers a set of steps of the accepted practices for a given point of reference, for example the human operator, and is either acceptable or unacceptable for another reference, for example the designer. It is less rigid than a procedure as it has to integrate the presence or absence of particular external elements such as safety barriers in the operational context.

This concept of protocol has to include, on the one hand, violations, or the different manners to obtain an identical result by following a voluntarily deviated behavior, and, on the other hand, added behaviors after observation on the field, or by inference from the functions of the system. As a result, these practices of possible use concern the new paths to obtain a given result; their study allows the evaluation of the motivation of human operators to aim for increased control of work situations.

The behavior of operators can then depend on various factors:

– factors external to the operator that can be separated into two categories. The first contains organizational factors linked to the environment or the organization of work such as recognition of the activity of an operator within a team, or free time. The second category contains technical factors linked to the performance of the piloted process such as speed or quality of production;

– factors internal to the operator such as those related to the perception of external factors, to experience, confidence, stress or mental workload.

The analysis of risk determines the conjunction of factors that can generate the occurrence of an undesirable event. The designer of a given machine can therefore anticipate behavioral deviations such as the crossing or the adding of barriers by the users of this machine, with regard to factors that can influence their occurrence or their impact on the human–machine system. This predictive approach can be based on retrospective analysis to identify risk factors.

An experimental study by simulation of a guided transport system showed the feasibility of such an evaluation [VAN 02, ZHA 04, VAN 09, POL 09]. The analysis took into account the immediate benefit, the possible cost and the potential deficit associated with removed barriers, and the factors that can affect four criteria of performance: the respect of the itinerary and the timing of trains, the respect of planned stops at stations, the respect of the safety of the traffic and the regulation of the workload caused by tasks of control and supervision of trains.

Nevertheless, removing or adding barriers can also be the result of errors of perception, diagnosis, prognosis or execution by human operators. As a result, the analysis of behavioral deviations in a given protocol must also be based on the concept of cognitive errors so as to optimize specification and the positioning of barriers and to reduce the occurrence of errors or to control their consequences. The management of errors by barriers must then contain the following stages [VAN 10a]:

– prevention to avoid the occurrence of undesirable events such as dangerous behavioral deviations;

– protection or confinement to deal with the consequences of these events;

– recover to control the occurrence and consequences of these events and return to a normal situation.

Methods of analysis of human reliability such as DYLAM, HITLINE or ACIH support the specification of these three levels of error management by taking into account the impact of dynamic aspects on human behavior. Moreover, faced with the specification of defenses in a human–machine system, the ACIH approach allows the evaluation of behaviors such as the adding, bypassing, avoidance, inhibition or modification of barriers by human operators. The removal of barriers, for example, is assimilated with particular violations: they can be precursors to incidents or accidents, but also indicators of the increase in control of a given situation in relation to a compromise between the level of safety and other criteria. New perspectives can therefore be envisaged. For example, they concern the combination of methods for the evaluation of behavior [VAN 10b], the application of principles of cooperation to make their control easier [ZIE 09], the development of learning support systems to increase human or technical knowledge [VAN 12] or the use of simulations to understand their origin [DE 07].

5.6. Conclusion

Even though current technology allows for the complete or partial automation of some processes, it remains insufficient for tasks of complex problem resolution, which require a human presence. We are still faced with the dilemma voiced by Bainbridge in his famous article "ironies of automation" [BAI 83]: the human remains invaluable, because an automated system is not flawless, and is incompetent when faced with unplanned dysfunction. However, human inactivity can lead to a decrease in vigilance and a deterioration of human competency that risks harming the safety of the global system. This safety is analyzed by the designers of machines who plan the normal modes of operation from the analysis of the associated risks. To these modes of operation, procedures or barriers are then designed to prevent accidents or limit their effects. However, observation in the field shows that these barriers are not always respected, which leads to modes that are not always taken into account during risk analysis. The occurrence of such deviations in behavior can be explained by different interests in terms of the performance to be reached.

This chapter presented the general process of analysis of human reliability and raises the problem of the validity and reusability of the current methods. These are insufficient because they do not detail in which conditions the evaluations are done. For example, faced with the definition of human error developed in this chapter, they only focus on the execution of the prescribed tasks without taking into account deviations such as added behaviors or violations of the human operator and are limited to the use of single model of the human–machine system. Three methods have thus been proposed: the DYLAM and HITLINE methods allow the evaluation of human reliability within a human–machine system by integrating several models such as behavioral models of the process, the human operator and human–machine interaction, with which a simulation system interacts. The

ACIH method enables the evaluation of erroneous, changed and added modes of behavior in terms of benefit, cost, potential deficit, control and loss of control of dynamic situations using different criteria and different references of analysis.

These methods serve as a base to extend knowledge on the human behavior during the design, analysis and evaluation of a given human–machine system. The concept of protocol is then evoked to assist the designer of the machine during this process of validation. Instead of specifying a set of procedures and studying possible deviations, it would seem important to define the references from which the evaluations are carried out and to compare the acceptability of behaviors between these references for a given protocol. The positioning of barriers as well as the act of removing them can then be studied so as to define adaptive barriers of prevention, protection and error recovery, affecting criteria such as safety, production, quality or even the workload of the human operator.

5.7. Bibliography

[AME 81] AMENDOLA A., REINA G., "Event sequences and consequence spectrum: a methodology for probabilistic transient analysis", *Nuclear Science and Engineering*, vol. 77, pp. 297–315, 1981.

[BAU 99] BAUMONT G., Modèle et méthode RECUPERARE – Une méthode d'analyse des incidents intégrant la fiabilité humaine, report DES 371, IPSN, February 1999.

[BAI 83] BAINBRIDGE L., "Ironies of automation", *Automatica*, vol. 19, no. 6, pp. 775–779, 1983.

[BEL 88] BELLO G.C., COLOMBARI V., "The human factors in risk analyses of process plants: the control room operator model, TESEO", *Reliability Engineering*, vol. 1, pp. 3–14, 1988.

[BIL 97] BILLINGS C.E., *Aviation Automation: The Search for a Human-Centered Approach*, Lawrence Erlbaum Associates, Mahwah, NJ, 1997.

[CAC 98] CACCIABUE P.C., *Modelling and Simulation of Human Behaviour in System Control*, Springer-Verlag, London, 1998.

[CAC 04] CACCIABUE P.C., *Guide to Applying Human Factors Methods*, Springer-Verlag, London, 2004.

[CAC 10] CACCIABUE P.C., "Dynamic reliability and human factors for safety assessment of technological systems: a modern science rooted in the origin of mankind", *Cognition, Technology & Work*, vol. 12, p. 119–131, 2010.

[DE 03] DE KEYSER V., "Les systèmes de report d'incident", in AMALBERTI R., FUCHS C., GILBERT C. (eds.), *Autour de la mesure du risque – Un questionnement multidisciplinaire*, Publications de la MSH-Alpes, Grenoble, France, pp. 41–71, 2003.

[DE 07] DE WINTER J.C.F., WIERINGA P.A., KUIPERS J., *et al.*, "Violations and errors during simulation-based training: objective "Reason"-ing", *Ergonomics*, vol. 50, no. 1, pp. 138–158, 2007.

[DOU 95] DOUGHERTY E., "'Violation' – does HRA need the concept ?", *Reliability Engineering and System Safety*, vol. 47, pp. 131–136, 1995.

[EMB 86] EMBREY D.E., *SHERPA: A Systematic Approach for Assessing and Reducing Human Error in Process Plants*, Human Reliability Associated Ltd, Lancashire, UK, 1986.

[FAD 97] FADIER E., La fiabilité opérationnelle des systèmes automatisés, Thesis, Institut national polytechnique de Lorraine, Nancy, France, 1997.

[GER 97] GERDES V., Identification and analysis of cognitive errors – Application to control room operators, Doctorate Thesis, University of Technology of Delft, The Netherlands, 1997.

[GOO 84] GOODSTEIN L.P., HEDEGARD J., SOE HEJBERG K., *et al.*, The GNP test-bed for operator support evaluation, RISØ report M-2460, Roskilde, Denmark, 1984.

[HAN 84] HANNAMAN G.W., SPURGIN A.J., LUKIC Y., "Human cognitive reliability model for PRA analysis", Electric Power Research Institute, Report no. 4531, Project RP2170-3, Palo Alto, CA, 1984.

[HOL 93] HOLLNAGEL E., *Human Reliability Analysis: Context and Control,* Academic Press, London, 1993.

[HOL 98] HOLLNAGEL E., *Cognitive Reliability and Error Analysis Method CREAM,* Elsevier, Amsterdam, The Netherlands, 1998.

[HOL 99] HOLLNAGEL E., "Accidents and barriers", in HOC J.M., MILLOT P., HOLLNAGEL E., CACIABUE P.C. (eds.), *Proceedings of the 7th European Conference on Cognitive Science Approach to Process Control (CSAPC '99)*, Presses Universitaires de Valenciennes, France, pp. 175–180, 1999.

[HUM 88] HUMPHREY P., "Human reliability assessors guide", in SAYERS B.A. (ed.), *Human Factors and Decision Making – Their Influence on Safety and Reliability,* Elsevier, Amsterdam, The Netherlands, pp. 71–86, 1988.

[KIR 97] KIRWAN M., "Validation of human reliability assessment techniques: part 2 – validation results", *Safety Science*, vol. 27, pp. 43–75, 1997.

[LAP 95] LAPRIE J.-C., FABRE C., GUILLERMAIN H., *et al.*, in LAPRIE J.-C. (ed.), *Guide de la sûreté de fonctionnement,* Cépaduès Editions, Toulouse, France, 1995.

[LEP 85] LEPLAT J., *Erreur humaine, fiabilité humaine dans le travail*, Armand Colin, Paris, 1985.

[MAC 92] MACWAN A., Methodology for analysis of operator errors of commission during nuclear power plant accidents with application to probabilistic risk assessment, PhD Thesis, University of Maryland, 1992.

[MAC 94] MACWAN A., MOSLEH A., "A methodology for modeling operators errors of commission in probabilistic risk assessment", *Reliability Engineering and System Safety*, vol. 45, pp. 139–157, 1994.

[MIL 87] MILLER D.P., SWAIN A.D., "Human error and human reliability", in SALVENDY G. (ed.), *Handbook of Human Factors*, John Wiley & Sons, New York, pp. 219–250, 1987.

[NAG 88] NAGEL D.C., "Human error in aviation operations", in WIENER E.L., NAGEL D.C. (eds), *Human Factors in Aviation*, Academic Press, San Diego, CA, pp. 263–303, 1988.

[O'LE 99] O'LEARY M., "British Airways human factors reporting programme", *Proceedings of the International Workshop on Human Error, Safety, and System Development*, Liège, Belgium, 7–8 June 1999.

[POL 00] POLET P., VANDERHAEGEN F., MILLOT P., "Modélisation du risque au travers de l'amélioration et la dégradation de situations", *Actes du 12e Colloque national de sûreté de fonctionnement*, Montpellier, France, pp. 213–220, 28–30 March 2000.

[POL 02a] POLET P., VANDERHAEGEN F., WIERINGA P., "Theory of safety related violation of system barriers", *Cognition Technology & Work*, vol. 4, no. 3, pp. 171–179, 2002.

[POL 02b] POLET P., Modélisation des franchissements de barrières dans l'analyse des risques des systèmes homme-machine, Doctorate Thesis, University of Valenciennes and Hainaut-Cambrai, 9 December 2002.

[POL 03] POLET P., VANDERHAEGEN F., AMALBERTI R., "Modelling border-line tolerated conditions of use (BTCUs) and associated risks", *Safety Science*, vol. 41, pp. 111–136, 2003.

[POL 09] POLET P., VANDERHAEGEN F., MILLOT P., "Human behaviour analysis of barrier deviations using a benefit-cost-deficit model", *Advances in Human-Computer Interaction*, pp. 10–19, 2009.

[RAN 98] RANKIN W., KRICHBAUM L., "Human factors in aircraft maintenance", *Integration of Recent HRA Developments with Applications to Maintenance in Aircraft and Nuclear Settings*, Seattle, WA, 8–10 June 1998.

[RAS 83] RASMUSSEN J., "Skills, rules and knowledge; signals, signs, and symbols, and other distinctions in human performance models", *IEEE Transactions on Systems, Man and Cybernetics*, vol. 13, no. 3, pp. 257–266, 1983.

[REA 90] REASON J., *Human Error*, Cambridge University Press, Cambridge, 1990.

[SAS 93] SASSEN J.M.A., Design issues of human operator support systems, PhD Thesis, Laboratory for Measurement and Control, Faculty of Mechanical Engineering and Marine Technology, Delft University of Technology, Delft, The Netherlands, 1993.

[SHA 97] SHARIT J., "Allocation of functions", in SALVENDY G. (ed.), *Handbook of Human Factors and Ergonomics*, John Wiley & Sons, New York, pp. 301–339, 1997.

[SHE 92] SHERIDAN T.B., *Telerobotics, Automation and Human Supervisory Control*, MIT Press, Cambridge, MA, 1992.

[SHE 99] SHERIDAN T.B., "Human supervisory control", in SAGE A.P., ROUSE W.B. (eds), *Handbook of Systems Engineering and Management*, John Wiley & Sons, New York, pp. 591–628, 1999.

[SIU 94] SIU N., "Risk assessment for dynamic systems: an overview", *Reliability Engineering and System Safety*, vol. 43, pp. 43–73, 1994.

[SWA 83] SWAIN A.D., GUTTMANN H.E., *Handbook of Reliability Analysis with Emphasis on Nuclear Plant Applications*, Nuclear Regulatory Commission, NUREG/CR-1278, Washington DC, 1983.

[SWA 90] SWAIN A.D., "Human reliability analysis: need, status, trends and limitations", *Reliability Engineering and System Safety*, vol. 29, pp. 301–311, 1990.

[VAN 97a] VAN DER SCHAAF T.W., "Prevention and recovery of errors in software systems", *Proceedings of the Workshop on Human Error and Systems Development*, Glasgow, UK, pp. 49–57, 19–22 March 1997.

[VAN 97b] VANDERHAEGEN F., "Multilevel organization design: the case of the air traffic control", *Control Engineering Practice*, vol. 5, no. 3, pp. 391–399, 1997.

[VAN 99] VANDERHAEGEN F., "Multilevel allocation modes – allocator control policies to share tasks between human and computer", *System Analysis Modelling Simulation*, vol. 35, pp. 191–213, 1999.

[VAN 01] VANDERHAEGEN F., POLET P., MILLOT P., "Modélisation duale des risques dans les conditions limites tolérées par l'usage", *3e Conférence francophone de Modélisation et Simulation – MOSIM'01 – Conception, Analyse et Gestion des Systèmes Industriels*, Troyes, France, pp. 869–876, 25–27 April 2001.

[VAN 02] VANDERHAEGEN F., POLET P., ZHANG Z., *et al.*, "Barrier removal study in a railway simulation", *Proceedings of the 6th International Conference on Probabilistic Safety Assessment and Management*, San Juan, PR, 23–28 June 2002.

[VAN 03] VANDERHAEGEN F., *Analyse et Contrôle de l'Erreur Humaine*, Hermès, Lavoisier, Paris, 2003.

[VAN 04] VANDERHAEGEN F., "The benefit-cost-deficit (BCD) model for human error analysis and control", *Proceedings of the 9th IFAC/IFIP/IFORS/IEA Symposium on Analysis, Design and Evaluation of Man-Machine Systems*, Atlanta, 7–9 September 2004.

[VAN 09] VANDERHAEGEN F., "A reinforced iterative formalism to learn from human errors and uncertainty", *Engineering Applications of Artificial Intelligence*, vol. 22, nos. 4–5, pp. 654–659, 2009.

[VAN 10a] VANDERHAEGEN F., "Human-error-based design of barriers and analysis of their uses", *Cognition, Technology & Work*, vol. 12, pp. 133–142, 2010.

[VAN 10b] VANDERHAEGEN F., CASSANI M., CACCIABUE P.C., "Efficiency of safety barriers facing human errors", *11th IFAC/IFIP/IFORS/IEA Symposium on Analysis, Design, and Evaluation of Human-Machine Systems*, Valenciennes, France, September 2010.

[VAN 11a] VANDERHAEGEN F., ZIEBA S., ENJALBERT S., *et al.*, "A benefit/cost/deficit (BCD) model for learning from human errors", *Reliability Engineering & System Safety*, vol. 96, no. 7, pp. 757–776, 2011.

[VAN 12] VANDERHAEGEN F., "Cooperation and learning to increase the autonomy of ADAS", *Cognition, Technology & Work*, vol. 14, no. 1 pp. 61–69, 2011.

[VIL 88] VILLEMEUR A., *Sûreté de Fonctionnement des Systèmes Industriels*, Octarès, Toulouse, 1988.

[ZHA 04] ZHANG Z., POLET P., VANDERHAEGEN F., *et al.*, "Artificial neural network for violation analysis", *Reliability Engineering and System Safety*, vol. 84, no. 1, pp. 3–18, 2004.

[ZIE 09] ZIEBA S., POLET P., VANDERHAEGEN F., "Using adjustable autonomy and human-machine cooperation to make a human-machine system resilient – application to a ground robotic system", *Information Sciences*, vol. 181, no. 3, pp. 379–397, 2009.

PART 3

Human–Machine Cooperation

6

Causal Reasoning: A Tool for Human–Machine Cooperation

6.1. Introduction

Supervision systems in control rooms must be a primary decision-making tool and therefore must aim to provide an online digital overview of the system status, which is necessary for a pertinent analysis of the situation. These systems must, of course, detect failures, locate them and filter them using more or less complex techniques and must not lose sight of the ultimate objective that is the human operator: they are in charge of what decision to make, which corrective measures to take and the timing of these, emergency protocols and stops; they generally have limited time to analyze and react to the malfunctioning situation. They also expect legitimate help for troubleshooting from the surveillance system and not simply new and increasingly sophisticated indicators irrelevant to the process's mental representation. Thus, artificial intelligence (AI) and specifically causal reasoning since the mid-1980s has had the aim of outlining models that allowed us to understand and explain systems, which had become too complex for control room operators, in order to assist operators' decision-making with higher level cognitive tasks [DUB 89].

In this chapter, we address causal reasoning through its application in the supervision of complex industrial applications. A quick description of the tasks performed by control room operators helps us to better understand the essential role of this cognitive mode in both normal situations (surveillance,

---

Chapter written by Jacky MONTMAIN.

observation) as well as incidental situations (fault detection, identification of malfunctions). The representation medium considered for causal reasoning in this chapter is an influence graph, which models the cause and effect relations between the variables of a process.

The different examples or illustrations provided originate from the reprocessing of nuclear waste. The interested reader can refer to [LEY 91c, LEY 91b, MON 92, PEN 93, MON 00a, MON 00b, CEL 07] for a description of the process, the basic components of which are liquid–liquid extractors.

6.2. Supervision

At the beginning of the 1980s, the first proposed support systems were exclusively for helping the human operator to diagnose faults [NEL 82, SAC 86], since this step was already considered crucial. Then, faced with the amplitude and difficulty of this task, other works were prioritized with the aim of providing assistance to the operator such that they would not find themselves in a critical situation [DE 89]. The role of such a system is therefore to provide a digital image of the process, in order to clearly explain its current situation, or the situation toward which it is evolving.

The state of a process is traditionally represented by the state of the different variables that define it. This leads to two distinct problems with the implementation of a pertinent supervision support system:

– The choice of variables to be kept for the creation of the interface for the operator, adapted to each situation.

– The choice of variables to be introduced at the modeling level.

Multiple tasks must be completed by a supervision support system: the detection of pertinent inputs and significant disturbances, the prediction of the process evolution, the early detection of malfunctions, the filtering of these faults through analysis of the propagation of a fault in the process, the delivery of advice, the locating of failures, etc. Supervision is therefore placed in the context of global observation. The pertinence of the provided decision support must be prioritized with regard to its precision: it therefore seems preferable to be able to provide the human operator with physical explanations of the analyzed behavior and behavioral predictions at any moment rather than systematically provide him with the precise values of every variable.

The reasonings required to successfully carry out the previous tasks call for the development of a supervision model used as a reference. The operation of a process conforms to the laws of physics. Therefore, an objective representation of this process does exist, independent of the reasonings we want to apply to it. The research for a model adapted to supervision purposes therefore consists of specifying this objective representation, which is able to support different types of reasonings (prediction, explanation, detection of failures, etc.). Nevertheless, the primary quality of a model is that it must be tailored to its objectives. The sought-after representation must first ensure a synthetic overview of the process; the specific and precise overviews of machines, for example, necessary in order to ensure good control or to provide a precise diagnosis, are not the main concern of such a model and must be managed elsewhere (traditional digital models of control, expert diagnosis systems). The main qualities of a supervision model have been mentioned in [GEN 97]:

– The model must be understandable: it must take into account the variables that are pertinent for the operator, be they measurable or not. Furthermore, the information provided by the monitoring does not need to be precise, since their essential goal is to provide a qualitative analysis of the operation of the process.

– The model must have multiple uses: it must allow for different reasonings able to help the operator, such as those mentioned above. The elementary steps of these reasonings are based on the analysis of the causal link between two variables, since at any moment each reasoning needs to be explainable to the operator. Consequently, the structure of a supervision model must be based on the causality of physical phenomena taking place in the process.

– The model must dynamically represent the process. In a continuously running process, considering the dynamic behavior is essential, since its duration is non-negligible (of the order of a minute, or even an hour). In addition, it is often during dynamic operation phases that situation perception is most difficult. The supervision model must therefore, in order to be efficient at any instant, represent the evolution of the working variables over time, and take into account the well-known simple temporal notions such as the delay and response time.

Process supervision is traditionally realized by advanced applications of industrial computing; however, it is not based on mathematical theory, unlike control. Communication with operators is achieved because of a human–machine interface. Supervision workstations are mostly made up of graphical screens on which the process supervision views are displayed; these views group together the set values of groups of variables to be monitored in analog and digital forms, the temporal evolution curves of these variables and the alarm thresholds. Supervision of the process state is often done through an alarm processing system, which can be connected to an emergency recovery computer system or to an interface with the operator to whom the assessment of the situation is allocated. Supervision covers a large field where precision is not a point of major interest but can require, in order to be completed successfully, the use of precision tools. Among these, there are signal processing algorithms, control algorithms and diagnosis algorithms.

The supervision tasks are shared between the computer systems and the human operators, which leads to a number of problems. The control/command systems are equipped with supervision systems that centralize the data from sensors in a control room. When the process is in a steady state and under normal operation, the computers ensure its running without the need for human intervention. On the other hand, when a state change or a malfunction takes place, the operators become essential for managing the decision-making tasks that we do not know how to automate (or at least not systematically).

Even when assisted by a digital control and monitoring system (DCMS) well adapted to the process, the operator remains the only one able to analyze the situation and make the important decisions. However, the operator only empirically understands the process, and rarely reasons in the long term. The evolution logs displayed do not always show sufficient information, particularly when the scale of the problem encountered exceeds the duration of the log. The operator occasionally has difficulty correlating certain physical phenomena that are linked but spread out over time. The logs present the variables independently of each other; it is up to the operator to understand their relations. Therefore, a mental image of the process is sometimes created that is completely different to its real functionality. Consequently, the responses of an operator to these problems are only adequate if the phenomena are repetitive. An occasional problem, even if it is known, will demand more thinking time and the solution runs the risk of being approximate, since lapses and deformations in human memory alter the responses. Worse still than the occasional problems that can be resolved through the use of simulators, there are problems that are unknown *a priori*. Faced with an unidentified malfunction, the response of the operator can only be valid if they perfectly understand the process and are under favorable conditions.

It is clear that in practice the actual response to supervision problems still essentially remains founded on the progress of industrial computing and is much more based on an increasing amount of information than on its pertinence. Traditional processing systems are capable of displaying a large number of rapidly evolving variables and this is ultimately what leads to the problem. Faced with this flood of information, the limits of humans become quickly apparent. The tendency for operators to overlook certain important pieces of information, not fully respond to a problem, react too slowly and panic as soon as the amount of information increases can lead to dangerous situations and the resulting untimely shutdowns cause the performances of the process to decline considerably from an economic perspective.

This information overload is due to the flexibility of current systems. Being able to adapt to different types of processes means that it must be possible to display all of the information at a given instant to the operator. For a nuclear reprocessing plant this is not simple. Field studies have proven that only 10% of alert messages were acted upon by operators and that their vigilance is very limited (around half an hour) for monitoring a given problem. In the case of an incident, they are quickly overwhelmed by the flood of events (cognitive overload). This overload is paradoxically due to the increasing complexity of operator support systems [BAI 83, BAI 91]. We therefore understand the need to invest in systems capable of detecting faults early enough and to filter false alarms and, in order to accomplish this, research general tools.

To successfully complete their work, the operator must link different types of reasonings (dynamic inference chain) based on the analysis of the process: for example, diagnoses, predictions and hypothesis tests. These reasonings with various aims and differing makeups are linked together in an order dependent on the objectives, the state of the process and the results of each of the previous reasonings.

In summary, supervision includes a more virtual overview of the process than control on its own, since all of the information available must be taken into account, along with all of the operation modes (therefore the limits of control algorithms); its study also includes humans who are at the forefront of the process, with their qualities and limits, and in this sense is not only interested in human–machine interfaces, but also in the cooperation between human and machine. One of the challenges of interactive computer systems lies in the selection, organization and dynamic presentation of information; the performance of the human–machine setup depends on the efficiency of the communication established by the interface (in the broad sense of the term). The creation of cooperative systems is based on the analysis of human–machine systems in order to establish the information requirements to define the goals, constraints and tasks to be completed. In [BEN 96, MON 00c, MON 05], it is explained that the specification of functions of a supervision system must be based on the analysis of the actions of the operator in the control room, particularly their cognitive perception actions, for the understanding of phenomena, diagnostics and decision-making or more specifically on human errors pertaining to these actions. The aid in the mental representation of the functionality of the process, the understanding of situations or even in diagnostics therefore seem to be essential components in an interactive system, the aim of which is to significantly reduce errors in the decisions of humans confronted with a situation requiring complex action.

This analysis on the needs of supervision operators in control rooms should be seen in the context of studies conducted at the end of the 1980s on *situation awareness* (SA: knowledge of the situation, sensitivity to context) by researchers into human factors. The question that was initially posed by SA was how do aircraft pilots manage to remain aware of the situation to filter pertinent information when the pilot's cockpit is overrun with increasingly numerous sensors and multiple complex events can occur during a flight? In addition, how is this information used to anticipate the most pertinent actions as soon as possible? The term SA has been adopted to describe the process of attention, perception and projection into the future, which together make up the mental representation of the situation the pilot is in for deciding on the pertinent actions to be taken [END 95a, END 95b]. SA was (and is still) an area of research prioritized for Human Factors in aviation ahead of it being extended to numerous other implementations (air traffic control, power plants, etc.) where the decision-makers are confronted with complex and quickly changing situations [BER 89].

Numerous definitions have been proposed for SA and their origins are mostly linked to applied fields [HAR 88, BER 89, CAR 92, END 95b, DAL 97, VID 95]. SA is the perception of environmental or contextual events (modifications of variables) relative to time and/or space, the understanding of the reasons for these events and the projection of their consequences in order to take pertinent and efficient action as soon as possible (see also Chapter 1). The absence of SA or a poor management of SA has been identified as one of the major factors in accidents linked to human error [HAR 91]. It therefore requires a mental model of the situation in a changing environment, which forces the operator to estimate the effects of their actions. Their SA reasoning is good if the diagnosis of the situation is correct and their capacity to anticipate allows for a pertinent estimation of the consequences of their actions with regard to the new objectives of the ever-changing situation. After the performances of these actions have been assessed, the operator re-evaluates the situation, and so forth. SA therefore involves cognitive mechanisms (perception, understanding, anticipation and projection into the future), which are part of a cognitive loop [END 95b]. In the field of industrial supervision, SA is necessary in decision-making since the operator first must have observed the perturbation, the anomaly or error that has affected the operation of the system, understood its type and origin and estimated the consequences before completing a piloting, reconfiguration or maintenance action that would be compatible with the evolution of the situation [PAR 08, SAL 09].

These preliminary observations allow us to better understand how AI has been able to provide interesting solutions to the supervision of industrial processes. The following two problems are more specialized and in the field of AI have played an important role:

– Qualitative Reasoning (QR) that aims to formalize the notions of magnitude or symbolic labels in order to be able to describe the behavior of physical systems in the same manner as a specialist who reasons over this behavior;

– the consideration of time that incorporates several aspects as follows:

- the resolution of problems in real time, which is focused on the execution speed of reasoning,

- the integration of the dimension of time in the knowledge being processed (it is necessary for the planning of actions and for the qualitative simulation of dynamic phenomena),

- the capacity to integrate a flow of information into the reasoning process: this is a question of updating information and previously processed inferences, or even interruptions in the reasoning process.

Causal reasoning, since the 1980s, has proven itself to be an important tool for tackling these problems.

It seems that the failures in the advanced automation of large industrial systems can be explained, among others, by the incompatibility between the "objective" model of the automation engineer and the "subjective" model of the user, the operator running the DCMS. AI can reduce this clash by aiming for an actual partnership between the operator and the supervision system (rather than a total substitution) because of an exhaustive knowledge of the human and what the machine does, which must be equipped with justification and explanation abilities. The explanation side of AI evidently favors human–machine cooperation; the declarative side of its programming enables it to implement different reasonings required in supervision (prediction, diagnosis, advice for actions, etc.) independently of the knowledge that underlies them [DUB 94]. Finally, a supervision system can also be perceived as a decision support system (DSS), which once again places the problem within the fields of research of AI.

6.3. Qualitative model

6.3.1. *The origins*

In the context of this supervision problem, AI focuses on QR. The development of qualitative representation techniques of the world is a very natural idea and therefore very old.

The qualitative model should be simpler than traditional physics. The commonly considered simplification method consists of reducing the digital model precision pushed to the extremes by De Kleer; this reasoning reduces the set of possible values of a variable to {negative value, null value, positive value}. The quantitative equations traditionally outlining the behavior of a system are then transformed into qualitative equations or confluences, which are functions of qualitative variables. The qualitative simulation program based on this model is called ENVISION. Starting from a system of confluences and an initial state, it provides an outline of the behavior of a system in terms of its possible future states.

After the works of Thom on a mathematical QR theory (catastrophe theory), the works of economists show that qualitative models can provide a good representation of certain economic systems and that significant conclusions can be made from purely qualitative information [THO 91].

All of the historical works that have been found pertaining to QR were traditionally motivated by the fact that quantitative information is not always available so as to question being able to obtain a digital model. This can be the case during the design phase where only the functions have been defined without allocating the digital values or when confronted with little-known systems. We must therefore construct a model of the system from an incomplete and imprecise set of information and know how to make use of it.

The motivation of the AI community toward qualitative methods was not linked to the absence of sophisticated models but rather elicited from the intent of reproducing the mental models used by man to reason over physical systems. Going beyond this original idea, it involves building models allowing for a behavioral description of a physical system over its qualitative features (significant changes of values, reversal of the direction of change, scale of phenomena, etc.).

What is interesting to note here is that the thought process resulting in QR in AI has given new perspectives to control engineers who had not considered until now the use of qualitative models as well as digital ones. Following the expert system-based approaches of AI built upon essentially associative knowledge, with QR it is a case of introducing more fundamental knowledge, termed "deep", and using models, the level of abstraction of which reflect significant phenomena, nevertheless, based on objective principles from established theories. Obtaining, followed by the analysis of such models is the goal of QR [TRA 97].

Qualitative physics aims at prepresenting a physical system and explaining its behavior, based on both the common meaning of causal reasoning used by humans to qualitatively analyze their environment, and scientific knowledge implicitly used by engineers [BOB 84]. The representation of continuous variables must follow the following principle: the distinctions made by quantification must be pertinent for the type of reasoning being carried out [FOR 84]. Qualitative simulation of the behavior of the physical system is distinguishable from the traditional digital simulation among others because time is sampled according to the occurrence of interesting events, rather than over a predetermined and regular increment [FOR 88].

These observations explain why qualitative physics seems appropriate for the supervision of continuous processes, one of the objectives of which is to explain the function of a process to operators, with more of a pertinence criterion rather than precision in order to assist them in the decision-making. Finally, we can say that QR is a more general term than that of qualitative physics, encompassing all reasonings based on pertinence criteria. QR has been considered as a field of AI since 1984.

6.3.2. *Current models*

The mathematical aspects of the qualitative model contributed significantly to the qualitative representation of physical measurements and the relations that can exist between these measurements and also to the mathematical structures, which allowed for calculatory or symbolic manipulations over qualitative measurements [DE 84, DOR 89, FOU 89, STR 88, MIS 89, RAI 89, DAG 88, DUB 89, DAG 86, DAG 87, TRA 88, KUI 86].

Causality is a concept often at the center of qualitative modeling. If qualitative approaches with mathematical models consider that the intervention of causality essentially presupposes an *a priori* on the behavior of the system, the causal-type models show that this knowledge of the phenomena is worth taking advantage of while it exists [IRI 79, IRI XX, DE 86, DE 84, IWA 86, SHI 85, OYE 88, CAL 88, FER 89a, LEY 91c, GUE 90, FER 89b, VES 90, VES 91, MON 92].

The *symbolic–digital interface* is concerned with the translation of digital observations into symbolic terms that are either directly explicit or are compatible with the qualitative models being used [FOU 90b, CHA 90, CHA 91, TRA 89a, TRA 89b, TRA 90, MON 92].

Qualitative simulation groups together the different methods that, starting from a qualitative model, allow for the deduction of the behavior(s) of the system. [KUI 84, KUI 85, KUI 86, KUI 87, KUI 88, BOB 84, FOR 84, FOR 88, DOR 88, CAL 88, FER 89a, LEY 91c, FOU 90a, CHA 92].

The main problem with qualitative simulation is the inherent ambiguity of the modeling: based on the initial given conditions, multiple future behaviors are proposed, certain of which may even prove to be incompatible with the physical system modeled. The inherent ambiguity of the description set $\{+,0,-\}$ renders simulations difficult to use in the case of complex systems, and even crippling in the context of supervision of industrial processes. Indeed, we cannot present the operator with different possible future behaviors for the process, including physically unrealizable ones. Furthermore, the model can be based on real knowledge of the behavior of the process, and the simulation aims to provide an image of this behavior under normal operating conditions, rather than proposing all of the possible outcomes, of which the distinction would not correspond to a significant phenomenon and would cause interpretation problems rather than resolving them. In diagnosis, this signifies that multiple interpretations of the observed faults can be proposed, the right one being among them; this is very rarely detected by the human operator, whose decision and reaction times are essential in incidental situations. Ambiguity or even incoherence are not concepts that we can think of introducing into the control room. It is regrettable to not be able to use all of the digital information that we have on a process in order to increase the resolution power of QR. Finally, traditional qualitative systems do not manage time, at least not explicitly; however, taking the dynamics of the process into account is vital at the supervision level.

6.3.3. *The evolution of qualitative reasoning (QR)*

From this analysis, QR takes on different forms, the common point of which is to refine the model or the simulation to obtain better descriptions of physical systems. Four of these forms stand out: the analysis of mathematical aspects, reasoning over orders of magnitude, the use of quantitative data and the consideration of time.

Mathematical analysis consists of transposing well-known quantitative theories into qualitative theories. Inspired by the reasoning commonly used for roughly calculating a result while disregarding certain numbers compared to others, the reasoning over orders of magnitude allows for the reduction in the number of ambiguous results of the "+ ⊕ –" operations. The easily conceivable solution when we have numeric data is to use this knowledge in case of ambiguities. The final form is temporal analysis.

6.3.3.1. *Analysis of mathematical aspects*

The first theoretical works clearly had the aim of reducing the number of incorrect behaviors generated by the simulation; filters are developed to eliminate incoherent behaviors and to make the simulation converge toward the actual behavior of the system. These works are based on the qualitative analysis of well-known qualitative theories: the analysis of the qualitative derivatives of order higher than 1, outlined by de Kleer, and followed through by Kuipers [KUI 87, KUI 89, TRA 86, DOR 87, DOR 88, TRA 88, TRA 89a].

Kuipers incorporates the analysis of qualitative derivatives of order higher than 1 into QSIM by filtering certain erroneous behaviors [KUI 87]. The transition of a qualitative variable X is ambiguous when the derivative equals zero, in other words from a qualitative state ([X], 0); the three possible transitions correspond to the three possible values of the derivative (–, 0, +). Nevertheless, the behavior of a system modeled by a linear differential equation is determined in a unique way at a critical point of the first-order non-zero derivative at this point. At a critical point, an algorithm then identifies the constraints where the first-order derivatives equaling zero intervene, and which provide an expression of the derivative of these constraints at the critical point, which allows us to calculate the sign of the second-order derivatives and to simply generate compatible transitions.

6.3.3.2. *Use of quantitative data*

Quantitative data was originally excluded from qualitative simulators. Subsequently, by analyzing real systems rather than case studies, it was revealed that the consideration of available digital information during a simulation was a simple method for refining the descriptions of the behaviors being carried out [GAL 86, CHA 90, CHA 91]. Kuipers also developed tools for processing quantitative information in QSIM, relative to the variables or constraints in the form of intervals [KUI 88].

6.3.3.3. *Reasoning over orders of magnitude*

In the analysis of electrical circuits, experts rarely have precise digital information available to them: certain variables are not measurable; others are no longer able to be estimated during breakdowns since a component, even remaining intact, can be placed in conditions such that it no longer satisfies its criteria for good operation. Nevertheless, an expert can efficiently make a diagnosis by approximating the value of variables according to their order of magnitude and by ignoring certain variables compared to others [DAG 86]. Qualitative ambiguities can therefore be resolved, when two opposing effects have differing orders of magnitude. Different formalisms have been proposed for managing reasonings over orders of magnitude [RAI 86, DAG 97, MAV 88, DUB 89].

6.3.3.4. *Temporal analysis*

The consideration of temporal aspects represents a fundamental stake in QR: explicitly reasoning over dynamic phenomena. In all of the works that have just been referenced, the notion of time is only present through the intermediary of the ordering of successive states in which the system finds itself. Although temporal reasoning is heavily studied in AI, there existed at the time, curiously, very few works on the explicit consideration of time in qualitative simulations; the approach proposed by Williams is one of the first of this type. Thus, for Williams, the analysis of qualitative ambiguity allows us to distinguish three typical scenarios [WIL 84]. Ambiguities appear and give way to multiple branches when:

– some effects are ambiguous (classic case $+ \oplus - = ?$);

– cycles appear in the reasoning (A cannot be deduced without knowing B, and vice versa); this is the problem of simultaneity;

– the order of the transitions of variables is unknown.

The third case is a result of the absence of an explicit representation of time: in representation by states, the behavior of a system requires a partial order in which the variables undergo a transition. For example, let A and B be two positive decreasing variables; the three possible cases must be considered: A reaches 0 first, B reaches 0 first, A and B reach 0 simultaneously. This leads to three different simulations, not only costly in terms of time and memory, but also corresponding to subtle distinctions that are not pertinent for describing the evolution of the system.

To resolve this, Williams built a temporal constraint propagator (TCP) [WIL 86, WIL 90], which takes into account an explicit representation of time based on Allen's works on interval-based temporal logic [ALL 83]. This representation of time is very close to that used by Caloud [CAL 88] and Feray-Beaumont [FER 89a] and is used as the basis for the event-based signal representation and Leyval's causal simulator [LEY 91c], which we will present in more detail in this chapter.

In the mid 1990s, the tendency was therefore to use *more or less simple* models, which are quantitative and over which we will reason qualitatively or in terms of orders of magnitude to avoid being deceived by the precision of numbers, the validity of algebraic differential equations, the legitimacy of statistical thresholds, the quality of the industrial measuring systems, etc. In the following, we will assume from now on that a human operator operates their system by causally interpreting the occurrence of phenomena and symbolically reasoning over the digital values of the measurements of the system, which are displayed on their control screen. The notions of approximated and causal reasoning therefore have a central role.

6.4. Causal graphs and event-based simulation

In the field of control theory, behavioral models are represented as block diagrams, where the blocks represent input–output relations. These types of relations are a reduced expression of causality: this means that the effects can never precede the causes. The simulation is easily interpretable as the propagation of the evolution of the inputs (disturbances or commands) to the evolution of the outputs. In the field of reliability, an abnormal behavior is characterized by its causes and consequences (failure modes and effects analysis (FMEA)). Systematic analyses of this type are local: the internal causes are referenced for each subsystem; however, the malfunction of a subsystem can be explained by an internal cause, such as the effects of a breakdown of another subsystem. Thus, a breakdown mode corresponds to the propagation of the initial effects throughout the entire system.

Be it a simulation or diagnosis, causality therefore acts as a founding principle for supervision. If we add to this the idea of propagation dynamics of a command, a disturbance or a failure mode, the notion of a dynamic causal graph for supervision is required: since the notable dynamics of coupled procedures complicates the understanding and explanation of abnormal operation and forms an additional cause for human operator interpretation errors, it is necessary to focus on the consideration of temporal and causal aspects in both the simulation and reasoning [FER 89b, LEY 94].

A causal structure is a description of the influences variables can have over each other. Consequently, this provides a conceptual tool for reasoning over the way that changes are propagated throughout a system. The behavior of any system can be described, at least partially, by a causal graph, made up of unidirectional relations between variables: a system can often be described by its structural equations, which are generally algebraic interpretations of the laws of physics governing the system; this causal sequencing provides a guide for identifying the asymmetries between the dependent and independent variables.

The most basic structure of the causal graph is that of the signed directed graph (SDG). The nodes of an SDG correspond to the state of the variables, and the directed arcs carry signs from the following influences: the sign is "+" when the variables corresponding to the arc evolve in the same direction, the sign is "–" when they evolve in opposite directions. The state of a variable is expressed within the set $\{+, 0, -\}$, such that the value is either normal (0), too high (+) or too low (–).

In addition to the causality given by the direction of the arc, other semantics can be attached to this also: thus, if the simplest version corresponds to the SDG, the functions related to these arcs can very well hold more quantitative information or even parameters that can inform as to the dynamics of these influences.

The simulator must allow for an artificial representation of the process behavior where the criterion of pertinence prevails over the precision of the information.

The supervision model has multiple objectives [MON 94a, MON 99]. It must be based on the notions of causality, the basic concept of simulation and also on the analysis of situations. Indeed, diagnostics results in causal reasoning, which has the aim of finding the causes of faults from observed effects. The explanation of the evolution of a variable is a very similar procedure: among the previous effectors of the variable, we look for those for which their own evolution has contributed to the analyzed behavior; the imprint of the reasoning carried out by the simulation provides an explanation of the system. Finally, prediction determines the future effects of a given phenomenon. The multiple uses of the same model for several supervision operations require the separation of the process knowledge and its operation, the knowledge pertaining to diagnosis and simulation reasoning, etc. Thus, the declarative nature of the model is essential.

6.4.1. *The causal graph*

Representation using a causal graph (Figure 6.1) is a dual representation of the one adopted by the functional automation diagram: the variables are nodes of the causal graph, whereas they are traditionally represented by flows in the functional diagram. Physical systems are the arcs of the causal graph, whereas they are black boxes in the functional diagram [LEY 91c, LEY 91a]. The arcs are parameterized by quantitative data that are easy to understand for the operative engineer, and which allow, in particular, for the consideration of the dynamics of the process (concept of delays, response time, etc.)

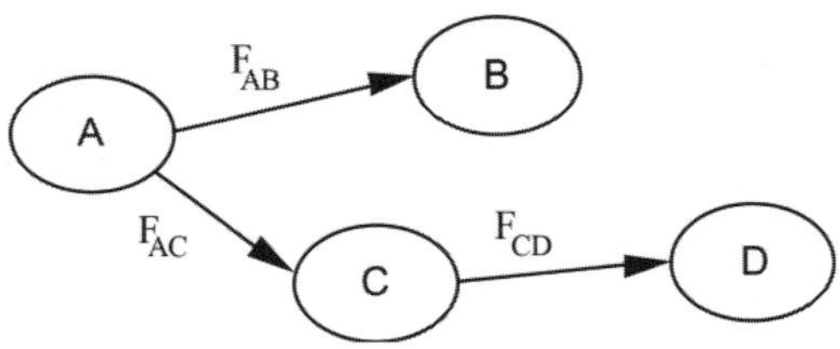

Figure 6.1. *A causal graph*

In the following, we have addressed the dynamics of the process through modeling of the time of events. This choice has the aim of illustrating the idea of a virtual description of the evolution of a variable, and emphasizes the notion of pertinence in the representation of phenomena rather than that of precision. This, therefore, requires the introduction of a formalism of specific calculation methods that can be restrictive, but allow us to place causal reasoning into a historical context, which is labeled as AI. Modeling and simulation using causal graphs are, therefore, presented in this formalism in the following sections. Obviously, considering the causal graph and functional diagram duality, we could have, for example, outlined the evolution of a variable through a sampled signal and the relations between variables through difference equations. This is how we will proceed in order to simplify the presentation of reasonings made over the causal structure other than that of simulation, and this is what we will discuss later in this chapter.

6.4.1.1. *Variables*

The variables considered in the model are the key variables used for process control and supervision. The majority of variables are measurable; however the non-measurable variables, the physical meaning of which is important to understand the workings of a system, can still be integrated. Each "variable" node is matched with the behavior of the variable over time, which we refer to as the history of the variable.

The simulated history of a variable is the result of the propagation of information collected at the inputs of the process (in other words, the sources of the graph). The use of this history allows us to understand and explain the behavior, which is present during a consultation; the future evolution follows from the prediction of the behavior of the process starting from the present state, and can help to anticipate phenomena and the recommended attitude to be taken up by the operator.

Since the goal is to provide the operator with a global image of the behavior of the process rather than the precise values of the evolutions, only the pertinent values of variable evolutions are considered, in other words those susceptible to modifying the global behavior of the process. They are described in the shape of a series of events.

The evolution of a variable is represented by a linear function of time in samples. A break in the slope corresponds to a significant change in the behavior of the variable; in order to take into account the instantaneous phenomena of the scale of the surveillance time, these functions are allowed to be discontinuous. An event is therefore characterized by the triplet: {time of occurrence t_i, γ_i the slope variation at t_i, δ_i the variation in amplitude at t_i}. An evolution is therefore seen as a series of chronologically ordered events.

6.4.1.2. *Arcs*

Each arc of the graph defines the causality relation that links two variables, with the notions usually used by automation engineers for dynamically describing influences (gain, pure delay, response time, etc.). By analogy with traditional control theory, the function carried by the arc is referred to as qualitative transfer function (QTF) [FER 89a]. The QTF must be an internal function in the vectorial subspace, which forms the evolutions: they allow for the calculation of the response of the downstream variable of an arc to the evolution of the upstream variable. The response of the QTF to an input signal is designed like the approximation through a sampled linear function of the response of a traditional transfer function with the same input signal: it is therefore an evolution, which allows us to ensure the stability of the operation for the chosen representation.

The graph is used as a linear model of the process around its operating point. When two variables A and B influence a variable C, the evolution of C is obtained by using the theorem of superimposition.

The influences represented in a causal superimposition graph must be derived from simple systems; this is why the QTF library must be limited to a small number of behavior types. For example, we find the pure QTF delay, the first-order QTF, etc. They are always defined with reference to the corresponding traditional transfer function.

The causal graph, which could be traced over the functional diagram of a close-loop regulation, with zero pure delay, is not directly useable by the simulation algorithm. Indeed, the good functioning of the propagation algorithm for the evolutions relies on the implicit hypothesis that every loop has a non-zero pure delay. A problem typically arises for regulation loops, of which the total delay, observed from a supervision point of view is almost always null. It is for these reasons that specific QTFs have been designed to model regulation loops in causal graphs, since they specifically allow us to consider the behaviors of regulated variables faced with a change in protocol and a disturbance [LEY 91a].

6.4.2. *Evolution and event*

The simulation is none other than the propagation of significant changes over the inputs of the graph from variable to variable with the help of QTFs.

Let $\sigma = \{t_0, t_1, ..., t_N\}$: σ is a subset of $[a,b[\Leftrightarrow (t_0 = a)$ and $(t_N = b)$ and $(t_0 < t_1 < ... < t_N)$.

$[t_0, t_1[, ..., [t_{N-2}, t_{N-1}[, [t_{N-1}, t_N[$ are intervals defined by the subset (the subset is not necessarily of fixed step).

A function E_σ defined over [a,b[is a sampled linear function if and only if there exists a subset σ of [a,b[such that E_σ is a linear function over each interval defined by the subset. Each such defined function is referred to as a segment, hence the term segmentation that defines the process transforming the gains from the process into a sampled linear function in real time [MON 92].

σ_{min} is the minimal subset with an associated function E_σ if and only if any other possible subset of E_σ includes σ_{min}. The minimal subset associated with E_σ is unique: it is the set of points of the interval [a,b[where E_σ is non-differentiable. This set is finite. To only take into account the significant changes intervening in the evolution of a variable, in other words obtaining a description of this evolution in a minimal number of segments, we will focus exclusively on sampled linear functions represented with the help of their minimal subset.

An evolution $E\sigma$ is therefore a discontinuous function of [a,b[in $\mathbb{R}$ with which its minimum subset σ is associated. The set of defined evolutions is labeled E. Figure 6.2 shows an evolution and its associated minimal subset.

An evolution $E\sigma$ can be discontinuous left of each point of its subset σ, since the intervals defined by σ are open to the right.

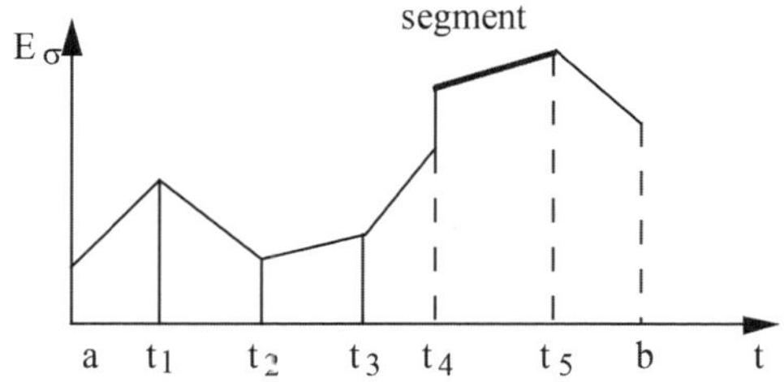

Figure 6.2. *An evolution*

Let (F, +, *) be the vector space of the functions of the interval [a,b[from $\mathbb{R}^+$ to $\mathbb{R}$; the + and * laws the addition of functions and multiplication by a real number, respectively. The set of evolutions E is a vector subspace of R of the vector space of functions of the interval [a,b[from $\mathbb{R}^+$ to $\mathbb{R}$.

An event e_i is a linear function, non-zero on the interval [t_i,b[and zero on the opposite interval [a,t_i[.

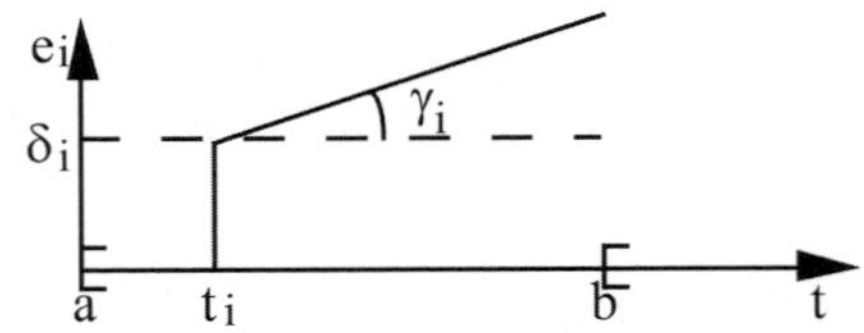

Figure 6.3. *An event*

Let σ = {a,t_i,b} be a subset of [a,b[in two intervals, and the function e_i such that:

$$(\forall\ t \in [a,t_i[),(e_i(t) = 0) \qquad [6.1]$$

$$(\forall\ t \in [t_i,\ b[),(e_i\ (t)\ =\ \delta_i + \gamma_i * (t - t_i) \text{ with } (\delta_i, \gamma_i) \neq (0,0)$$

is an event (Figure 6.3). This is an evolution.

Let $E\ (t - t_i)$ be the unit step at $t = t_i$ defined over [a,b[:

$$\forall t \in [a,t_i[, E(t) = 0 \text{ and } \forall t \in [t_i,b[, E(t) = 1$$

Let $R\ (t - t_i)$ be the ramp of unit value from $t = t_i$ defined over [a,b[:

$$\forall t \in [a,t_i[, R(t) = 0 \text{ and } \forall t \in [t_i,b[, R(t) = t$$

$E\ (t - t_i)$ and $R\ (t - t_i)$ are events and $R\ (t - t_i) = (t - t_i) * E\ (t - t_i)$.

Every event e_i is written as: $e_i\ (t) = \delta_i * E\ (t - t_i) + \gamma_i * R\ (t - t_i)$.

The attributes of e_i are:

– its date is the initial echelon time: Date (e_i) = t_i;

– its amplitude is its value at the start time: Ampl (e_i) = δ_i;

– its associated minimal subset is {a, Date (e_i), b} = {a, t_i, b};

– its slope is the slope of the linear function: Slope (e_i) = γ_i.

Leyval has shown that any event has a unique representation in the form of a linear combination of the components of the set { E (t – z), R (t – z); z ∈ [a,b[} [LEY 91c]. This set is therefore a basis of the evolution space vector. This decomposition is less intuitive than through episodes as adopted by Feray-Beaumont [FER 89a] and Caloud [CAL 88]; however, it is particularly adapted to linear propagation. Indeed, when a significant event is detected, it must be taken into account without knowing its duration, an implicit parameter required for the breakdown into episodes [LEY 90].

The absolute references being zero, every event from e_i to t_i has as coordinates the variation in amplitude of the evolution at t_i labeled Δa_i, and the variation of the segment slope relative to the preceding segment labeled Δp_i: $e_i(t) = \Delta a_i * E(t - t_i) + \Delta p_i * R(t - t_i)$.

An evolution is therefore written as:

$$E_\sigma(t) = \sum_{i=0}^{N-1} e_i(t) = \sum_{i=0}^{N-1} \Delta a_i * E(t - t_i) + \sum_{i=0}^{N-1} \Delta p_i * R(t - t_i) \qquad [6.2]$$

Figure 6.4 illustrates the breakdown of an evolution into events.

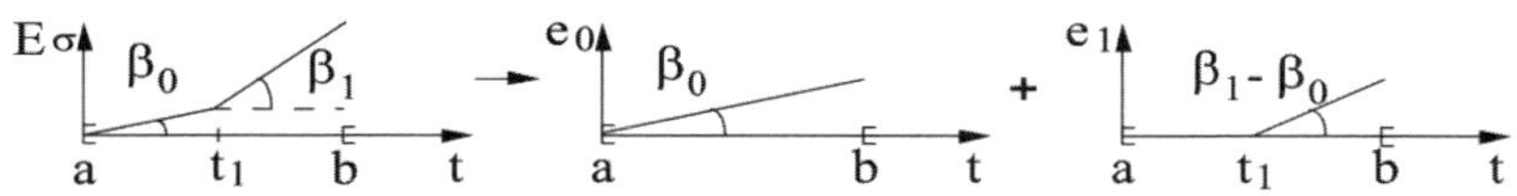

Figure 6.4. *Breakdown of an evolution into events*

6.4.3. *Simulation*

The image of an evolution through an QTF is once again an evolution labeled as a qualitative response (RQ).

According to [6.2], the evolution of the upstream variable can be written as:

$$E_{\sigma}(t) = \Delta a_0 * E(t-t_0) + \Delta p_0 * R(t-t_0) + \sum_{i=1}^{N-1}(\Delta a_i * E(t-t_i) + \Delta p_i * R(t-t_i)) \quad [6.3]$$

The evolution of the downstream variable, taking the initial conditions into account, is obtained as such:

$$E_{\sigma'}(t) = \Delta a'_0 * E(t-t_0) + \Delta p'_0 * R(t-t_0) + \mathrm{FTQ}(\sum_{i=1}^{N-1}(\Delta a_i * E(t-t_i) + \Delta p_i * R(t-t_i))) \quad [6.4]$$

where $\Delta a'_0$ and $\Delta p'_0$ are the initial values of the downstream variable.

The linear QTF in the evolution space vector enables the calculation of a response to an arbitrary evolution from the responses to the basis elements.

$$\mathrm{FTQ}(E_{\sigma}(t)) = \mathrm{FTQ}(\sum_{i=1}^{N-1}(\Delta a_i * E(t-t_i) + \Delta p_i * R(t-t_i))) \quad [6.5]$$

(Initial zero conditions: $\Delta a_0 = 0$ and $\Delta p_0 = 0$).

The basis of the evolution space vector is made up of step functions and ramps. The responses of traditional systems to these inputs are well known, and serve as a model for the design of QR.

For a linear system that is to be modeled, the problem consists of determining the significant instants of the response of the traditional transfer function, which make up the subset of the QR. The first important point is, of course, the starting instant. All of the QTFs have a parameter corresponding to the pure delay between the input evolution and the output evolution. For the majority of systems, when the input becomes constant, the output becomes equally constant after a certain amount of time. In this case, modeling the output behavior through a differential equation solution asymptotically tends toward a line. It would therefore not be pertinent to endlessly choose response points to construct the QR, since the segments would indicate insignificant distinctions. We can rely on the frequently used criterion, which states that the steady state is achieved when the response differs by less than 5% of the asymptote. The second pertinent point therefore corresponds to the response time [LEY 91c].

The number of notable points chosen to describe the QR influences the precision of the approximation. Modeling the whole dynamics of a system using only a single line segment quickly proved to be insufficient, especially when the processes to be modeled were of a slow dynamic that, for example, is the case for liquid–liquid extractors used in nuclear waste reprocessing [LEY 91c, LEY 91b, MON 93a, PEN 93, MON 00a, MON 00b]. Research for a better description of the " first-order QR" therefore comes down to outlining its behavior over [0,3τ] in two intervals: [0,θ] and [θ,3τ]. The point θ can be chosen such that the sum of errors between the digital response and the QR is minimal, in the sense of least squares:

$$\varepsilon = y_{\text{digital}} - y_{\text{qualitative}}$$

$$J(\theta) = \int_0^{3t} \varepsilon^2 dt$$

is the criterion to be minimized.

In fact, in [LEY 91c] it is shown that we can choose two other notable points in the interval [0,3τ], such that the approximation carried out always remains within a 5% distance of the curve. Figure 6.5 illustrates the response, defined by four notable points, at an echelon input. The precision is limited, since we consider that an approximation of less than 5% would lead to non-pertinent segments: in the case of "first-order QR", it is therefore unnecessary to choose a smaller subset.

We can finally tackle the question of QR to an evolution. The responses at an echelon and a ramp are still evolutions, made up of K segments, thus:

$$\forall\ 1 \leq i \leq N\text{-}1 \text{FTQ}\ (E\ (t - t_i)) = \sum_{j=0}^{K-1} (\alpha_j * E(t - t_j) + \beta_j * R(t - t_j))$$

$$\text{FTQ}\ (R\ (t - t_i)) = \sum_{j=0}^{K-1} (\gamma_j * E(t - t_j) + \delta_j * R(t - t_j))$$

with $t_j \geq t_i$ (causal system).

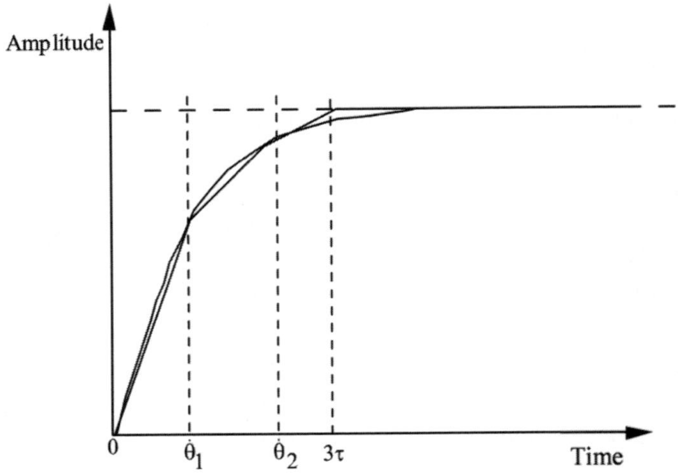

Figure 6.5. *First-order qualitative response*

Thus:

$$FTQ\ (E_{\sigma}(t)) =$$

$$\sum_{i=1}^{N-1} \Delta a_i . \sum_{j=0}^{K-1} (\alpha_j * E(t-t_j) + \beta_j * R(t-t_j)) + \sum_{i=1}^{N-1} \Delta p_i . \sum_{j=0}^{K-1} (\gamma_j * E(t-t_j)$$

$$+\delta_j * R(t-t_j)) \qquad [6.6]$$

Consider the simple case where two events at the input are to be propagated, and where the QR at a ramp is an evolution made up of only two events: B (t) = RQ (A(t)). The calculations performed are shown in Figure 6.6. The calculation of the response to an evolution therefore simply consists of multiplying the coefficients of the evolution (Δp1, Δp2) by the coefficients associated with the responses to the base components (Δk1, Δk2), and then chronologically ordering the obtained events according to their date.

6.4.3.1. *Contribution and explanation*

The QTF and the parameters enable the calculation of the QR to an event. The contribution is a history archiving the evolution having been propagated by the QTF of the arc; this contribution is updated by the algorithm for each propagation along the arc. It is later used to causally explain the influence of a node by its preceding nodes in the graph: the simulated evolution of a node is the sum of the contributions carried by the input arcs, under the linearity hypothesis used in this model; it could, however, be obtained using a more complex function of the contributions. The explanation of an evolution is therefore naturally linked to the notion of contributions from predecessors.

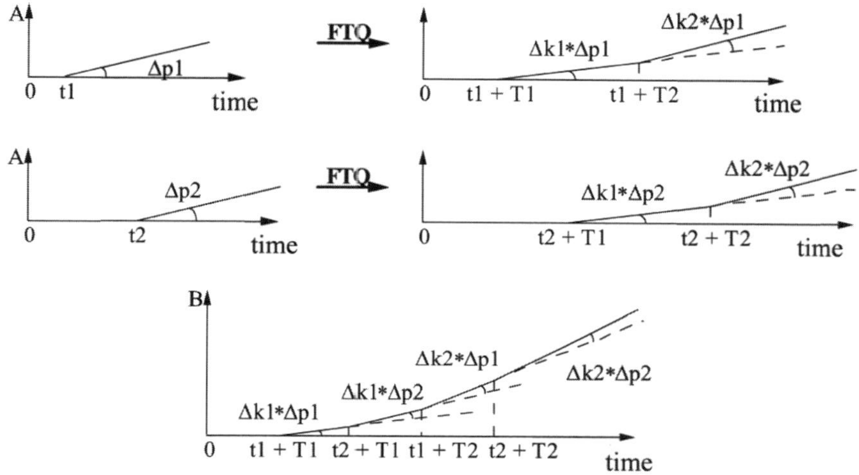

Figure 6.6. *Calculation of the image of an evolution using a qualitative transfer function*

6.4.3.2. *Granularity of the simulation*

The working of the simulation algorithm requires the implementation of a control over the duration separating two consecutive events in the evolution of a variable. Indeed, the sampling induced by the QR can lead to the introduction of distinctions in the history, which are in a sense insignificant, but more penalizing for execution time and memory used. The consequence of this phenomenon on the propagation is the generation of non-pertinent histories, since they are made from events with indistinguishable dates.

It is therefore necessary to define a "minimal distance" between the dates of the two events, referred to as temporal granularity g; this is commonly expressed in the form of a temporal distinguishability condition:

$$t_1 \text{ is not distinguishable from } t_2 \Leftrightarrow /t_1 - t_2/ < g;$$

where *g* is the temporal granularity perceived by the observer of the process.

Although it is easy to fulfill this condition for an offline operation of the simulator, using compacting techniques, it causes problems with an online simulation [LEY 91c]. This is why granularity management must be based on an equivalence relation between the instants.

Constructing an equivalence relation between the instants consists of choosing a function *f*(*t*) such that:

$$t_1 \equiv t_2 \Leftrightarrow f(t_1) = f(t_2)$$

f can, therefore, be fixed such that the previous temporal distinguishability condition becomes a particular case of it; for example, in the context of a propagation of events, *f* can be chosen as such:

$$(\forall t \in \mathbb{R}^+)\,(f(t) = g * \text{Integer\_part}(t/g))$$

Time management becomes much more simple, since all of the instants are expressed as a multiple of the granularity (supposing that the starting time is zero): any new event disrupting an evolution will therefore be placed at a pertinent date (a pre-existing date in history or a new date at a distance of at least g from any other existing date). A duration control mechanism, separating two consecutive events of an agenda representing the evolution to be propagated from a node, must be put in place: this is compacting. Compacting consists of replacing *N* consecutive events of an agenda by a single equivalent event. Compacting was developed to follow the temporal distinguishability relation.

6.4.3.3. *Segmentation*

The aim of this functionality is to detect the input variables of the process or the measurable disturbances, which are sources of the causal graph, the significant events disrupting their evolution.

When a segment other than the initialization segment is detected on a source variable of the graph, the segmentation module provides the event propagation module with the corresponding event. This event is immediately processed, since its date is less than the current time of the simulation.

By noticing that the phenomena of short duration compared to the dynamics of the process have no effect on its behavior, as long as they remain at normal amplitude, the selection criterion of pertinent evolutions has been defined relatively to the amplitude of phenomena and their duration. The dynamics of an evolution is then estimated by observing the connected recording components over a window of duration L_f, chosen to be equal to g in this case (generally, $L_f \geq g$).

If the amplitude is quantified in the space {negligible, significant, abnormal} and the duration in the space {insignificant, significant}, the threshold of which corresponds to the granularity g, the evolutions of negligible amplitude and the evolutions of significant amplitude and negligible duration are ignored, and the evolutions of significant amplitude and significant duration are taken into account, as well as abnormal evolutions regardless of their duration. Hence, the notations of Figure 6.7 over the scales of perception in amplitude and time.

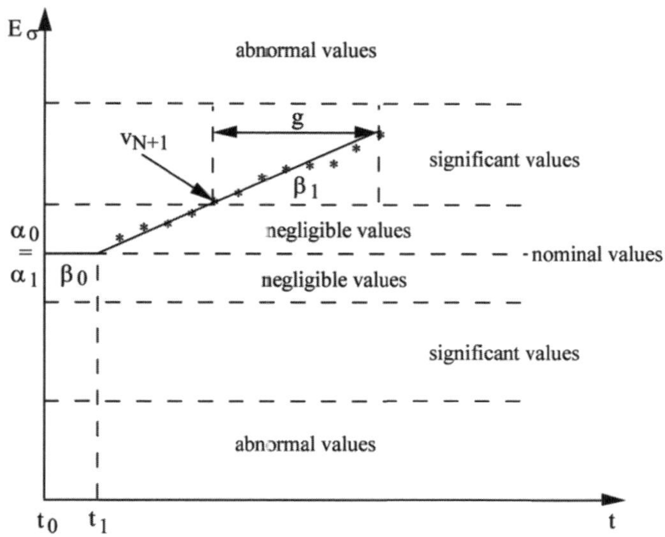

Figure 6.7. *Segmentation of an evolution*

A new measurement v_{N+1} is disregarded if it is located in the field of negligible values, and immediately considered if it is an abnormal value. In the case where it is found in the "significant" zone of the scale, the evolution of the process will only be pertinent if it lasts at least g ($g = k \cdot Te$). It is modeled by a line of slope β_1, β_1 being the best slope approaching the significant measurements, using the smallest squares, for example: this line intersects the reference evolution at one point (t_1, α_1), and the event with attributes $(t_1, 0, \beta_1)$ becomes the new reference when the position of t_1 relative to $[t_{current} - g; t_{current}]$ permits it. The event provided by the simulator is $(t_1, 0, \beta_1)$. If t_1"is too far in the past", or "later then $t_{current}$", the start date of the new event chosen is not t_1 but $t_{current} - g$ and the event has a non-zero amplitude $\alpha_1 - \alpha_0$ (α_0 being the value of the previous event at the date $t_{current} - g$) [MON 92]. An analysis of the results of the segmentation in terms of g and the maximal interpolation error permitted is provided in [MON 92] in order to make the comparison with the more traditional notions in signal processing.

The segmentation algorithm takes measurements as inputs and, according to the qualification zones previously described, produces segments for the storage of evolutions as its output and the events to be supplied to the simulation. The qualitative timescale is determined in an empirical manner. For the implementation of the DIAPASON process [PEN 93] in the field of liquid–liquid extraction for the treatment of nuclear waste, measurements from sensors were carried out every 300 ms, and the digital regulators had a sampling period of 300 ms. The visualization on the control panels was set at a scale of 10 times the size, or 3s; the archiving of data was carried out every 30 s. Only the phenomena lasting more than 300 s or 5 min were considered significant: the dates of the pertinent events in the evolution of a variable were separated by at least 5 min from each other.

The detection of rupture points in the slope guarantees:

– that these are at a distance of at least g from each other; the temporal distinguishability condition is therefore verified. The consideration of quick phenomena is carried out with the help of dicontinuities. The supply and modeling of signals as well as the simulation form a coherent entity;

– beyond its use for supplying the causal simulation, this event-based representation of the signal still remains relevant to this day, since in the fields where we are only equipped with expert knowledge, which cannot be modeled with the help of differential equations over dynamic systems, surveillance methods without a formal model consist of extracting pertinent characteristics of signals, before fusing them with the help of a decision-making model emulating human expert knowledge.

The implementations particularly concern hospitalized patients in intensive care units as well as the recognition of physiological state [CHA 04, CHA 05, CHA 06].

6.4.3.4. *Conclusion*

The dynamic causal graph has proven to be a useful model for the simulation and explanation of evolutions. The illustration of this "multiuse" characteristic which the model of a supervision system must be equipped with is completed, in the following, by the reasoning over localization and fault detection. Its use in the guidance of actions has been addressed in [FER 89a].

However, the main reproach we can make against such a representation is the uniqueness of the graph [LEY 93, DZI 93b]. Its use in supervision exposes the need to have several levels of representation depending on the context: the choice and the number of variables that we wish to follow in a given system depend on the situation, its importance, its complexity, etc. [DZI 93a, DZI 94]. The operator must be able to adapt the representation level depending on the depth of the analysis, which he wishes to carry out, and for this we must focus on the construction of a hierarchy of causal graphs [DZI 96, GEN 05].

6.5. Hierarchy of behavior models

With the duty of the operator having evolved toward a supervision role, they have had to include in their activities information seeking, recognition of scenarios, diagnoses and planning of actions over the process, with all of this being accomplished from their supervision station. The operator must therefore be able to switch from a global view to a detailed view of the system [IWA 86], from a qualitative interpretation to a precise report of the measurements and for this, the operator is generally only equipped with overviews all showing the same level of detail, with the most detailed overview at the level of acquisitions. It is obvious that this fixed representation cannot be the best for each of the varied analyses, which the operator must accomplish during his job. To favor human–machine cooperation at any instant, the operator must be assisted in their various reasonings with a model adapted to each situation. The temporal and QRs are fundamental processes for decision-making [CAC 92].

In [DZI 96], the idea is to take from the principle that a causal interpretation of phenomena, in terms of temporal precedence, is a vital tool for the representation of cognitive mechanisms in the human understanding of physical phenomena [RAS 93], to which it is useful to add that one way to address complex systems is to structure the information via an abstraction hierarchy [RAS 85, LIN 11a, LIN 11b] (see also Chapter 1). On the other hand, the models of this hierarchy must remain homogeneous to develop a coherent description of the system behavior.

We must therefore establish the creation of causal models, represented by graphs and a method for developing the hierarchy of graphs based on the model presented in the previous section. To simplify the presentation of these works without interfering with event-based modeling, traditional transfer functions (instead of QTFs) are assigned to the arcs of the graph.

6.5.1. *Definition of a graph hierarchy*

We start from the example of a graph, which determines a level of the hierarchy. The graph in Figure 6.8 represents a report of the flows that enter and exit through an extraction column of a nuclear combustibles reprocessing plant. The (free) overflow throughput QG600 is seen logically as a terminal node, the value of which is fixed by the value of other throughputs (Figure 6.8).

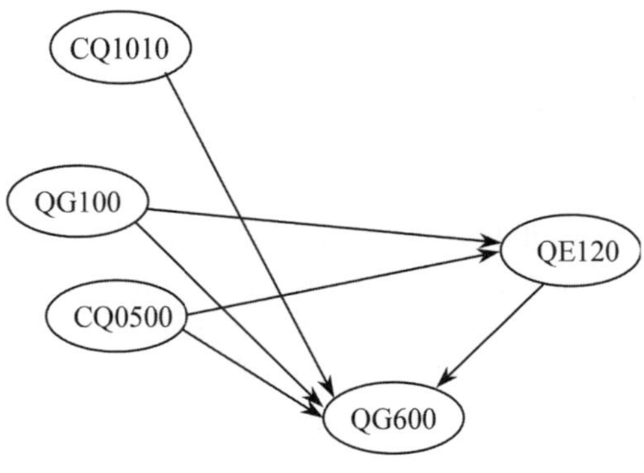

Figure 6.8. *Throughput balance graph*

Operators generally work with only a few variables being constantly observed; these variables will be retained to construct the graph of the highest level of the hierarchy. As soon as an abnormal phenomenon occurs, the operator will tend to turn his interest to a particular subsystem to verify the hypotheses or identify a course of action. In other cases, the operator will simply want several variables at their disposal to understand and follow the evolution of the situation.

A graph G2 at a level in the hierarchy lower to that of a graph G1 must also include all of the variables of graph G1, in addition to other nodes used to take other phenomena into account. Consequently, the detailed graph can contain new sources (regulation instructions), new sinks (additional measurements) or intermediate variables (for a more detailed view); in this latter case, an arc of the highest level will be broken down into a path at the level below.

As arcs represent transfer functions, this hierarchical decomposition consists of highlighting the internal variables by dividing a complex transfer function into elementary transfer functions.

6.5.2. *Creation of the hierarchy*

A top-down approach allows us to construct an *a priori* hierarchy of causal graphs. The construction is initialized by the graph G_0, containing the set of variables that are essential to the operation of the reprocessing plant. These variables correspond to the minimal information needed to monitor the usual operation. The highest level graph for the column can correspond to the graph in Figure 6.8. Additional variables are introduced. A complex function is broken down into elementary functions; an arc is thus replaced by a path. The development can be built upon a particular function, such as a regulation. In this case, at an abstraction level, regulated variables and instructions share the same meaning. To take the changing values into account and diagnose the regulation malfunctions, the arcs linking instructions and disturbances that are measurable at the regulated variable and at the actuator must be created.

In this example, the second level of the hierarchy includes retention (BETAE) and pressure (CPRE801). This second level therefore includes 7 nodes and 11 arcs (Figure 6.9). The third level of the hierarchy this time includes the column regulated interphase (NIRE). CNIRE is the instruction, and QE120 is the actuator. The variations in entering fluid throughput of the column interfere with QE120.

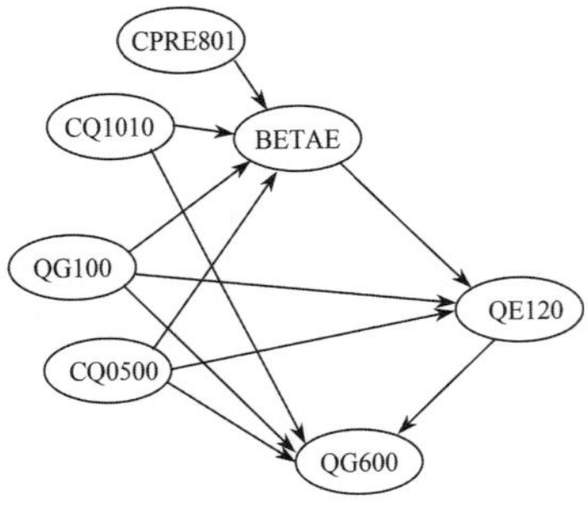

Figure 6.9. *Level 2 graph*

Variables are then introduced into a new model (DELTAQE) in order to take into account these disturbances (Figure 6.10). The addition of further variables to obtain the adequate level of detail finally leads to the more detailed graph including the set of variables displayed on the supervision interface.

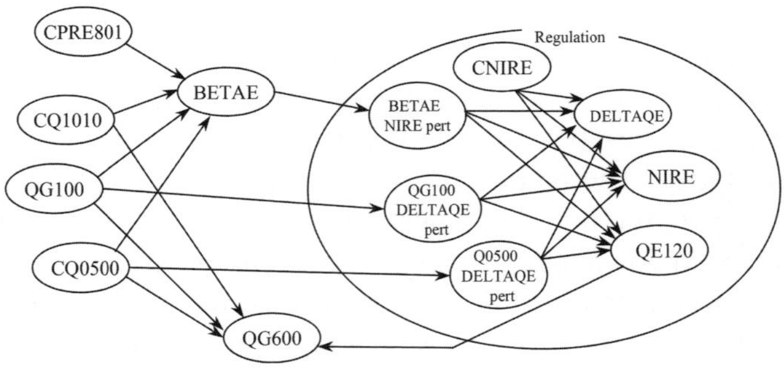

Figure 6.10. *The third level of the hierarchy with the regulation of the interphase level*

6.5.3. *Online construction of graphs*

The situation is very different if it involves creating hierarchical graphs for the online needs of an operator. To create a graph online, it is necessary to identify the paths of the low level graph that will become arcs at a higher level. This method must also guarantee that no pertinent information of the detailed graph is lost.

The operator generally begins with a graph G_0 and has the intention of focusing on a particular subsystem depending on the context: either by adding new variables or by outlining functions. It is therefore impossible to specify *a priori* pertinent hierarchical abstraction levels. The construction of the detailed graph depends on the context. This approach is at the center of human–machine cooperation, and aims to promote a responsive behavior from the operator. The required graph is created online using the variables available on the display and additional variables needed by the operator, and it is obtained online from the lowest level graph (the most detailed).

Let us then consider the opposite situation where the operator wishes to have a more global overview of the situation. After having validated the structure of the desired graph, by identifying eventual circuits, all of the non-pertinent nodes must be eliminated. When several nodes are eliminated from a graph, they must be destroyed one by one through an iterative process [DZI 96].

Path algebra [GON 95] allows us to resolve a large number of pathfinding and calculation problems in graphs. This method uses a matrix representation of graphs as well as the appropriate algebraic structures. By equipping this set of matrices S with an appropriate sum $\oplus$ and a product $\otimes$, we obtain a semi-ring structure. This allows us to efficiently resolve numerous operations over graphs by simply defining the nature of the elements of the matrix and the operations. Thus, it is possible to find the number of paths of a given length between two variables of the graph through a simple matrix multiplication.

The enumeration of paths is also possible by associating a name with each arc as shown in Figure 6.11. The name can be a symbol, a character or even a transfer function depending on the semantics associated with the arc. The enumeration allows us to obtain, for each path length, a chain of characters corresponding to the aggregation of the arcs of the path.

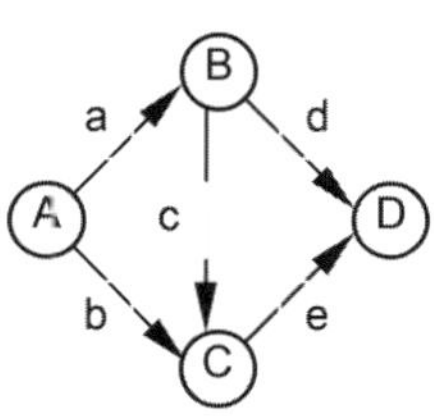

Figure 6.11. *Example of labeled arcs*

The graph in Figure 6.11 is represented by the matrix N_V when the variables are ordered according to the vector V [6.7]. The matrix of names N_V contains all of the paths of length 1 in the graph. For example $N_V(2,1) = a$ represents the semantic arc *a* from the node A to the node B:

$$N_V = \begin{bmatrix} 0 & 0 & 0 & 0 \\ \boxed{a} & 0 & 0 & 0 \\ b & c & 0 & 0 \\ 0 & d & e & 0 \end{bmatrix} \qquad V = \begin{bmatrix} A \\ B \\ C \\ D \end{bmatrix} \qquad [6.7]$$

The paths of length *n* are obtained by calculating the *n*th power of the matrix N_V. By calculating the square and the cube of N_V, four paths containing two arcs of length 2 and one of length 3 can be identified (equations [6.8] and [6.9]). There are no paths of length 4 or more, since all of the elements of the fourth power of N_V are equal to zero. For example, the path of length 3 between *A* and *D* is "ace":

$$N_V^2 = \begin{bmatrix} 0 & 0 & 0 & 0 \\ 0 & 0 & 0 & 0 \\ ac & 0 & 0 & 0 \\ ad + be & ce & 0 & 0 \end{bmatrix} \qquad [6.8]$$

$$N_V^3 = N_V \cdot N_V^2 = \begin{bmatrix} 0 & 0 & 0 & 0 \\ 0 & 0 & 0 & 0 \\ 0 & 0 & 0 & 0 \\ \boxed{ace} & 0 & 0 & 0 \end{bmatrix} \qquad [6.9]$$

A circuit is identified when a matrix includes a non-zero diagonal element. The circuits must be treated because it is not possible to remove all the nodes they contain without losing the sense of causality [DZI 96].

The total number of paths is obtained by calculating the limit N_V^*, sum of all the (non-zero) powers of the matrix N_V [6.10]. The element $N_V^*(4,1)$ indicates three possible paths between *A* and *D*:

$$N_V^* = \sum_{n=1}^{\infty} N_V^n = \begin{bmatrix} 0 & 0 & 0 & 0 \\ a & 0 & 0 & 0 \\ b+ac & c & 0 & 0 \\ \boxed{ad+be+ace} & d+ce & e & 0 \end{bmatrix} \qquad [6.10]$$

This method allows us to find the semantics of all of the paths linking the two variables of the graph, by respecting the direction of the arcs and subsequently the causality of the influences.

The elimination of a node from the graph is based on the multiplication of the matrix defining the graph and a matrix made from the only influence of the variable to be removed (column matrix of the graph matrix associated with the variable to be removed added to the identity matrix of the corresponding dimension). This matrix product is used to calculate a matrix in which the influence of an eliminated node remains by considering the links between the remaining nodes (Figure 6.12). An algebraic iterative method has been outlined [DZI 96, GEN 05].

The transfer functions associated with the arcs generally correspond to simple transfer functions of limited rank in order to provide an approximate description of the behavior of the process, which we consider as satisfactory in the context of systems supervision. An algorithm was proposed in [DZI 96, GEN 99] to control the complexity of functions when combined.

This approach provides the operator with the possibility of constructing and adapting models online at his own will, by selecting the variables pertinent for a given situation and task.

Path algebra and the aggregation of functions are only mathematical subtleties for this preliminary approach to the hierarchical modeling of systems for intelligent supervision. This supposes the operator to be capable of a considerable capacity of autonomy since he is the only one responsible for the choice of the supervision abstraction level.

This hypothesis is without a doubt to be questioned in incidental situations where the exceptional and stressful conditions run the risk of affecting the analytical ability of the operator; it is therefore clear that there remains work to be done involving a guide for the automatic online selection of the most pertinent variables. In addition to causal and dynamic readings, the choice of the abstraction level provides the representation with a dimension, which is relevant with an approach centered on the human–machine cooperation.

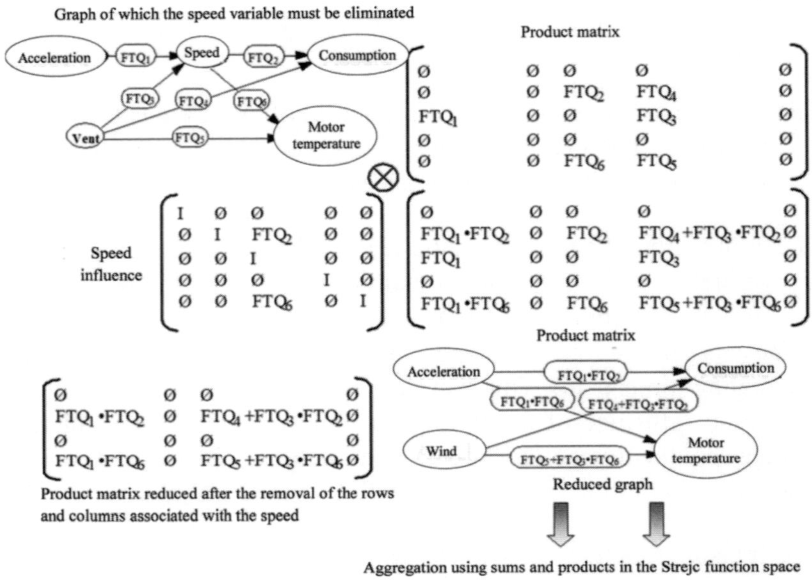

Figure 6.12. *Principle of graph reduction: matrix calculation and aggregation of transfer functions*

6.6. Fault filtering

6.6.1. *Causality and digital simulators*

In the large chemical or nuclear industries, the available simulators are not specifically adapted to the fault detection and isolation (FDI) methods used by control engineers and it generally requires a huge effort in additional modeling in order to benefit from this sort of tool. Using a simulator directly for supervision leads to the creation of an open loop residual[1] that can enable the detection of faults, but not isolate them. The idea of minimizing the effort of additional modeling and almost directly reusing the simulators already available is therefore an important requirement. In the following, we

1 We refer to residual in a very simplified manner as the distance between the expected value and the observed value, for example a residual in open loop will be the distance between the output of a simulation and the associated measurement.

will show how causal reasoning, based on simulation errors, enables the design of a structured set of residuals, which supply the process of isolation without having to resort to algebraic manipulations over the available models or initial residuals.

Now, let us return to the idea of a causal model of the process for use in FDI. Diagnosis is typically a causal process, since it involves determining which components are failing, explaining the malfunctions observed. Thus, the question of "why" has been answered.

If one aims to apply QR to diagnosis, and if the definition of a fault remains "the modification of the relations between cause and effect in a process", the approach used by a causal graph seem well suited to the tasks of understanding the behavior of the process, through the implicit graphical representation of the proposed causal reasoning [IRI 79, IRI XX, YU 91, SHI 85, KRA 87, KRA 89, MOS 97, MOS 98]. The notion of causality is linked to that of explanation and one such prioritized implementation of it is in diagnosis [DAV 84, POO 89].

Current alert processing systems provide the operators with an unordered list of simultaneous alarms that affect the process; afterward, the operators must interpret this and extract the source alarm; in other words, the one that helps to explain the set of the other triggered alarms. Causal graph-based methods aim at helping operators in their diagnosis task, by creating from the set of alarms a tree the root of which is the source alarm and the branches of which illustrate the propagation of alarms from one to the other.

Davis had noted that an important part of the knowledge required for the analysis of disrupted systems is the understanding of mechanisms in terms of causality [DAV 83]. Bandekar [BAN 89] has shown that an explicit representation of such relations is directly useful for the diagnosis: in a reasoning based on a model, this signifies that the knowledge of causal dependencies can be used in the search for the first deviation in the graph as we will see in this section.

Alarm filtering with a causal graph allows for a switch from a set of alarms to an ordered list of causes (Figure 6.13). In this example, a simple signed directed graph is used. A variable is said to be abnormal if it has a variation index close to 1 [KRA 87]. Thus, this example contains two significant deviations, X4 and X8. As X8 can be explained by the propagation of the positive deviation of X4, we make the prediction of a unique fault.

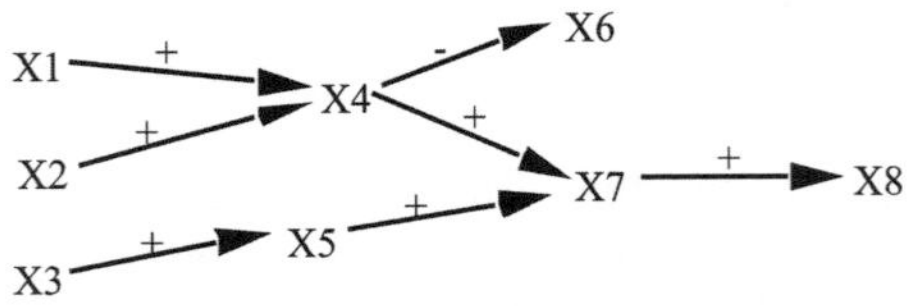

X1	X2	X3	X4	X5	X6	X7	X8
0.3	-0.4	-0.1	1.0	0.3	-0.7	0.8	1.1

Figure 6.13. *Signed directed graph (SDG) and alarm filtering*

A *backwards chaining* therefore starts with X4; the algorithm eliminates X1 and X2 since the indices are not significant; thus, the system identifies X4 as the most probable first-order deviation. A *forwards chaining* on X7 and X8 reinforces this hypothesis, as well as the index of X6. We note that if the index of X6 had a value of –0.3, we could have made the hypothesis that the fault propagation has not yet reached X6 (for example in the case of a pure delay).

Kramer and Finch [KRA 89] distinguish two kinds of knowledge: behavioral knowledge, which describes the behavior of a system, and diagnosis knowledge, which corresponds to reasoning over the system malfunctions. Diagnosis knowledge is generally not available during the creation phase; it must be acquired through experience or due to a conversion of behavioral knowledge. For reasoning in diagnosis, behavioral knowledge must be combined with a diagnosis method; this is the usual procedure in model-based diagnosis approaches. In our case, it also involves introducing causality into the behavioral model of the process.

Causal graphs are adequate tools for model-based diagnosis, since they provide a good dynamic support for the behavioral model and the implicit diagnosis methodology. The most usual form of causal graph for diagnosis was the directed signed graph with new isolation algorithms proposed in 1997 [MOS 97]. Through the enrichment of causal graph information of the causal graph (the quantification and introduction of time to the functions carried by the arcs or the definition of description space of non-variables), it is easy to see that the causal graph is a reasoning structure, which is easy to make evolve given the knowledge we have of a system during its operation.

We cannot only use algorithms with a simple SDG and process simple alarms with the values of variables, but we can also introduce nonlinear dynamic functions over the arcs and generate residuals over the variables in order to turn toward legitimate dynamic model-based diagnosis approaches [MON 00a, MON 00d].

Yu and Lee [YU 91] have used fuzzy sets to show how to enrich and make the causal structure evolve according to the knowledge available on the process over time. The basic principle of this work is to introduce quantitative data into an SDG. Let us consider, for example, the significance of the qualitative gain "+" between two variables A and B. The binary relation between A and B can be described by the ratio ΔB/ΔA and by taking the "+" value from 0 to infinity. In terms of membership functions, μ_{BA} (ΔB/ΔA) = 1 (see Figure 6.14(a)). As soon as we obtain quantitative information, the membership function can be "remodeled". For example, if the gain between A and B is between 5 and 10, the membership function can be modified as in Figure 6.14(b).

Finally, when the gain between two variables is known precisely (7.5), μ_{BA} is readjusted according to Figure 6.14(c). The membership function from fuzzy set theory provides a simple way to integrate quantitative knowledge into an initially qualitative representation.

The history of these types of approaches has been demonstrated in [MON 94c, MON 94a, MON 00d]. Our DIAPASON [PEN 93] and SALOMON [MON 96] prototypes fit perfectly into the frame of such systems.

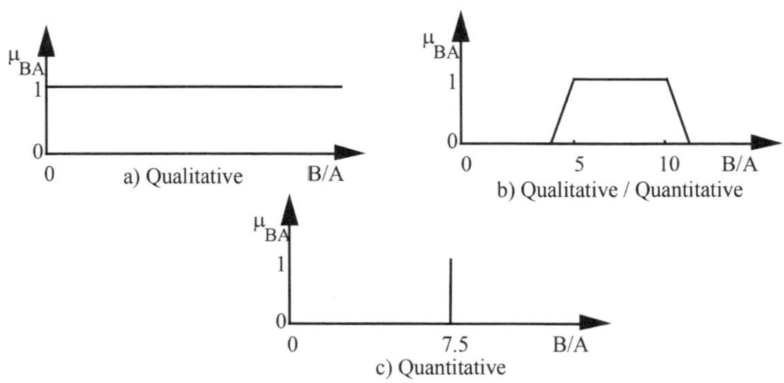

Figure 6.14. *Precision of the gain knowledge between A and B*

The causal structure is a representation of the process at a high abstraction level. The causal graph is a structure that can absolutely support a more traditional simulation than that mentioned in the relative works of [LEY 91c, DZI 96, MON 92]. Therefore, since each QTF represents a concept of gain, delay and response time, it can be replaced, for example, by a differential equation. The obtained graph is thus the dual of the block diagram used in the control theory.

For a process in which we know the *a priori* model described by a system of differential equations, the causality links between variables are implicit: the effects in the output cannot be prior to the variations of the inputs. The causality that is then defined becomes equivalent to the notion of computability and is linked to the implicit discretization of differential equations through simulation algorithms. This means that the system of differential equations can be represented in the form of a causal graph possibly containing loops:

$$\begin{cases} \dot{x}_1 = g_1(u_1, u_n, x_1) \\ \dot{x}_2 = g_2(u_i, x_2) \\ \dot{x}_3 = g_3(u_1, x_2, x_n, x_3) \\ \dots \\ \dot{x}_n = g_n(x_3, x_4, x_i, x_n) \end{cases} \quad [6.11]$$

The output of an equation becomes an input of the following equation corresponding to the choice of a partial strict order relation between variables. The graph in Figure 6.15 does not show anything other than this order relation.

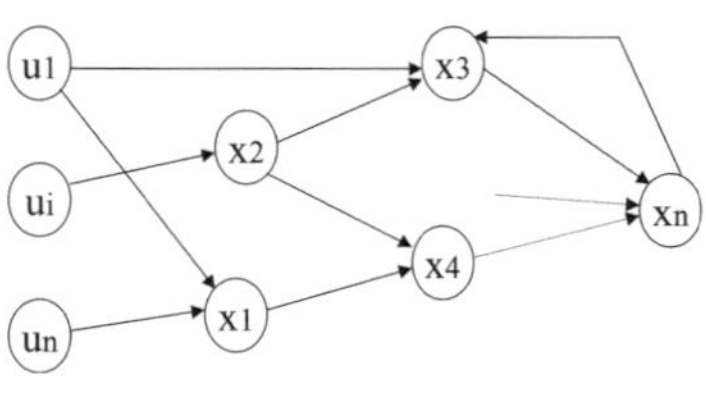

Figure 6.15. *Causality, partial strict order relation*

The simulator therefore calculates the evolution over time $x_i(k)$ of the variables X_i from the evolution $x_j(k)$ of its antecedents $X_j, j \in P_i$ where P_i is the set of predecessors X_i, and the evolution $u_j(k)$ of the inputs of the process $U_j, j \in Q_i$ acting directly on X_i.

6.6.2. *Generation of residuals and causal structure*

The method proposed here allows us to obtain the power of digital isolation methods without resorting to algebraic manipulation in order to obtain a set of residuals structured by using only the structure of the causal graph. It allows for more diverse process representations than those proposed by traditional diagnosis methods. Its formal link with the more traditional diagnosis approaches has been explained in [GEN 04]. In particular, it is flexible enough to be easily applied to the previously mentioned industrial simulators without additional modeling effort [CEL 07].

The real behavior $\overline{x}_i(k)$ of X_i is itself a function of the evolution $\overline{x}_j(k)$ of the antecedents $X_j, j \in P_i$ and the evolution $u_j(k)$ of the inputs of the process $U_j, j \in Q_i$.

Let ε_i be the open loop residual:

$$\varepsilon_i(k) = \overline{x}_i(k) - x_i(k) \qquad [6.12]$$

The use of only these residuals only allows us to carry out detection. Indeed, it is clear that the residual ε_i is stimulated either by a local fault f_i or by a fault in a variable that is higher from the source of ε_j, which will be propagated to ε_i. If we ban the use of a model other than the simulator to carry out the isolation, we must equip the diagnosis system with another set of fault indicators by acting on the potential inputs of this simulator and by using causal reasoning and QR.

A second type of residual is therefore introduced: the predictive residuals that are based on the previously mentioned causal graph and do not involve parameters other than those from the simulator equations. To this effect, we define $x_i^p(k)$, which is the prediction of X_i for the antecedent $X_p, p \in P_i$ of X_i: the simulated evolution $x_p(k)$ of X_p is replaced by its measured evolution $\overline{x}_p(k)$ to obtain the predicted evolution $x_i^p(k)$ of X_i due to the simulator. This is the prediction X_i *over reconfiguration* of the antecedent X_p in the simulator.

We can therefore define the prediction error *over reconfiguration* of X_p:

$$\varepsilon_i^p(t) = \overline{x}_i(t) - x_i^p(t) \qquad [6.13]$$

Generally, it is possible to reconfigure several antecedents at a time. Let P_i^r be the set of indices of the reconfigured antecedents of X_i. Therefore, if $\varepsilon_i^{P_i^r}$ designates the prediction error *over the reconfiguration* of antecedents associated with P_i^r:

$$\varepsilon_i^{P_i^r}(t) = \overline{x}_i(t) - x_i^{P_i^r}(t) \qquad [6.14]$$

In the linear case, it is easy to show that it is possible to compute $\varepsilon_i^{P_i^r}$ in terms of the threshold ε_i^p: the unit reconfigurations give access to the prediction errors obtained when combining reconfigurations. The limiting case is $\varepsilon_i^{P_i}$, the error of the prediction: the calculated evolution of X_i is only obtained from the measurements of its antecedents. Finally, if we take away the prediction error *over the reconfiguration* of the X_p of P_i^r at the simulation error over X_i, $\varepsilon_i(k) - \varepsilon_i^{P_i^r}(k)$, we show that this difference is not more susceptible to simulation errors $\varepsilon_p(k)$ over the antecedents X_p of P_i^r [MON 00a, MON 00d].

This result means that in practice, the difference between the simulation error and the prediction error *over the reconfiguration* of X_p allows us to evaluate the contribution of the simulation error ε_p propagated from X_p to the simulation error observed over X_i.

6.6.3. *Interpretation of the errors for the isolation and filtering of faults*

By using the different predictive residuals with the different reconfigurations, it is therefore possible for a variable X_i to say whether the observed error over X_i has a local cause (primary fault) or is simply the consequence of a fault that is higher (secondary fault). This test is the fundamental isolation procedure in a diagnosis diagram based on the causal graph.

The proposed method, due to an evaluation and a qualitative analysis of different simulation errors of the graph's variables, allows us to decide whether these errors are dynamically linked or not. This is not simply the case of, having detected multiple faulty variables in a causal graph, declaring that the path linking them is a propagation path from the highest detected variables toward the others, but rather deciding, arc by arc, whether it is suspect or not. Establishing the link between the two errors is based on the knowledge of the causal graph's structure and the local dynamic analysis of the behavior.

Let us take the simple example in Figure 6.16 to illustrate this principle. The simulated evolution of X_3 is computed with the simulated evolutions of X_1 and X_2. X_3 is the detection variable. A significant error ε_3 has been detected on X_3. The aim is to isolate the cause of this fault: it can be due to a fault directly affecting X_3, or be the consequence of a malfunction affecting X_1 or X_2. We can note that X_1 and X_2 are not necessarily detection variables. We can assume that the error on X_1 (or X_2) was not significant enough (as defined by the chosen detection criteria) to be detected; however, it could be significant enough to be held responsible for the error on X_3.

The use of local reconfigurations allows us to test it. A local prediction where the real evolution of X_1 is replaced with its simulated evolution is carried out as if X_1 was an input of the simulator.

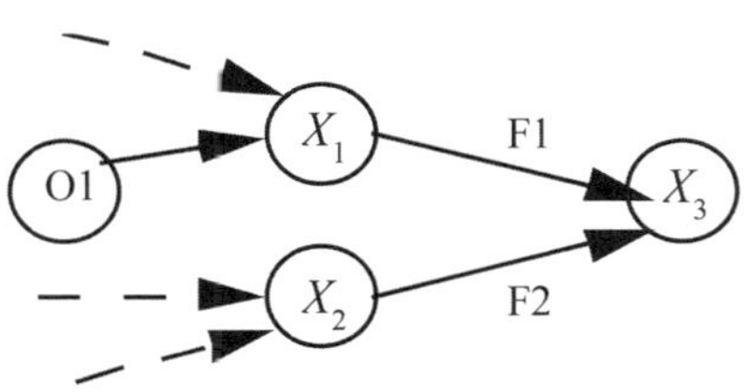

Figure 6.16. *A causal graph*

The reconfiguration of X_1 and/or X_2 provides three prediction errors: ε_3^1, ε_3^2 and $\varepsilon_3^{\{1,2\}}$ with the notations of equation [6.14]. Thus, according to the conclusion of section 6.4.2, we can note that for the following cases [MON 00a, MON 00d]:

– if ε_3^j and ε_3 are near identical, which will then be written as $\varepsilon_3^j \cong \varepsilon_3$, then $\varepsilon_j \cong 0$ and X_j cannot be considered as responsible for the error detected on X_3;

– if ε_3^j is negligible compared to ε_3, which will be written as $\varepsilon_3^j \ll \varepsilon_3$, then ε_3 is essentially stimulated by ε_j and X_j is incriminated as responsible for ε_3;

– if ε_3^j and ε_3 are of the same order of magnitude, which will then be written as $\varepsilon_3^j \approx \varepsilon_3$, we can draw the unique conclusion that ε_3 is only partially explained by ε_j; in this case, ε_3 is explained by the multiple faults and if, in addition to this, $\varepsilon_3^{\{1,2\}} << \varepsilon_3$, then X_1 and X_2 are both incriminated.

This can be summarized in the following way:

– $\varepsilon_3^1 << \varepsilon_3$ and $\varepsilon_3^2 \cong \varepsilon_3$: secondary fault, X_1 is incriminated;

– $\varepsilon_3^1 \cong \varepsilon_3$ and $\varepsilon_3^2 << \varepsilon_3$: secondary fault, X_2 is incriminated; [6.15]

– $\varepsilon_3^1 \cong \varepsilon_3$ and $\varepsilon_3^2 \cong \varepsilon_3$: primary fault, X_3 is incriminated;

– $\varepsilon_3^1 \approx \varepsilon_3$ and $\varepsilon_3^2 \approx \varepsilon_3$ and $\varepsilon_3^{\{1,2\}} << \varepsilon_3$: secondary fault, X_1 and X_2 are incriminated.

The residuals which are structured for the isolation of the fault are obtained because of the causal decomposition without the need for the slightest algebraic manipulation. The proposed test generalizes the coherence test between the initial nodes and the final node [SHI 85, KRA 87b, MOS 97] in the most complicated case where the causal relations are quantitative and dynamic, and the test is applied on a computed entity (the residual) and not directly on the deviations in the measurements associated with the variables. A more theoretical presentation of the method is available in [MON 00d, GEN 04, CEL 07].

We note that X_1 and X_2 are not necessarily detection variables; the error on X_1 (or X_2) may not be significant enough, from the point of view of the criteria used by the detection, but can be significant enough to explain the error on X_3. This means that even if errors X_1 and X_2 have not been detected, X_1 and X_2 can be proposed as source variables through the isolation process, and this result allows us to render the diagnostics reasoning more robust.

The generic rules corresponding to the example of equation [6.15] and which provide the analysis in terms of the orders of magnitude of the residuals are the following:

– ($\forall k \in P_i$, $\varepsilon_i^k \cong \varepsilon_i$): primary fault, X_i incriminated;

– ($\varepsilon_i^k << \varepsilon_i$) and ($\forall$j $\in$ Pi, j $\neq$ k, $\varepsilon_i^j \cong \varepsilon_i$): secondary fault, X_k incriminated;

– ($\ni P_i^r \subset P_i$) such that ($\varepsilon_i^{P_i^r} \ll \varepsilon_i$) and ($\forall k \in P_i^r$, $\varepsilon_i^k \approx \varepsilon_i$): secondary fault, X_k, $k \in P_i^r$ incriminated. [6.16]

The relations with orders of magnitude for interpreting the coherence represented by [6.16] for the isolation illustrate the idea that reasoning qualitatively over real numbers is often enough in the case of practical resolutions. However, the implementation of simple rules as proposed in [6.16] requires an appropriate formalism. The mathematical processing of relations with orders of magnitude $\ll$, $\approx$ and $\cong$ is taken on in [MON 00a]. It is inspired by the formalism *O(M)* in QR [MAV 88].

To generalize the previous example, it is enough to note that due to causality, the local analysis of a variable X_i can be included in a recursive pathfinding strategy of the graph. The test on X_i is followed by the testing of its antecedents and so on and so forth. The complete propagation graph of the effects of the fault can thus be recovered, the source(s) of the graph therefore being identified as the cause(s). The other nodes of the propagation graph are only variables that have been affected by the effects of the primary fault(s). At date t, we can therefore generate a subgraph of faults that exposes the causal chain of the detected symptoms: this is the isolation at date t. Supervision over time of the evolutions of this subgraph therefore corresponds to the fault filtering functionality (Figures 6.17(a) and (b)). In [MOS 97, MOS 98], the notion of *fault detection* is used.

6.6.4. *Advantages for supervision*

We emphasize that a variable which would have not been detected as a fault, because the detection threshold of this would have been chosen in an inadequate fashion, can however, after this filtering procedure, be declared as a variable at the origin of the fault and therefore belongs to the fault propagation subgraph: the consequences of a bad parameterization of the detection procedure are therefore lessened. Furthermore, the source of the graph can very well be revised and modified during a filtering stage later than t. This offers more flexibility in the choice of detection thresholds – quantitative analyses are no longer necessarily required – and assures a greater robustness of the detection process with its own parameters [MON 92, MON 00a].

Let us take the evolution in time of the fault subgraph (*fault detection*). When a propagation subgraph has been identified, the faults which could appear later will be tested with the influence of the same process to determine if they correspond to the appearance of a new fault or if they are the consequences of faults detected and explained previously: this is no longer an isolation step and rather one of fault filtering. Indeed, if the effects of the fault are propagated, new simulation errors will appear, and they will become new sinks in the fault propagation graph. Each time measurements are taken, the fault graph evolves according to the dynamic propagation of the effects of the fault. The dynamic monitoring of the fault trace allows us to observe at each instant the evolution of the consequences on the process (Figure 6.17).

If diagnosis is perceived as the essential phase of a maintenance strategy, this dynamic monitoring is therefore not a fundamental process: only the source of the fault, found through isolation, must be provided so that the maintenance or the adequate corrective action can be executed at the earliest moment. On the other hand, to carry out supervision, it is required that the new faults be filtered and mapped to the problem that was previously highlighted. The filtering of faults is therefore an additional functionality that allows us to dynamically follow the evolution of the trace of the fault, and therefore remain in control of the intervention process. The visualization of this fault propagation subgraph on the operating interface makes up an explanation process that is popular with operators [PEN 93, EVS 98b].

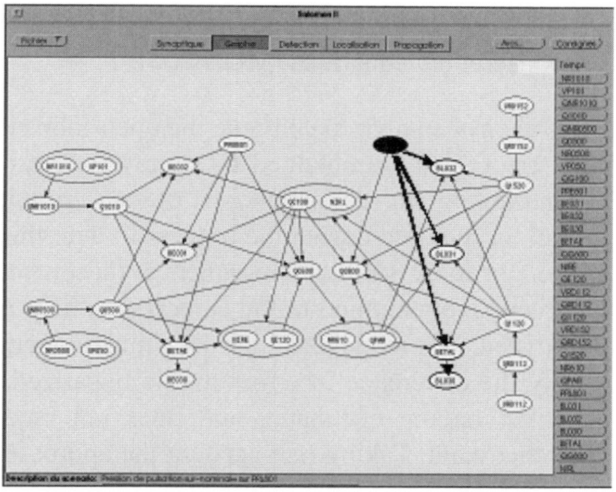

a) Detection at an early stage of a pulse problem

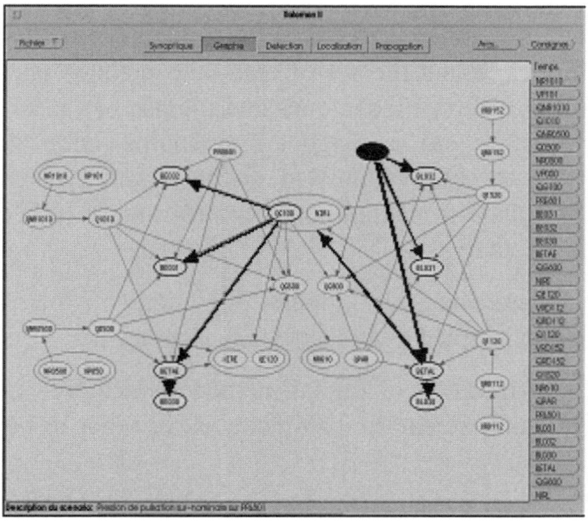

b) The same fault at a later stage,
the hydraulics of the column are affected

Figure 6.17. *Illustration of the propagation of a fault*

The principle of fault filtering that is summarized here is not linked with a given representation or with a particular resolution method; it remains valid whether we use QTFs [MON 93b], linear difference equations

[EVS 97a, EVS 97b, EVS 00] or continuous [VAD 97a, VAD 97b] or discrete nonlinear differential equations [MON 00d].

In practice, it is always enough to compute the open-loop simulation error and the prediction errors that are obtained by simply successively replacing each variable intervening in the equation of the simulator with its measurement, which is acquired during the process to determine if it is from a primary fault or from a simple consequence (secondary fault). The algorithm is then recursively applied to the antecedents held responsible for the error being analyzed. The model used for the simulation error is therefore nonlinear; it is only the difference in errors that is linearized, which is still much less restrictive, on the one hand, and does not require additional modeling, on the other hand. Taking into account the approximations, which are due, among others, to the research being used, the layer of approximate reasoning to be added for a pertinent interpretation of the different errors in play justifies this even more [MON 00a].

The basic principle and the set of resulting methods take advantage of both the precision of digital techniques and the use of causal knowledge on the process to carry out a diagnosis reasoning adapted to industrial processes. If the use of manipulated residuals can be compared to the generation of residuals from traditional control theory [GEN 04], we must remember however that:

– in these approaches, the global modeling of the system is not necessarily the best adapted to a diagnosis reasoning;

– the power of isolation of digital methods is acquired in the proposed method without using algebraic manipulations in order to obtain the set of structured residuals.

Causal knowledge is used to apply concepts, although traditional, of model-based diagnosis – simulation, prediction, coherence of measurement – to local submodels, which are extremely reduced. Isolation is therefore obtained naturally by testing the coherence of local models with the measurements. Contrary to an approach with a group of observers where the state is usually reconstructed globally, all that is reconstructed in this type of approach – reconfiguration and simulation again – is within the scope of a backwards chaining of the causal graph where the only variables necessary for the current diagnosis reasoning are examined.

The method allows for the creation of a complete detection, isolation and fault filtering system for mathematical representations of the process which are less constraining than a single state representation. If the simulator based on differential equations is already available, using, for instance, a few computer properties such as a *switch* of the inputs of the differential equations, it is not even necessary to know the analytical structure of the model to carry out the isolation: once the causal structure is associated with the established simulator, it can simply be seen as a black box with commutable inputs. In this configuration, the method allows for the fault isolation without extensions or refinement of the available industry systems models.

The algorithms based on the principal of causality remain simple, disassociated from the model to and allow us to avoid all of the algebraic manipulations that allow us, lose sight of the progression of the diagnosis reasoning. The fact that a digital command is based on a mathematical model that does not have any sense for the operator is not fundamentally important, since it is in this case a closed loop in the process. However, in the case of computerized supervision, the operator remains within the decision loop, and it is therefore required that the mathematical subtleties used by the system find a cognitive transcription adapted to the representation and reasoning of the operator in situations involving malfunctions. We can consider the method proposed here as dedicated to monitoring as well as diagnosis: it takes advantage of the causal decomposition process to minimize the calculations to be executed, equips the diagnosis system with an essential explanatory characteristic for a good overview in the control room and allows for monitoring of the evolution of the failure mode over time.

An FDI system in a complex setup is primarily an operating assistance tool. The concept of causal model-based operation seems to head in this direction: the visualization of the behavior of the model and the process in a history constitutes a primary tool for understanding the diagnosis. The visualization of the graph and its fault propagation paths is another of these, which is very popular with operators working with nuclear procedures [PEN 93, EVS 98b, LAM 98, LAM 99]. Thus, when the failure modes are simple or monophasic, in other words only concerning the aqueous or organic phase of the process, the operators use the graph more or less as a flow chart. However, when the problems are biphasic, and the hydraulic and chemical problems are mixed or the counteraction of a regulation intervenes, the operators resort to *the deep knowledge made explicit through causal representation.*

6.7. Discussion and conclusion

The introduction of an operating support system to the control room has inevitably (and must have) had an impact on the operation. Indeed, a human operator in charge of procedure execution creates his action through representations constructed by the "specialized artifact", the "physical symbols system" generally referred to as the control station. This explains the fact that every other project aiming to transform this station, for example by introducing advanced supervision functionality, inevitably leads to an evolution of the operating process itself, the operator adapting "intelligently" to the newly proposed representations.

Intervention decisions, particularly in an abnormal situation, depend on the sense (operationalized knowledge) and the value (assessment of the information utility) that the observer accords to the perceived phenomena. To intervene at the chosen level of reality, the situation must take shape and make sense regarding the displays, for the operator confronted with understanding, anticipating or practicing difficulties (DSS).

The supervision of a system operated by an operating team therefore defines a complex cognitive situation. Within this, the human operator must adapt to the situation, in other words apply their cognitive structures to real perceived circumstances in order to make pertinent decisions to achieve the goal. Faced with the complexity of the situation, a representation of the situation in intelligible and communicable terms is required: the model. This model (or models) is an information medium and can be more or less useful depending on:

– its representation power, in other words, its capacity to relate to the real situation, the goal of the act of knowledge;

– its argumentation power, in other words, its capacity to accompany the process of intervention.

To see through their actions, the operator must be capable of combining:

– knowing how to observe the object: the human subject perceives it through descriptive characteristics that give it a particular form;

– know-how: the human subject creates representations of the object that are appropriate to the situation;

– a practical knowledge: knowing how to behave through actions depending on the acquired knowledge.

A cognitive situation therefore involves three distinct processes (Figure 6.18). For the first, the object takes shape in the project. This process of information produces the basic data, the process of confrontation of the object with its environment and the concepts that are formed throughout the history of the object. For the second, the project mobilizes the subject. The object is mentioned in the project–human subject relation, in other words, it is both designated and interpreted by the human subject, a procedure which we qualify as representation, through knowledge, expressed around symbol systems and incorporated through competencies. For the third, the object gains meaning. We give meaning to the object–human subject relation through the process of signification, which exposes the elementary (actionable) knowledge developed by the human, the strategic reasonings, which are founded on calculations that are applied to the object and rely on reference principles.

Which lessons can we take from this general analysis of a cognitive situation in order to understand the different approaches needed for supervision, and in particular how are the control theory approaches placed (directed toward an automation of the cognitive functions and excluding the human from the loop) compared to those of AI (directed toward supporting human decision-making)? For this, we return to the values mentioned in the general diagram in Figure 6.18 applied to supervision. As we will choose either a *control* or *AI* point of view, the effort of modeling does not apply in the same way to the situation.

Thus, from the control theory perspective of supervision, using the principles of the techniques from FDI, it is clear that the crux of the modeling is found within the frame of the system (object) representation: the model is inspired from a representation from state variables or which can return, "revisited" for a supervision activity. In a very schematic manner, for the FDI community, the behavioral model of a system expresses the constraints or relations that exist between its description variables. An analytical redundance relation (ARR) is a relation where only known variables appear (for example measured): the non-accessible orders of magnitude were eliminated through substitution, projection, etc. from the initial relations of the model [CAS 94, GER 00]. ARR allows us to verify the coherence of variable values and cannot be verified while the behavior of the system corresponds to that from the model. A failure manifesting itself in the measurement of a variable or the change of a parameter value of an ARR will lead to its non-verification. For a failure, it is therefore possible to determine the set of ARRs affected during the occurrence of this failure. The list of affected ARRs with the measurements acquired online allows us to know which relations are respected or not at the current instant. The comparison of a set of non-respected relations at the current instant with theoretical signatures of the failures therefore allows us to identify the theoretical signature that is closest to the observations, and thus to determine the current failure.

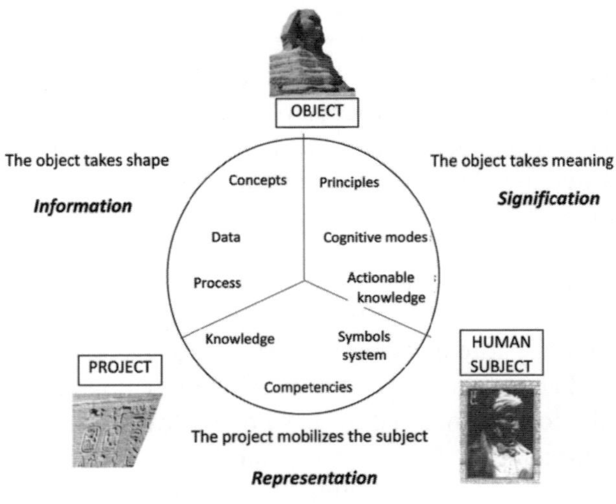

Figure 6.18. *The cognitive situation*

This type of approach is based on a very strong axiomatic structure [COR 00, COR 04]:

– "Exoneration" type axiom: a component that behaves normally is in a normal state.

– Non-interaction or non-compensation axiom: extension of the previous axiom (in the case of multiple faults).

– Fault uniqueness axiom: only one component (or no more than k) can be faulty.

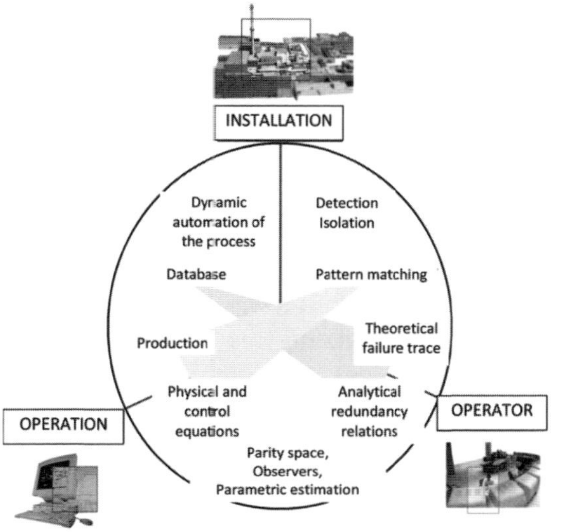

a) Modeling commitment from a cognitive automation point of view

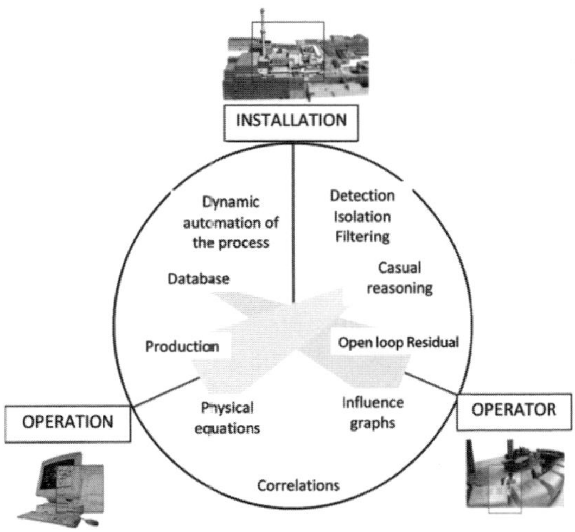

b) Modeling commitment from an operator support point of view

Figure 6.19. *The commitment of the modeling differs according to the different supervision approaches*

Thus, when the representation of the system can be perfectly outlined by a mathematical model in a strong axiomatic form (the principles), the actionable knowledge and the cognitive modes have little added value: the actionable knowledge (here, the cognitive mechanisms for the act of diagnosis) are in a way "precompiled" from the representation phase through the notion of failure. For the theoretical failure trace, created *a priori*, it is simply a case of recognizing the current trace observed among the different theoretical traces online. The theoretical failure trace is therefore a very powerful actionable knowledge, "completely computable", which according to the theory of the previous axioms reduces the diagnosis cognitive mode to a simple logical operation of *pattern matching* (Figure 6.19(a)).

From the control theory point of view, we can therefore consider that the "representation" type of modeling is mostly prioritized over the "signification" type of modeling: once the failure traces are established based on the ARR, the cognitive act of diagnosis strictly speaking appears to be significantly reduced.

From the operator support point of view, we find the AI approach (Figure 6.19(b)), as we have seen in this chapter,"AI being used in the sense of a knowledge representation methodology directed toward the mechanization of reasoning, specifically in the common sense, by explicitly taking into account the symbolic character of certain pieces of information" [DUB 94]. The human operators bring with their problem solving approach, capacities of reasoning, which allow them to determine the difference between different contexts, take shortcuts, use and accumulate experience, manage imprecision and even uncertainty. However, their reasoning is founded neither on a mathematical model of the process nor on the precision of numeric data, which they are presented with, but on the symbolic interpretation of these, which is the crux of the explanations that a support system must provide. These ideas gave rise to causal reasoning and QR and since the end of the 1980s approximate reasoning, which we have highlighted throughout this chapter. AI proposes algorithms that simulate or accompany the reasoning capacities of a human being. From this second point of view, the idea is evidently to prioritize the metric of significance and not that of the representation in a cognitive supervision situation (Figure 6.19(b)). The models are more qualitative, or at least less precise in general, the diagnostic hypotheses less drastic, and subsequently the decision cannot be automated: the assistance for reasoning becomes essential and the human–machine synergy must be prioritized in order to resolve supervision problems. The main quality expected from models is no longer precision, but pertinence, the compatibility between the representation used and the cognitive modes of the operator. In this chapter, a causal reasoning (the cognitive mode of the significance metric) supports the analyses of the operator in the control room, confronted with complex decision-making situations, and therefore favors human–machine cooperation.

6.8. Bibliography

[ALL 83] ALLEN J.F., "Maintaining knowledge about temporal intervals", *Communications of the ACM*, vol. 26, no. 11, pp. 832–843, November 1983.

[BAI 83] BAINBRIDGE L., "Ironies of automation", *Automatica*, vol. 19, no. 6, pp. 775–779, 1983.

[BAI 91] BAINBRIDGE L., "Les systèmes experts résoudront-ils tous les problèmes des opérateurs", *Introductory lecture: Facteurs humains de la fiabilité et de la sécurité des systèmes complexes*, Editions INRS, Vandœuvre, 1991.

[BAN 89] BANDEKAR V., "Causal models for diagnostic reasoning", *Artificial Intelligence in Engineering*, vol. 4, no. 2, pp. 79–91, 1989.

[BEN 96] BENKHANNOUCHE S., Aide à la supervision des processus industriels: vers une méthodologie de conception, Thesis, Pierre and Marie Curie University, Paris, 1996.

[BER 89] BERINGER D.B., HANCOCK P.A., "Exploring situational awareness: a review and the effects of stress on rectilinear normalization", *Proceedings of the 5th International Symposium on Aviation Psychology*, Ohio State University, Columbus, vol. 2, pp. 646–651, 1989.

[BER 11] BERRAH L., MONTMAIN J., CLIVILLÉ V., *et al.*, "Management decision-aiding with a fuzzy integral based performance measurement system", *European Society for Fuzzy Logic and Technology (Eusflat)*, Aix-les-bains, France, 2011.

[BOB 84] BOBROW D.G., "Qualitative reasoning about physical systems: an introduction", *Artificial Intelligence*, vol. 24, pp. 1–5, 1984.

[CAC 92] CACCIABUE P., DECORTIS F., DRODZDOWICZ B., *et al.*, "COSIMO: a cognitive simulation model of human decision making and behavior in accident management of complex plants", *IEEE Transactions on Systems, Man and Cybernetics*, vol. 22, no. 5, pp. 1058–1074, 1992.

[CAL 88] CALOUD P., Raisonnement Qualitatif: Application à l'aide à la supervision des procédés continus, Thesis, National Polytechnic Institute of Grenoble, 1988.

[CAR 92] CAROL L.A., "Desperately seeking SA", *TAC Attack (TAC SP 127-1)*, vol. 32, no. 3, pp. 5–6, 1992.

[CAS 94] CASSAR J.P., LITWAK R., COCQUEMPOT V., *et al.*, "Approche structurelle de la conception de systèmes de surveillance pour des procédés industriels complexes", *Revue européenne. Diagnostic et Sûreté de Fonctionnement,* vol. 2, 1994.

[CHA 90] CHARBONNAUD P., BERTIN R., "A technical diagnosis expert system using qualitative physics", *COGNITIVA 90,* Madrid, Spain, pp. 20–23, November 1990.

[CHA 91] CHARBONNAUD P., BERTIN R., "A quasi-qualitative formalism for symbolic diagnosis as a solution to numeric-symbolic interface problems", *IMACS International Workshop sur les systèmes d'Aide à la décision et le Raisonnement Qualitatif,* Toulouse, 1991.

[CHA 92] CHARLES A., Aide à la décision d'anomalies de fonctionnement de systèmes dynamiques: une approche fondée sur des modèles qualitatifs et quantitatifs, Thesis, University of Technology of Compiègne, 1992.

[CHA 04] CHARBONNIER S., BECQ G., BIOT L., "On line segmentation algorithm for continuously monitored clinical data in Intensive Care Units", *IEEE Transactions on Biomedical Engineering*, vol. 51, pp. 484–492, 2004.

[CHA 05] CHARBONNIER S., "On line extraction of temporal episodes from ICU high-frequency data: a visual support for signal interpretation", *Computer Methods, Programs in Biomedicine*, vol. 78, pp. 115–132, 2005.

[CHA 06] CHARBONNIER S., "Surveillance de systèmes continus à l'aide de méthodes n'utilisant pas de modèles formels: Application aux systèmes médicaux", Thesis, INPG, 2006.

[CEL 07] CELSE B., GENTIL S., MONTMAIN J., "Raisonnement causal", in GENTIL S., (ed.), *La supervision des systèmes industriels complexes*, Hermès, Lavoisier, Paris, pp. 353–416, 2007.

[COR 00] CORDIER M.-O., DAGUE P., LÉVY F., *et al.*, "A comparative analysis of AI and control theory approaches to model-based diagnosis", *ECAI'2000*, Berlin, Germany, 2000.

[COR 04] CORDIER M.-O., DAGUE P., LÉVY F., *et al.*, "Conflicts versus analytical redundancy relations: a comparative analysis of the model based diagnosis approach from the artificial intelligence and automatic control perspectives", *IEEE Transactions on Systems, Man, and Cybernetics – Part B: Cybernetics*, vol. 34, no. 5, pp. 2163–77, November 2004. DOI:10.1109/TSMCB.2004.835010.

[DAG 86] DAGUE P., DEVES P., RAIMAN O., "Raisonnement Qualitatif dans le diagnostic de pannes", *6e Journées Internationales sur les Systèmes Experts et leurs Applications*, Avignon, 1986.

[DAG 87] DAGUE P., RAIMAN O., DEVES P., "Troubleshooting: when modeling is the trouble", *American Association for Artificial Intelligence (AAAI) Conference*, Seattle, WA, pp. 600–605, 1987.

[DAG 88] DAGUE P., "Order of magnitude revisited", *Qualitative Physics Workshop*, Paris, France, 1988.

[DAL 97] DALRYMPLE M.A., SCHIFLETT S.G., "Measuring situational awareness of AWACS weapons directors. Situational awareness in the tactical air environment", *Augmented Proceedings of the Naval Air Warfare Center's First Annual Symposium, CSERIAC SOAR Report# 97-01*, WP-AFB, Ohio, 1997.

[DAV83] DAVIS R., "Diagnosis via causal reasoning: paths of interaction and the locality principle", *American Association for Artificial Intelligence*, pp. 88–94, 1983.

[DAV 84] DAVIS R., "Diagnostic reasoning based on structure and behaviour", *Artificial Intelligence*, vol. 24, pp. 347–410, 1984.

[DE 84] DE KLEER J., BROWN J.S., "A qualitative physics based on confluences", *Artificial Intelligence*, vol. 24, pp. 7–83, 1984.

[DE 86] DE KLEER J., "An assumption-based truth maintenance system", *Artificial Intelligence*, vol. 28, pp. 127–162, 1986.

[DE 89] DE LAPPARENT D., CALOUD P., DESCOTTES Y., *et al.*, *Le projet ALLIANCE – Systèmes experts et conduite de processus*, *OFTA*, Masson, Paris, pp. 81–94, 1989.

[DOR 87] DORMOY J.L., Résolution Qualitative : Complétude, Interprétation Physique et Contrôle Mise en œuvre dans un langage à base de règles: BOOJUM, PhD Thesis, University of Paris 6, 1987.

[DOR 88] DORMOY J.L., RAIMAN O., "Assembling a device", *Proceedings of 7th National Conference on Artificial Intelligence (AAAI-88),* Saint-Paul, MN, pp. 330–335, 1988.

[DOR 89] DORMOY J.L., "Nouvelles méthodes de calcul qualitatif", *Revue d'Intelligence Artificielle, numéro spécial sur le Raisonnement Qualitatif*, vol. 3, no. 4, pp. 39–53, 1989.

[DUB 89] DUBOIS D., "Order of magnitude reasoning with fuzzy relations", *Revue d'Intelligence Artificielle, numéro spécial sur le Raisonnement Qualitatif*, vol. 3, no. 4, 1989.

[DUB 91] DUBOIS D., GENTIL S., "Intelligence Artificielle: un outil pour l'automatique? " Rapport du GR Automatique, in GENTIL S., DUBOIS D. (eds.), *Automatique et Intelligence Artificielle*, CNRS, 1991.

[DUB 94] DUBOIS D., GENTIL S., "Intelligence Artificielle pour l'Automatique", *Revue d'Intelligence Artificielle*, vol. 8, no. 1, pp. 1–27, 1994.

[DZI 93a] DZIOPA P., GENTIL S., "Hierarchical organisation of causal graphs", *2nd IMACS International Symposium on Mathematical and Intelligent Models in System Simulation,* Bruxelles, Belgium, 1993.

[DZI 93b] DZIOPA P., LEYVAL L., GENTIL S., "Qualitative reasoning for supervision", *International Joint Conference on Artificial Intelligence, Qualitative Reasoning Workshop*, Chambéry, France, 1993.

[DZI 94] DZIOPA P., LEYVAL L., GENTIL S., "Hierarchical causal modelling for large industrial plants supervision", *ISE Workshop,* Hambourg, Germany, 1994.

[DZI 96] DZIOPA P., Représentation multi-modèle pour la supervision des procédés continus, Thesis, National Polytechnic Institute of Grenoble, 1996.

[END 95a] ENDSLEY M.R., "Measurement of situation awareness in dynamic systems", *Human Factors*, vol. 37, no. 1, pp. 65–84, 1995.

[END 95b] ENDSLEY M.R., "Toward a theory of situation awareness in dynamic systems", *Human Factors*, vol. 37, no. 1, pp. 32–64, 1995.

[EVS 97a] EVSUKOFF A., MONTMAIN J., GENTIL S., "Les systèmes flous pour l'aide en ligne aux opérateurs", *6e Congrès français de Génie des Procédés*, Paris, France, 1997.

[EVS 97b] EVSUKOFF A., MONTMAIN J., GENTIL S., "Dynamic model and causal knowledge-based fault detection and isolation", *IFAC Symposium on Fault Detection, Supervision and Safety for technical Processes, Safeprocess' 97*, Hull, UK, 1997.

[EVS 98a] EVSUKOFF A., Le raisonnement approché pour la surveillance de procédés, Thesis, National Polytechnic Institute of Grenoble, 1998.

[EVS 98b] EVSUKOFF A., MONTMAIN J., GENTIL S., "Causal model based supervising and training", *IFAC Workshop on Supervision of Chemical Industries*, Lyon, France, 1998.

[EVS 99] EVSUKOFF A., COMBASTEL C., GENTIL S., "Qualitative and neural decision for fault detection and isolation", *IFAC'99, 14th World Congress of International Federation of Automatic Control*, Beijing, China, 1999.

[EVS 00] EVSUKOFF A., GENTIL S., MONTMAIN J., "Fuzzy Reasoning in co-operative supervision systems", *Control Engineering Practice*, vol. 8, pp. 389–407, 2000.

[FER 89a] FERAY-BEAUMONT S., Modèle qualitatif de comportement pour un système d'aide à la supervision des procédés, Thesis, National Polytechnic Institute of Grenoble,1989.

[FER 89b] FERAY-BEAUMONT S., LEYVAL L., GENTIL S., "Declarative modelling for process supervision", *Revue d'Intelligence Artificielle*, vol. 3, no. 4, pp. 135–150, 1989.

[FLE 08] FLEUREY F., DELHEN V., BENCOMO N., *et al.*, "Modeling and Validating Dynamic Adaptation", *3rd International Workshop on Models @Runtime (MRT'08) at MoDELS'08*, Toulouse, France, 2008.

[FLE 09] FLEUREY F., SOLBERG A., "A domain specific modeling language supporting specification, simulation and execution of dynamic adaptative systems", *12th International Conference Model-Driven Engineering Languages and Systems (MoDELS '09)*, Colorado, pp. 606–621, 2009.

[FOR 84] FORBUS K.D., "Qualitative process theory", *Artificial Intelligence*, vol. 24, pp. 85–168, 1984.

[FOR 88] FORBUS K.D., *Qualitative Physics: Past, Present, and Future – Exploring Artificial Intelligence*, Morgan Kaufmann Publishers, San Francisco, pp. 239–296, 1988.

[FOU 90a] FOUCHÉ P., KUIPERS B., "An assessment of current qualitative simulation techniques", *4th International Workshop Qualitative Physics*, Lugano, Switzerland, 1990.

[FOU 89] FOULLOY L., "Using qualitative reasoning to write expert controllers", *Revue d'Intelligence Artificielle, numéro spécial sur le Raisonnement Qualitatif*, vol. 3, no. 4, pp. 151–173, 1989.

[FOU 90b] FOULLOY L., Du contrôle symbolique, Démarche, Outils, Exemples, Doctoral Thesis, University of Paris-Sud, 1990.

[GAL 86] GALLANTI M., GILARDONI L., GUIDA G., *et al.*, "Exploiting physical and design knowledge in the diagnosis of complex industrial Systems", *ECAI*, Brighton, UK, pp. 335–349, 1986.

[GEN 97] GENTIL S. *et al.*, (ed.), Filtrage d'alarmes en temps réel, Convention de recherche scientifique et technique no. 93.34.098.00.470.75.01, Laboratoire d'automatique de Grenoble, 1997.

[GEN 99] GENTIL S., DZIOPA P., MONTMAIN J., "Hierarchical causal Model for complex plant supervision", *3rd IMACS-IEEE International Multiconference on Circuits, Systems, Communications and Computers, CSCC'99*, Athens, Greece, 1999.

[GEN 04] GENTIL S., MONTMAIN J., COMBASTEL C., "Combining FDI and AI Approaches within Causal – Model-based Diagnosis", *IEEE Transactions SMC-Part B*, vol. 34, no. 5, pp. 2207–2221, 2004.

[GEN 05] GENTIL S., MONTMAIN J., "Complex systems hierarchical representation supporting human decision-making", *Advanced Engineering Informatics,* vol. 18, pp. 143–159, 2005.

[GER 00] GERTLER J., "All linear methods are equal – and extendable to nonlinearities", *Safeprocess'2000, IFAC Symposium on Fault Detection, Supervision and Safety for Technical Processes*, Budapest, Hungary, 2000.

[GIO 02] GIORGINI P., MYLOPOULOS J., NICCHIARELLI E., *et al.*, "Reasoning with Goal Models", *21st International Conference on Conceptual Modeling (ER02)*, Tampere, Finland, pp. 167–181, 2002.

[GON 95] GONDRAN M., MINOUX M., *Graphes et Algorithmes*, Eyrolles, Paris, 1995.

[GON 04] GONZALEZ-BAIXAUL B., DO PRADO LEITE J.C.S., MYLOPOULOS J., "Visual variability analysis for goal models", *IEEE International Conference on Requirements Engineering (RE)*, Washington, DC, pp. 198–207, 2004.

[GUE 90] GUERIN F., "Raisonnement Qualitatif pour l'aide au contrôle des processus biologiques: interprétation en écologie", *10es journées internationales sur les systèmes experts et leurs applications*, Avignon, France, 1990.

[HAR 88] HARWOOD K., BARNETT B., WICKENS C., "Situational awareness: a conceptual and methodological framework", Technical Report no. USAFA-TR-88-1, pp. 316–320, CO: US Air Force Academy (AD-A198723), *Proceedings of the Psychology in the Department of Defense Eleventh Symposium*, Colorado Springs, CO, 1988.

[HAR 91] HARTEL C.E., SMITH K., PRINCE C., "Defining aircrew coordination: searching mishaps for meaning", *6th Symposium on Aviation Psychology*, Colombus, OH, 1991.

[IMO 11] IMOUSSATEN A., MONTMAIN J., TROUSSET F., *et al.*, "Multi-criteria improvement of options", *European Society for Fuzzy Logic and Technology (Eusflat)*, Aix-les-Bains, France, 2011.

[IRI 79] IRI M., O'SHIMA E., MATSUYAMA H., "An algorithm for diagnosis of system failure in the chemical process", *Computer and Chemical Engineering*, vol. 3, pp. 489–493, 1979.

[IRI 80] IRI M., AOKI K., O'SHIMA E., *et al.*, "A graphical approach to the problem of locating the origin of the system failure", *Journal of the Operations Research Society of Japan*, vol. 23, no. 4, pp. 295–311, 1980

[IWA 86] IWASAKI Y., SIMON H.A., "Causality in device behavior", *Artificial Intelligence*, vol. 29, pp. 3–32, 1986.

[IWA 90] IWASAKI Y., "Reasoning with multiple abstraction models", *4th International Workshop on Qualitative Physics*, Lugano, Switzerland, pp. 183–194, 1990.

[KRA 87] KRAMER M., PALOWITCH B.L., "A rule-based approach to fault diagnosis using the signed directed graph", *Artificial Intelligence in Chemical Engineering*, vol. 33, no. 7, pp. 1067–1078, 1987.

[KRA 89] KRAMER M., FINCH F.E., in TZAFESTAS S.G., (ed.), *Fault Diagnosis of Chemical Processes, Knowledge-Based System Diagnosis and control*, Plenum Press, New York, pp. 247–263, 1987.

[KUI 84] KUIPERS B., "Commonsense reasoning about causality: deriving behaviour from structure", *Artificial Intelligence*, vol. 24, pp. 169–203, 1984.

[KUI 85] KUIPERS B., "The limits of qualitative simulation", *IJCAI*, Los Angeles, pp. 128–136, 1985.

[KUI 86] KUIPERS B., "Qualitative simulation", *Artificial Intelligence*, vol. 29, pp. 289–388, 1986.

[KUI 87] KUIPERS B., CHIU C., "Taming intractible branching in qualitative simulation", *IJCAI* 87, Milan, Italy, pp. 1079–1085, 1987.

[KUI 88] KUIPERS B., BERLEANT D., "Using incomplete quantitative knowledge in qualitative reasoning", *AAAI*, Saint-Paul, MN, pp. 324–329, 1988.

[KUI 89] KUIPERS B., "Qualitative reasoning: modeling and simulation with incomplete knowledge", *Automatica*, vol. 25, no. 4, pp. 571–585, 1989.

[LAM 98] LAMBERT M., RIERA B., MARTEL G., *et al.*, "Conception et évaluation d'interfaces d'aide à la décision pour la supervision", *Colloque ERGO'IA, Ergonomie et Informatique avancée*, Biarritz, France, 1998.

[LAM 99] LAMBERT M., Conception centrée sur l'homme d'un système de supervision avancée – application à un procédé de retraitement de combustibles nucléaires, Thesis, University of Valenciennes and Hainaut-Cambrai, 1999.

[LEY 90] LEYVAL L., GENTIL S., "Interval versus event based simulation", *Mathematical and Intelligent Models in System Simulation Symposium*, Brussels, Belgium, 1990.

[LEY 91a] LEYVAL L., GENTIL S., "On line event based simulation through a causal graph", *Qualitative Reasoning & Decision Support Systems Workshop*, Toulouse, France, 1991.

[LEY 91b] LEYVAL L., LEDOUX A., "Simulation for a nuclear retreatment facility", *IFAC Symposium on Fault Detection, Supervision and Safety for Technical Processes*, Baden-Baden, Germany, 1991.

[LEY 91c] LEYVAL L., Raisonnement causal pour la simulation de procédés industriels continus, Thesis, National Polytechnic Institute of Grenoble, 1991.

[LEY 93] LEYVAL L., DZIOPA P., MONTMAIN J., "Simulation Causale et Interprétation Qualitative pour la Supervision des Procédés Continus", *Colloque sur les Nouvelles Avancées en Automatique et en Productique, 1er Congrès biennal de l'AFCET*, Versailles, France, 1993.

[LEY 94] LEYVAL L., GENTIL S., FERAY-BEAUMONT S., "Model based causal reasoning for process supervision", *Automatica,* vol. 30, no. 8, pp. 1295–1306, 1994.

[LIN 11a] LIND M., "An introduction to multilevel flow modeling", *International Journal of Nuclear Safety and Simulation*, vol. 2, no. 1, pp. 22–32, March 2011.

[LIN 11b] LIND M., "Control functions in MFM: basic principles", *International Journal of Nuclear Safety and Simulation*, vol. 2, no. 2, pp. 1–8, June 2011.

[MAV 88] MAVROVOUNIOTIS M., STEPHANOPULOS G., "Formal order-of-magnitude reasoning in process engineering", *Computer and Chemical Engineering*, vol. 12, no. 9/10, pp. 867–880, 1988.

[MIS 89] MISSIER A., PIERA N., TRAVE-MASSUYES L., "Order of magnitude qualitative algebra", *Revue d'Intelligence Artificielle, numéro spécial sur le Raisonnement Qualitatif*, vol. 3, no. 4, pp. 95–109, 1989.

[MON 92] MONTMAIN J., Interprétation qualitative de simulation pour le diagnostic en ligne de procédés continus, Thesis, National Polytechnic Institute of Grenoble, France, 1992.

[MON 93a] MONTMAIN J., LEYVAL L., "L'apport du Qualitatif et du Flou pour le Diagnostic de Procédés Industriels: Application dans le projet DIAPASON", *4e Congrès français de Génie des Procédés*, Grenoble, France, 1993.

[MON 93b] MONTMAIN J., GENTIL S., "Interprétation qualitative pour le diagnostic en ligne", *Revue Européenne Diagnostic et Sûreté de Fonctionnement*, vol. 3, no. 1, pp. 23–45, 1993.

[MON 94a] MONTMAIN J., LEYVAL L., "Causal graphs for model based diagnosis", *IFAC Symposium on Fault Detection, Supervision and Safety for Technical Processes*, Espoo, Finland, vol. 1, pp. 347–353, 1994.

[MON 94b] MONTMAIN J., LEYVAL L., GENTIL S., "Qualitative analysis for decision making in supervision of industrial continuous processes", *Mathematics and Computers in Simulation*, vol. 36, pp. 149–163, 1994.

[MON 94c] MONTMAIN J., "Raisonnement approximatif et graphes causaux pour la détection et la localisation de défaillances", *Journées d'étude S3 "Sûreté, Surveillance, Supervision" organisées par le GdR Automatique*, Paris, France, 1994.

[MON 96] MONTMAIN J., GENTIL S., "Operation support for alarm filtering", *Symposium on Modelling, Analysis and Simulation, CESA'96 IMACS Multiconference*, Lille, France, 1996.

[MON 00a] MONTMAIN J., Des modèles pour la supervision, Thesis, National Polytechnic Institute of Grenoble, 2000.

[MON 00b] MONTMAIN J., "Supervision applied to nuclear fuel reprocessing", *AI Communications, European Journal of AI*, vol. 13, no. 2, pp. 61–81, 2000.

[MON 00c] MONTMAIN J., "Un superviseur pour les apprentis opérateurs?", *NimesTic'2000, Ingénierie système et NTIC*, Nîmes, France, 2000.

[MON 00d] MONTMAIN J., GENTIL S., "Dynamical causal model diagnostic reasoning for online technical process supervision", *Automatica,* vol. 36, pp. 1137–1152, 2000.

[MON 05] MONTMAIN J., "La supervision homme-machine", *Techniques de l'ingénieur*, Paris, France, 2005.

[MON 10] MONTMAIN J., CLIVILLÉ V., BERRAH L., *et al.*, "Preference and causal fuzzy models for manager's decision aiding in industrial performance improvement", *FUZZ-IEEE 2010*, Barcelona, Spain, 2010.

[MOS 97] MOSTERMAN P., BISWAS G., NARASIMHAM S., "Measurement selection and diagnosability of complex physical systems", *Eight International Workshop on Principles of Diagnosis, DX'97*, Mont Saint-Michel, France, pp. 79–86, 1997.

[MOS 98] MOSTERMAN P., BISWAS G., MANDERS E., "A comprehensive framework for model-based diagnosis", *Ninth International Workshop on Principles of Diagnosis, DX'98*, pp. 86–93, Cape Cod, 1998.

[NEL 82] NELSON W.R., "Reactor: an expert system for diagnosis and treatment of nuclear reactor accidents", *AAAI*, Pittsburgh, PA, pp. 296–301, 1982.

[OYE 88] OYELEYE O.O., KRAMER M.A., "Qualitative simulation of chemical process systems: steady state analysis", *AIChe Journal*, vol. 24, no. 9, pp. 1441–1454, 1988.

[PAR 08] PARASURAMAN R., SHERIDAN T.B., WICKENS C.D., "Situation awareness, mental workload, and trust in automation: viable, empirically supported cognitive engineering constructs", *Journal of Cognitive Engineering and Decision Making*, vol. 2, no. 2, pp. 140–160, 2008.

[PEN 93] PENALVA J.M., COUDOUNEAU L., LEYVAL L., *et al.*, "DIAPASON: a supervision support system", *IEEE Expert Intelligent Systems and their Applications*, vol. 8, no. 5, pp. 57–65, 1993.

[POO 89] POOLE D., "Normality and faults in logic-based diagnosis", *IJCAI*, Detroit, MI, pp. 1304–1310, 1989.

[RAI 86] RAIMAN O., "Order of magnitude reasoning", *Fifth National Conference on Artificial Intelligence, AAAI,* Philadelphia, PA, 1986.

[RAI 89] RAIMAN O., "Le raisonnement sur les ordres de grandeur", *Revue d'Intelligence Artificielle, numéro spécial sur le Raisonnement Qualitatif,* vol. 3, no. 4, pp. 55–67, 1989.

[RAS 85] RASMUSSEN J., "The role of hierarchical knowledge representation in decision making and system management", *IEEE Transactions on Systems, Man and Cybernetics*, vol. 15, no. 2, pp. 234–243, 1985.

[RAS 93] RASMUSSEN J., "Diagnostic reasoning in action", *IEEE Transactions on Systems, Man and Cybernetics*, vol. 23, no. 4, pp. 981–991, 1993.

[SAC 86] SACHS P.A., PATERSON A.M., TURNER M.H.M., "Escort – an expert system for complex operations in real time", *Expert Systems*, vol. 3, no. 1, pp. 22–28, 1986.

[SAL 09] SALMON M.P., STANTON N.A., WALKER G.H., et al., "Measuring situation awareness in complex systems: comparison of measures study", *International Journal of Industrial Ergonomics,* vol. 39, pp. 490–500, 2009.

[SHI 85] SHIOZAKI J., MATSUYAMA H., TANO K., *et al.*, "Fault diagnosis of chemical processes by the use of signed, directed graphs: extension to five-range patterns of abnormality", *International Chemical Engineering*, vol. 25, no. 4, pp. 651–659, 1985.

[STR 88] STRUSS P., "Mathematical aspects of qualitative reasoning", *Artificial Intelligence in Engineering,* vol. 3, no. 3, pp. 156–173, 1988.

[THO 91] THOM R., *Prédire n'est pas expliquer, entretiens d'Emile Noël avec René Thom,* rédigés par Yves Bonin et illustré par Alain Chenciner, Editions Eshel, 1991.

[TRA 86] TRAVÉ-MASSUYÈS L., KASZKUREWICZ E., "Qualitative controllability and observability of linear dynamical systems", *Proceedings of IFAC/IFOR Symposium on Large Scale Systems*, Zurich, Switzerland, pp. 964–970, 1986.

[TRA 88] TRAVE-MASSUYES L., DORMOY J.L., "Qualitative calculus and applications", *Congrès mondial IMACS sur le calcul scientifique*, Paris, France, pp. 267–270, 1988.

[TRA 89a] TRAVE-MASSUYES L., PIERA N., "Order of magnitude models as qualitative algebras", *11e IJCAI*, Detroit, MI, 1989.

[TRA 89b] TRAVÉ-MASSUYÈS L., PIERA N., MISSIER A., "What can we do with qualitative calculus today ?", *IFAC/IMACS/IFORS Workshop on Advanced Information Processing in Automatic Control*, Nancy, France, 1989.

[TRA 90] TRAVÉ-MASSUYÈS L., MISSIER A., PIERA N., "Qualitative models for automatic control process supervision", *IFAC World Congress*, Tallinn, Estonia, 1990.

[TRA 97] TRAVE-MASSUYES L., DAGUE P., GUERIN F. (eds.), *Le Raisonnement Qualitatif pour les sciences de l'ingénieur*, Hermes, Paris, 1997.

[VAD 97a] VADAM O., MONTMAIN J., CASSAR J.-P., "Fault detection using paralell simulations", *IFAC Symposium on Fault Detection, Supervision and Safety for technical Processes, Safeprocess' 97*, Hull, England, 1997.

[VAD 97b] VADAM O., MONTMAIN J., CASSAR J.-P., "Fault isolation using parallel simulations", *DX'97*, Mont Saint-Michel, France, 1997.

[VES 90] VESCOVI M., "A new model for qualitative reasoning representing time explicitely", *Mathematical and Intelligent Models in System Simulation Symposium*, Brussels, Belgium, 1990.

[VES 91] VESCOVI M., La représentation des connaissances et le raisonnement sur les systèmes physiques, Thesis, University of Savoy, Chambéry, France, 1991.

[VID 95] VIDULICH M.A., "The role of scope as a feature of situation awareness metrics", *Proceedings of the International Conference on Experimental Analysis and Measurement of Situation Awareness,* Embry-Riddle Aeronautical University Press, Daytona Beach, FL, 1995.

[WIL 84] WILLIAMS B.C., "Qualitative analysis of MOS circuits", *Artificial Intelligence*, vol. 24, pp. 281–346, 1984.

[WIL 86] WILLIAMS B.C., "Doing time: Putting qualitative reasoning on firmer ground", *AAAI*, Philadelphia, PA, pp. 105–112, 1986.

[WIL 90] WILLIAMS B.C., "Temporal qualitative analysis: explaining how physical systems work", in WELD D.S., DE KLEER J. (eds), *Qualitative Reasoning about Physical Systems*, Morgan Kaufmann, San Mateo, CA, pp. 133–177, 1990.

[YU 91] YU C., LEE C., "Fault diagnosis based on qualitative/quantitative process knowledge", *AIChE Journal*, vol. 37, no. 4, pp. 617–627, 1991.

7

Human–Machine Cooperation: A Functional Approach

7.1. Introduction

The use of the term "human–machine cooperation" could be considered, at worst, as an advertising slogan for machine sellers and, at best, as a metaphor for the concept generally reserved for the description of human relations. A team of humans and machines might seem fantastic like the mythological chimera that saw the monstrous joining of a lion, a goat and a dragon. However, science, even the most serious, more or less always makes use of metaphors. This computational "metaphor" has led to models of human intelligence using concepts usually reserved for machines: processor, storage unit, encoding, etc. The human metaphor has led to "humanoid" or biological models of the machine: neural networks, genetic algorithms, communication, cooperation, etc. All of these models, with regard to the complex objects they represent, evidently remain simplifications, the pertinence of which must be evaluated by the added value they produce.

Metaphors are more useful when used with a certain level of precision and, of course, precaution. It is naturally the cognitive sciences that spread these metaphors. A few decades ago, we talked about human–machine interaction, which would avoid any misunderstanding, such was the vagueness of the term. The artificial intelligence industry first aimed to "mimic" the human. When the machine was deemed to be gifted with an adequate capacity for processing natural language (notably speech

Chapitre written by Jean-Michel HOC.

recognition or emulation), the terminology of "communication" pervaded the human–machine interaction community. Linguistic theory, mostly pragmatic, has modeled these approaches to human–machine relations. For some time, the term "cooperation" has been integrated into this community. This term recognizes a potential characteristic of an autonomous agent within the machine. Beyond autonomy, such an agent can show signs of a certain intelligence; in other words, a certain capacity to adapt to unforeseen situations or to optimize its interactions with its environment to best achieve its goals.

However, finding metaphors for the the human–machine relationship also requires a sense of realism, since humans will be confronted with this. Telling them that they are faced with a machine that knows how to communicate induces an attitude of communication in them, such that it is developed to meet their expectations. Therefore, if the communication properties between humans are not accomplished by the operations of the machine, misunderstandings and rejections may occur. Cooperation should therefore proceed as follows: if we can show that humans have a tendency to develop a cooperative attitude toward machines [NAS 96], we must take care not to deceive them. Since the initial ideas of Hollnagel and Woods [HOL 83], which intentionally introduce the point of view of a human–machine system, seen as the reunion of two joint cognitive systems, the notion of human–machine cooperation was an eventuality.

In short, we want to apply the model of cooperation between humans (when it is successful, since this is not always the case) to human–machine cooperation; however this undertaking has many requirements. In this chapter, we recall the major points of a minimal cooperation model to highlight the creation of human–machine cooperation, which we will approach in a functional manner [HOC 03]. Structural approaches will be discussed later in this book (see Chapter 9). These two types of approaches clearly complement each other [MIL 11]. The relational structure between the agents that cooperate constrains the way in which they can cooperate. However, this structure does not resolve everything: the agents still require the necessary "know-how-to-cooperate" [MIL 98].

First we recall our definition of cooperation by placing it in the context of related concepts and by introducing the notions on which it relies. Subsequently, we will outline the cooperative activities according to three levels of abstraction, corresponding to three time spans, and we will derive a few implications regarding the conception of cooperation.

7.2. A functional approach to cooperation

The term "cooperation" is used in a number of fields: economics, sociology, psychology, computer science, etc. Suffice to say that it is not easy to find a consensus over the definition. Our concept of it will more likely resemble the psychological point of view, in the sense that it concerns small work collectives, which we can qualify as teams. However, even here, the definitions vary. With the exception of larger, more macroscopic work collectives such as organizations (for example companies) or groups (for example workshops), which justify more global approaches (for example sociology and social psychology), a team is only made up of a small number of individuals. Moreover, the activity of the members of a team is very centered on active goals. Everyday language is not short of very typical examples of this property, such as a football team. Nevertheless, we will refer here to technical areas, such as air traffic controllers or to an airplane's cockpit setup.

In the dynamic situations with which we will concern ourselves here, and which are mainly used in this book, the concrete roles and contributions of each agent of a team are not fully determined in advance. This is because we will be confronted with exceptions. Thus, the notion of cooperative activity makes even more sense. We can, in fact, imagine situations where the behavior of agents is precisely defined in advance, such that only individual activities are carried out, even though we can refer to them as *collective activities*, from the point of view of an observer of the entirety of the agents' activities. It is true that this absence of cooperative activity, because of perfect planning, is an engineer or planner's dream more than it is reality. Even in the field of planning and work organization, real-time adjustments are necessary [MCC 01]. We will reinforce the idea here that this is precisely what cooperation is useful for. Cooperation, therefore, appears to be a particular case of cooperative activity where the agents ensure the integration of their individual activities themselves, in real time.

Among the activities of integration, *coordination* plays a major part by allowing the sequential or parallel (synchronized) scheduling of individual activities. In cooperation, coordination is assured by the agents in real time, similar to a meta-operation dealing with individual operations [PIA 65]. When a particular agent is exclusively in charge of coordination, it will be appropriate for us to say that the agents do not cooperate and the situation resembles a preplanned integration. It is enough for each agent to develop its individual activity without taking into account the activities of others. However, cooperation requires the development of activities other than coordination. The different agents in the situation can also carry out diagnoses between each other or keep their representations consistent.

Communication is another component of cooperation. It plays a major part in the development of cooperative activities, such that it allows an observer to identify them [LOI 01]. Nevertheless, it is useful to not restrict communication to only the use of symbolic (or verbal) codes. In dynamic situations with strong time constraints, non-verbal communication can play a major part. This could be similar, for example, to a football team when playing or the authorized communications in a card game.

Consequently, we can therefore say that cooperation is a particular case of collective activity where the agents ensure the integration of each other's activities in real time. Cooperation makes use of coordination and communication with no restraints. Specifically, we can give the following definition of cooperation, inspired by Piaget [PIA 65]:

> "We consider that two (or more) agents are in a situation of cooperation given the following two minimal conditions:
>
> – They each pursue goals that can interfere with one another, with regard to the goals themselves, resources, procedures, etc.
>
> – They make sure to address these interferences in order to facilitate the individual activities (and) or the task they have in common where it exists." [HOC 03, MIL 97]

Two key notions are included in this definition: interference and facilitation. The concept of *interference*, used frequently in artificial intelligence and in the field of planning, is particularly adapted to a more specific context of cooperation. On this note, our concept of interference is similar to that of Castelfranchi [CAS 98]. Interference between goals reflects the fact that the goals of one agent are pertinent to the goals of other agents, in which they can facilitate the achievement of others (positive interference) or put them in jeopardy (negative interference). The most frequent forms of interference [HOC 03] are interference of precondition, interaction, mutual control, and redundancy. The first two are present in individual planning activities. Mutual control interference is a good example of a frequently positive form of interference: one of the agents verifies the work of another and provides it with an evaluation. The management of redundancy interference between agents is susceptible to achieving similar functions and is at the center of the delegation of functions [HOC 13]. However, interference can always be considered negative if it disrupts individual activities. This would be neglecting the cost/quality ratio of efficiency: the cost of managing a certain interference can considerably improve the quality of the individual activity (mutual control), or even reduce the cost (redundancy). This would also be neglecting an eventual common task, the successful completion of which supposes that each agent has a certain cost, including the one being reevaluated. Nevertheless, in our concept of cooperation, the existence of a common task (taken into consideration in the activities of each agent) is not a necessary precondition: agents can cooperate to manage common resources in the pursuit of very different goals. For example, if several people share the same printer, they will need to cooperate to avoid one of them starting a very long session when another has an urgent form to be printed and sent away.

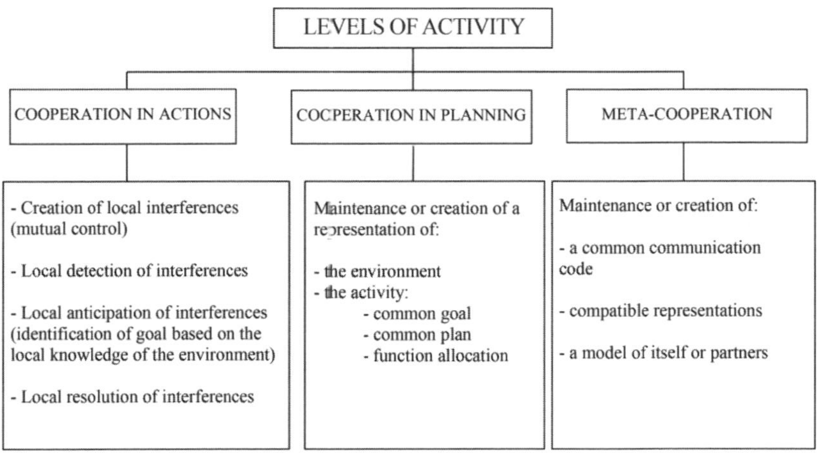

Figure 7.1. *Cooperative activities*

To manage these interferences, cooperative activities can be defined at three levels of a hierarchy of abstraction and time span of the control (Figure 7.1):

– Cooperation in actions manages local interferences, without any deep abstraction, limiting itself to a brief analysis of the situation. When a profound reconsideration of the common plan is required, this form of cooperation only deals with the symptoms of the problem.

– Cooperation in planning tackles the more extensive problem of negotiation of the common plan or facilitates this process through the explicit creation or maintenance of consistent representations of the situation between the agents (common reference). Such a maintenance also allows for the facilitation of activities of the previous level by anticipating the more pertinent adjustments [HOC 02].

– Meta-cooperation capitalizes on and makes use of more abstract knowledge, which is based on experience in cooperation, such as models of partners or oneself.

7.3. Cooperation in actions

Cooperation in actions includes cooperative actions that aim to manage interferences with a minimal level of abstraction in the short term.

– *Creation of local interference:* for this to be an activity in its entirety and not a side effect of another activity, this creation is deliberate. It is typically the introduction of a mutual control interference. In human–machine cooperation, the role of the machine in mutual control, which it can have over a human operator, has often been seen as a positive aspect. For example, we have shown the benefits in driving through haptic feedback on the steering wheel, which prompts corrections that are twice as efficient which doubles the driver's trajectory accuracy compared to when the feature is absent [DER 12, NAV 07]. On the other hand, reverse control is too often implicitly considered when its benefits could be substantial. It is usually a testament to the maintenance of human operator supervision in the work environment that guarantees good working conditions for the role of adjustment and adaptation of the human–machine system delegated to the human operator. A good design rule in this field would be to avoid creating too much of a separation in the monitoring between human and machine. Such a separation has been related to instances of "contentment" (where the human is satisfied with mediocre performance from the machine which could be improved) or loss of skill (lack of practice), beyond the reduced adaptive capacity of the human–machine system.

– *Detection of local interference:* it seems trivial to say that the machine must at least allow the human to detect the interferences it produces with its activity. However, the recommendations made following several flying accidents have precisely mentioned this functionality. During the air incident of a Boeing 747 in San Fransisco in 1985 (vertical descent of 9,000 m in 2.5 min with a maximum acceleration of 5 g), among other things, the pilot had not realized that the autopilot had compensated for a fault when the switch was made to manual control at the moment the aircraft stalled [NTS 86]. It is desirable for the machine to provide feedback of its own activity. For example, automatic devices in airplanes, which act upon the controls, must give feedback over the controls (for example, moving the power control as if the pilot had executed the action himself/herself). The feedback can also be produced for the effect of the machine's actions.

– *Local interference resolution:* the resolution of interference is completed through individual activities, but with the ultimate goal of cooperation. The objective can be reached in different ways. It can be a correction in the coordination, for example the carrying out of precondition interference. An operator who has forgotten to transmit crucial information to their colleague can remember and send it at the last moment after having noticed their colleague's inability to respond. It is also possible to inhibit an action or direct it toward a more appropriate one. However, at this level of cooperation, it is impossible to completely reconsider the plan. On the other hand, identifying a failure of resolution can activate this higher level.

– *Identification of goals and subgoals:* the anticipation of a problem, when it is valid, always facilitates the identification or resolution of this problem. When somebody has an expertise (through experience of performing the action) in a field, it will be easier for him to anticipate the goal pursued by an agent by noticing certain aspects of its actions. In many cases, the identification of the goal is no interest in itself, but only with the creation of a possibility of identifying the subgoals another agent will interfere with. Knowing that someone will exit a room through a door that we will not use is not interesting in itself, but it allows us to anticipate the schedule that will be adopted and an eventual clash with one's own schedule. It is, therefore, desirable for the machine to provide the necessary information for the identification of its goals and subgoals.

7.4. Cooperation in planning

To facilitate the execution of activities of cooperation through actions, a superior level of abstraction, also covering a larger time span, might be required. It allows for the creation or maintenance of a representation of the controlled process and the control activity at the center of the team. This representation is a common reference between every member of the team. It is not, strictly speaking, a sharing of identical representations of the situation. Each agent can (and must) be equipped with a pertinent representation of its own action for reasons of efficiency. However, it is convenient that the representations of each one be consistent between one another. For example, the representations of the same change in direction of a motorway for the passenger working on a map and for the driver are distinct. In countries where the cars driving on a motorway must follow the right lane, if the passenger wishes to turn left, the driver must make it his goal to turn right. The representations are different, but compatible with each other to achieve the goal. These works are numerous and highlight the importance of a common reference, but few are willing to reiterate that its representation for another agent must adopt a format that remains pertinent to its or his action. The machine must provide the common reference with a form that remains pertinent to its use by its human partner.

Designers of support for cooperation between humans (for example remote cooperation) or between humans and machines have introduced the notion of "common workspace" to support the common reference (see Chapter 8). This is an important initiative for the furthering of human–machine cooperation. Nevertheless, the common workspace would not constitute the common reference as an internal representative structure. It is only an external reference, the presence of which is not sufficient to ensure its use. Furthermore, if it is shared by several human agents, it is not certain that it would be operational. Indeed, as mentioned earlier, each agent might require its own specific format for external activity support. On the other hand, it is important that these personalized supports be compatible with each other in order to equally support a common reference between the agents.

In a common reference, we often think of ranking the representations of the external situation (the controlled process, in the dynamic situations we are involved in). However, when all the agents have direct access to the external situation, it is rather the representations of the team's activities that they require. These are, in particular, the representation of the objective, the goals, the common plans and the allocation of functions. In addition, as in operational representations necessary for the support of individual activities, the common reference must also comprise resource representations (notably cognitive) available within the team.

Speaking of common references, we focus mainly on its creation, in particular, because of linguistic communications. We have too much of a tendency to neglect the important role of frequent updates, without requiring complex work, such that we can see them in highly dynamic time-constrained or changing situations. It is difficult for a machine to enter into negotiations with humans, over a common plan, over the allocation of functions, etc. It is easier to provide a common reference based on observations of the external situation and its (or his) own activity, or even that of other agents.

7.5. Meta-cooperation

Moving on to a superior level of abstraction, we arrive at meta-cooperative activities that facilitate the execution of those mentioned earlier. These activities contribute toward a more long-term structure:

– *Creation of a common communication code*: in the majority of professional activities, communications are based on technical languages or jargon that save time and allow for better precision. This does not mean that the use of language is no longer necessary; however, it is limited to problematic situations and therefore quite rare [FAL 89]. In fact a machine will not be able to cooperate with professionals in a field if we do not equip it with the applicable communication code of the field.

– *Creation of compatible representations*: as was seen with common references, the agents of a team do not necessarily use strictly identical representations. Nevertheless, to ensure a certain consistency, it is preferred that each one be interpretable in terms of the activity of the agent to which it is transmitted. In other words, if a machine must cooperate with multiple agents, it is preferred that it addresses each one in operational terms when dealing with its assigned activity. The notion of a common workspace must therefore be adopted with caution, and in any case must be used in situations where all of the concerned agents can benefit from the same form of representation.

– *Creation of a model of oneself and partners:* insofar as cooperation aims to manage interferences to facilitate tasks, the anticipation of the occurrence of this interference, if it is correct, can only reduce the cost of this management. However, to anticipate interference in cooperation, in other words non-independence between goals or subgoals of its own action and those of other agents, it is preferable to outline two types of model: a model of oneself (which we often qualify as "meta-knowledge" or knowledge of its own functionality) and models of partners. To cooperate with human agents, a machine must of course outline minimal models of its partners. Similarly, human agents themselves must have a minimal model of the machine to best manage the interferences between their activities and their own. It is on such models that the confidence between partners is established. Nevertheless, it is preferable to note that the human must not only trust the machine, but also trust its interaction with the machine. Therefore, it helps if the human builds, often through experience, a model of its interaction with the machine. For example, when using adaptive cruise control (ACC) that regulates both the cruising speed of the vehicle and its relative distance to the vehicles located in front of it (because of a radar), the driver gradually builds trust in its interaction with the device when their car must overtake another vehicle [RAJ 03]. For instance, the driver learns that from a relative distance of around 150 m, the radar of the device will detect the slow vehicle and slows down their overtaking vehicle when the latter must accelerate. He/she is then capable of making a coherent decision, for instance anticipating a change of lane of his vehicle.

7.6. Conclusion

A cooperative machine should be capable of completing the above-described cooperative activities. In certain specific and well-constrained contexts, solutions have been proposed to accomplish, through the use of cooperative machine activities such as the identification of the partner's goals, the negotiation of common plans, the use of a human operator model, etc. It is needless to say, however, that certain attempts are quickly overwhelmed by the complexity of the situation. Yet, more modest objectives remain within reach with considerable gains in the quality of human–machine interaction, seeking to support the human in its undertaking of cooperative human–machine activities. In this sense, this is why we have proposed a few recommendations for each of the classes of cooperative activities.

Two questions remain difficult, however partial solutions are possible. First, it is preferable to equip oneself with evaluation and cooperation tools. This evaluation can be very extrinsic, starting with the results of the activity: we compare the obtained performances with different cooperation modes (see the work carried out in the field of air traffic control and presented in Chapters 8 and 9). Evaluation can (and must) also be intrinsic to allow the evaluation of the cost of the cooperative activities, but it is always preferable to compare this cost with the achieved benefits. For example, a long dialogue for establishing a common reference can be very costly, but can represent a desirable benefit for a long period of activity by considerably facilitating cooperation. Second, we must note that the work on cooperation would be exclusively restricted to the processing level of symbolic information (which can be made discrete, often verbal and requires a shift in attention), as it is undertaken by supervisory control. However, in the progress of research in robotics, for example, the human interacts with the machine at the motor-sensory coordination level, which equally implicates human–machine cooperation activities. The work tackling these questions still needs to be consolidated before reaching justified recommendations. A few advances in this direction have been accomplished in the field of secure driving assistance (e.g. [HOC 09]).

7.7. Bibliography

[CAS 98] CASTELFRANCHI C., "Modelling social action for agents", *Artificial Intelligence*, no. 103, pp. 157–182, 1998.

[DER 12] DEROO M., HOC J.-M., MARS F., "Influence of risk expectation on haptically cued corrective manoeuvres during near lane departure", *Ergonomics*, vol. 55, pp. 465–475, 2012.

[FAL 89] FALZON P., *L'ergonomie cognitive du dialogue*, Presses Universitaires de Grenoble, Grenoble, 1989.

[HOL 83] HOLLNAGEL E., WOODS D.D., "Cognitive systems engineering: new wine in new bottles", *International Journal of Man-Machine Studies*, vol. 18, pp. 583–600, 1983.

[HOC 02] HOC J.-M., CARLIER X., "Role of a common frame of reference in cognitive cooperation: sharing tasks between agents in air traffic control", *Cognition Technology and Work*, vol. 4, pp. 37–47, 2002.

[HOC 03] HOC J.-M., "Coopération humaine et systèmes coopératifs", in BOY G. (ed.), *L'ingénierie cognitive: IHM et cognition*, Hermès, Lavoisier, Paris, pp. 139–187, 2003.

[HOC 09] HOC J.-M., YOUNG M.-S., BLOSSEVILLE J.-M., "Cooperation between drivers and automation: implications for safety", *Theoretical Issues in Ergonomics Science*, vol. 10, pp. 135–160, 2009.

[HOC 13] HOC J.-M., "Human-machine cooperation", in LEE J.D., KIRLIK A. (eds.), *The Oxford Handbook of Cognitive Engineering*, Oxford University Press, Oxford, pp. 395–403, 2013.

[LOI 01] LOISELET A., HOC J.-M., "La gestion des interférences et du référentiel commun dans la coopération: implications pour la conception", *Psychologie Française*, vol. 46, pp. 167–179, 2001.

[MCC 01] MAC CARTHY B.L., WILSON J.R. (eds.), *Human Performance in Planning and Scheduling*, Taylor & Francis, London, 2001.

[MIL 97] MILLOT P., HOC J.-M., "Human-machine cooperation: metaphor or possible reality ?", *Workshop on Human-Machine Cooperation (ECCS 97)*, Manchester, UK, April 1997.

[MIL 98] MILLOT P., "Concepts and limits for human machine cooperation", *IEEE SMC CESA 98, Computational Engineering in System Application Conference*, Hammamet, Tunisia, April 1998.

[MIL 11] MILLOT P., DEBERNARD S., VANDERHAEGEN F., "Authority and cooperation between humans and machines", in BOY G. (ed.), *Handbook for Human-Machine Interaction*, Ashgate Publishing Ltd, Surrey, 2011.

[NAS 96] NASS C., FOGG B.J., MOON Y., "Can computers be teammates ?", *International Journal of Human-Computer Studies*, vol. 45, pp. 669–678, 1996.

[NAV 07] NAVARRO J., MARS F., HOC J.-M., "Lateral control assistance for car drivers: a comparison of motor priming and warning systems", *Human Factors*, vol. 49, pp. 950–960, 2007.

[NTS 86] NTSB, Aviation accident report, Boeing 747, 19 February 1985, Washington, 1986. Available at http://www.rvs.uni-bielefeld.de/publications/Incidents/DOCS/ComAndRep/ChinaAir/AAR8603.html.

[PIA 65] PIAGET J., *Etudes sociologiques*, Droz, Geneva, 1965.

[RAJ 03] RAJAONAH B., ANCEAUX F., ESPIE S., *et al.*, "A study of the link between trust and use of adaptative cruise control", in VAN DER VEER G.C., HOORN J.F. (eds.), *Proceedings of CSAPC'03*, Rocquencourt, France, pp. 29–35, 2003.

8

The Common Work Space for the Support of Supervision and Human–Machine Cooperation

8.1. Introduction

Why and when to automate? The technically centered approach promotes automation from the moment it is possible to do so. The humanist approach consists of automating difficult, tedious, troublesome and even risky tasks for a human being. The economical approach encourages automation when it is less expensive. Finally, the "human"-centered approach considers that automation does not consist of simply and exclusively removing humans from the control/command loop. Thus, automation consists of conceiving a system that partially or entirely carries out a job which was or could have been considered partially or entirely executed by a human operator [PAR 00]. This definition implies that automation can vary on different levels, from a manual mode to fully automated, and that an automated system must, in some cases, work with one or more human operators. The consideration of the latter therefore requires a different approach to automation.

In this context, Figure 8.1 [RIE 01a] shows that the approach of human engineering[1] revolves around five points:

Chapter written by Serge DEBERNARD, Bernard RIERA and Thierry POULAIN.

1 Human engineering is often used in place of human-centered automation.

– The main objective is the global improvement of the human–machine system (HMS) studied.

– We are not only interested in technical performances, but also in effects caused by the installation of an artifact. For this, the HMS must be studied while taking into account the characteristics of humans (required for decisional models) as well as those of the technical system.

– Human engineering proposes alternatives to technically centered automation. Human–machine cooperation (HMC) and the dynamic task allocation (DTA) are original solutions proposed and evaluated by human engineers.

– Human engineering attaches great importance to the evaluation phase of the human–machine system. The evaluation criteria must take into account not only the global performances of the HMS, expressed for example in terms of the distance between the real production and the goals, but also the criteria allowing for the evaluation of the challenges encountered by the operators in terms of resolving problems and communicating with the technical tool, such as workload for example. To this effect, problems exist considering the impossibility of directly observing the cognitive behavior of operators.

– The human engineer considers automation according to a global approach involving humans, by focusing particularly on interactions between humans and the technical system.

The supervision of automated systems is a particularly interesting field for the human engineer, since humans play a very fundamental role that is dependent on the level of automation. This chapter sets itself in this frame and presents different research, the main objective of which is to assist human operators with the introduction of tools automating part of their activities. In order to keep these operators in the control "loop", a strict cooperation between the different actors involved, human or artificial, must be put in place.

The first section of this chapter therefore presents the technical means necessary for sustaining the cooperative activities between a human operator and a technical system. In this regard, the idea of a common work space (CWS) has been developed.

The other two sections of the chapter present two different applications, each with different levels of automation. The first involves air traffic control, a process with a low level of automation. The second is a process of high-level automation: the reprocessing of nuclear combustibles. In the two cases, the experiments were carried out with expert operators and the advantages of the "human engineering" approach were shown.

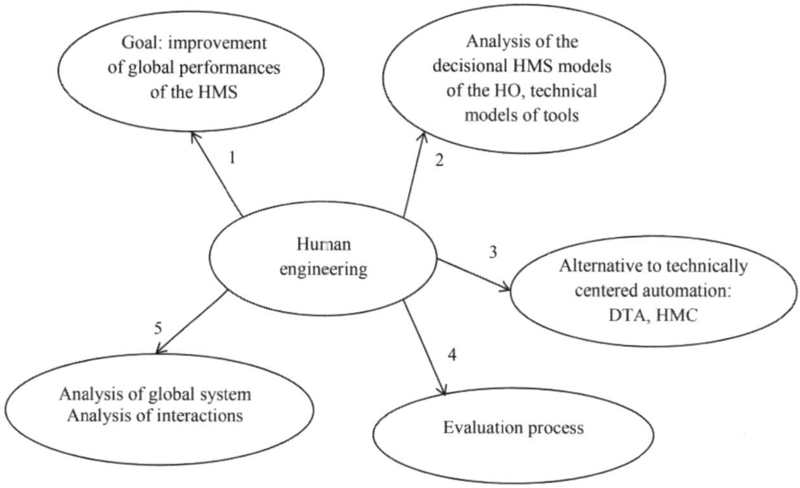

Figure 8.1. *Human engineering approach*

8.2. Human–machine cooperation

8.2.1. *Definitions of human–machine cooperation*

During the creation of a complex system, the choice of the level of automation is a difficult step, in which it is necessary to precisely take the needs of human operators into account [PAC 11]. Indeed, an overly forced automation can lead to a long-term deterioration of the competencies of the supervision operators. By contrast, an overly weak automation of the process cannot react to the needs for assistance of these operators and can lead to deterioration in the safety and productivity of systems [GAR 91]. This automation can be executed through two approaches. The first consists of defining the static allocation of different tasks or functions to be undertaken between the one or more operators and the control systems [HOL 00]. The second approach, which can work in addition to the first, is focused on human–machine cooperation.

Different research areas are dedicated to this problem. Giving a definition to human–machine cooperation which satisfies the different approaches is therefore difficult. Nevertheless, one has been proposed by Hoc, inspired by the previous work of Piaget [PIA 65] on human–machine cooperation [MIL 97, HOC 01].

"Two agents are in a situation of cooperation if the two following conditions are met:

– The agents pursue goals that can interfere with those of others, at the level of their own goals, the sub-goals, processes or resources.

– Each agent attempts to generate these interferences to facilitate their task, those of others or the common task."

In this definition, the notion of the goal does not correspond to the global goal to be achieved in order to monitor and control a process, but the goals of different tasks to be completed to achieve the global goal. The notion of interference corresponds to the different interactions between the activities of the different agents which can be normal or conflicting, for example if the activities of the agents lead to the production of different results.

In order to best characterize human–machine cooperation, for each agent Millot defines his *know-how* (KH) and *know-how-to-cooperate* (KHC) [MIL 98]. The *know-how* corresponds to his knowledge, competences, etc., which allow it to complete the tasks he has been given, whereas the *know-how-to-cooperate* allows him to manage his own activities by taking into account the activities of others in order to facilitate their work.

It is clear that in the frame of human–machine cooperation, it is necessary to favor the interactions between agents which facilitate the correct processing of different tasks and the research of methods that help to minimize the possible conflicts. However, it is clearly a fundamental difficulty if we want to integrate the capacities for facilitation, detection and management of interferences into the machine, competencies with which human operators are naturally equipped. Human–machine cooperation is therefore not (or no longer) a simple coordination problem of actions between the human operators and the support we provide them with. Royer insists on this by pointing out that a system is more cooperative if it incorporates difficult levels of interactions, these levels being linked to the perception and analysis of the situation to be managed, the decision-making and the implementation of actions, and the interactions allowing for the fusion of information, confirmation of analyses and the convergence of decisions [ROY 94].

In order to model these interactions, it is possible to adopt the following forms of cooperation defined by Schmidt [SCH 91]:

– In the augmentative form, agents have similar know-hows. These agents are simply "duplicated" to tackle a workload that is too heavy for a single one. The global supervision task is therefore broken down into many similar tasks that are allocated to different agents.

– In the *integrative* form, the agents have different but complementary know-hows. The contribution of each is necessary to carry out a particular task.

– Finally, in the *debative* form, the agents confront the results of their activities with the aim of obtaining the "best" solution. The agents therefore have identical know-hows.

According to Grislin and Millot, every situation of cooperation can be described as a combination of these three basic forms [GRI 99]. However, the authors suppose that the tasks (or sub-tasks) are processed autonomously by agents, and the interactions between them happen before or after the processing of the task, in accordance with the three applied forms mentioned above. This reasoning can be expanded upon further and we can assume that every situation of cooperation can be described, in part, by a combination of these three applied forms, not through a decomposition of the tasks, but through a decomposition based on the human activities required for the completion of these tasks (seen, for instance, as functions) [DEB 01].

8.2.2. *Characterization of cooperation activities*

To characterize cooperation activities, human–human or human–machine, Rasmussen's model has been used [RAS 83, RAS 85]. This problem-solving model shows four main activities: the collection of information, identification of the situation, decision-making and execution; and all of these can also be broken down into sub-activities. Although this model is simplistic in terms of psychology, it allows for the symbolic information handled by human operators to be pointed out during the processing of tasks that they need to complete [PAC 02]:

– Information, from activities of information collection.

– Problems, from identification activities.

– Strategies, from schematic decision-making activities.

– Solutions, from precise decision-making activities.

– Commands corresponding to solution implementation activities.

This list is of course non-exhaustive and depends on the tasks to be completed and the process being monitored. However this information provides the operator with a frame of reference that allows them to properly manage their work, mostly by better planning their activities according to their internal resources and the time and stress constraints they can be subject to [HOC 96].

When several operators monitor and control the same process, they create and maintain a COmmon Frame Of Reference (COFOR) on top of their own frame of reference [DEC 94]. This COFOR is therefore a representation of the environment and resources of the team and "contains", in addition to the representations specific to the process that we qualify as tactics, the goals and common plans of the team and the allocation of activities within this team. In order to cooperate, these operators communicate this information, problems, strategies, etc. to share and construct this COFOR, which of course has only a virtual existence [HOC 01]. Figure 8.2 gives a schematic representation of the COFOR, by including data only from problem-solving activities.

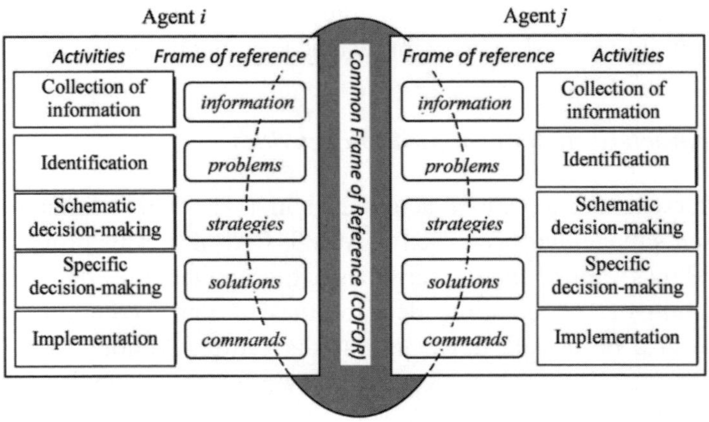

Figure 8.2. *Notion of COmmon Frame Of Reference (COFOR)*

The cooperative activities led by the different operators therefore facilitate:

– The updating of the COFOR, which will be part of the frames of reference of each participant.

– The control of the COFOR through the comparison and detection of differences with the frame of reference of each of the participants; these activities correspond in fact to mutual control and facilitate the detection of interferences between one or more pieces of information "contained" in each of the frames of reference.

– The management of the interferences which consists, on the one hand, of diagnosing the one or more differences between the frames of reference and, on the other hand, resolving these interferences.

Three forms of interference resolution can be used by the operators [PAC 02]: *acceptance*, *imposition*, and *negotiation*. These three forms naturally have cognitive costs and communication costs that differ and are therefore used according to the workload of the different agents. *Acceptance* consists of adding data to its own frame of reference, for example a strategy, from another agent. This acceptance can be chosen to facilitate the activities of the other agent because the cost of negotiation is too great in a particular situation. In contrast to acceptance, we have *imposition*, by which one agent forces another to accept particular data into its frame of reference, for example a solution to a particular problem. *Negotiation*, on the other hand, tries to reduce the differences between the respective frames of reference of the agents by starting, for example, an explanation activity centered on the data from the problem-solving activity that led to the interference (collection of information, definition of the problem, strategy and/or solution), which results in an adjustment of the individual representations [KAR 95, KAR 00]. The process of negotiation goes through the formulation of a proposition which is analyzed by the other agent who accepts or refuses this proposition, or even proposes an alternative [WAN 98]. According to Wang [WAN 02], this negotiation comes down to changing some constraints and looking to satisfy them.

It is obvious that the cost induced from cooperative activities can be relatively important. When the agent is a human operator, it is not uncommon for him to construct, in order to minimize the cost, an interpretation of the frame of reference of the other agent. This interpretation can be, for example, based on the detection of intentions of the other agent, according to a certain situation and his actions [MUN 01a, MUN 01b, TAH 05]. Providing a physical medium to operators representing this COFOR, provided that it is easily accessible and properly informed, avoids a misinterpretation of the frame of reference of the other agent. Furthermore, this medium would facilitate the desynchronization of the cooperative activities between operators and, therefore, avoid having an operator wait while the other is available to cooperate, rendering the information persistent [ZOU 98]. Finally, the information integrated into this medium can also be recovered through a machine and, therefore, sustain the human–machine cooperation.

The implementation of this common medium is referred to as common work space (CWS), which is presented bellow.

8.2.3. *Common work space: human–machine cooperation medium*

The definition of interactions between operator(s) and the assistance tools(s) remains a problem in the design of complex systems, particularly when it is necessary to include forms of human–machine cooperation. For example, Hoc describes the different possible failures that must be found out and avoided: the loss of expertise, the act of complacency toward assistance tools, confidence issues and the loss of adaptability [HOC 00].

The integration of human–machine cooperation in a system must therefore be executed by defining, on the one hand, the allocation of the activities required for the completion of tasks between artificial and human agents, and, on the other hand, the interactions between these agents.

Extensive research has been carried out on concepts similar to the notion of CWS [BEN 92, DEC 94, ROY 94, JON 97]. For example, in the field of air traffic control, Bentley considers that the set of handled information by the air traffic controllers represents a "shared workspace", which provides a representation of the air traffic adapted to the different users on different machines [BEN 92]. Decortis and Pavard define "the shared cognitive environment" as a set of manifest facts and hypotheses, which agents have in common and make up a sub-set of the cognitive universe of each agent [DEC 94].

It is obvious that the creation of a cooperative human–machine system requires the exact definition of the CWS, which depends on the field and the tasks to be completed. Furthermore, the use of such a medium by an operator generates additional cooperative activities, the usage cost of which will be compensated by the activities or rather by the functions carried out by the machines [DEB 08].

In order to model these cooperative activities between the agents, which will therefore depend on the allocation of activities/functions facilitating the control of the process, it is possible to use the three forms suggested by Schmidt [LEM 98] (see Figure 8.3).

In the debative form (case *a* in Figure 8.3), the agents "feed" the workspace with new data, with this being for an activity/function corresponding to a particular task. If interferences are detected by the agents, they must be resolved through acceptance, imposition or negotiation. In the integrative form (case *b* in Figure 8.3), one of the agents feeds the CWS with data (a strategy in the example of the figure) and the other agent carries out, using this, the following activity/function. Finally, in the augmentative form (case *c* in Figure 8.3), the agents carry out the same activity/function, for different tasks however, depending on their allocation, and inform the CWS.

The implementation of such a CWS, in the form of a human–machine interface between human and artificial agents, will nevertheless bring large constraints depending on the competencies of the machine, particularly in what concerns the cooperative activities of negotiation. Indeed, when two human operators negotiate, they can exchange very effective high-level symbolic information. An artificial agent requires operational information and it is currently very difficult to provide a machine with the capacity for real negotiation.

In order to illustrate these concepts, two research projects, conducted on processes with different levels of automation, are now presented.

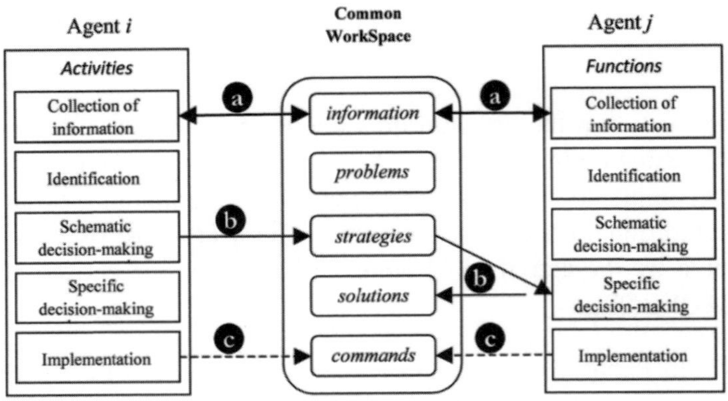

Figure 8.3. *Interactions between the agents through the common work space (CWS)*

The first concerns air traffic control, which consists of a low degree of automation. In order to mitigate the regular increase of traffic, different approaches can be considered. The technically centered approach aims for the improvement of control methods, or even the automation of human activities. The human-centered approach provides the operator with assistance tools, allowing them to correctly understand the different traffic situations, but without necessarily automating the different tasks [LER 96]. An intermediate approach consists of varying the level of automation of the human–machine system depending on the needs of humans, by establishing an augmentative cooperation and more specifically a DTA between controllers and assistance tool.

The second research project concerns a process with a high degree of automation, namely a nuclear waste reprocessing plant (also see Chapter 6 of this book). If we consider that it is not possible to remove the supervision operator, the creation of adapted supervision tools requires the establishment of debative and integrative tools for cooperation. These are presented next.

8.3. Application in air traffic control

This section deals with the concepts of DTA, which are used in lightly automated processes where the operators need support. Extensive research conducted in the field of air traffic control is presented. The main results obtained lead to the establishment of a new principle – that of the delegation of functions – which necessitates the definition of a CWS supporting the cooperative activities of different agents, human and artificial. These studies were carried out within the framework of research conventions with the *Centre d'études de la navigation aérienne* (CENA-DGAC, Centre for aerial navigation studies).

8.3.1. *Dynamic allocation of tasks*

Dynamic task allocation (DTA) is an augmentative form of cooperation. It was studied within the context of the supervision processes for the first time in France toward the end of the 1980s by Millot [MIL 88a]. It involves assisting a human operator with an automatic system capable of completing certain tasks, referred to as sharable by dynamically distributing them between the agents. This automatic system must therefore integrate all of the necessary functions in order to fully complete the tasks it can be allocated, from the recording of information to the implementation of solutions.

An optimal allocation of tasks would be a division of the tasks that optimizes the performance of the process, all the while taking into account the competencies and capacities of the different agents (see Chapter 9 of this book). The DTA therefore facilitates the adaptation of the level of automation for the human in order to keep them active or "in the loop"; this is why DTA is also referred to as adaptive automation. The main difficulty of this type of approach is to define the set of sharable tasks online which correspond to the intersection of the set of tasks relevant to the competencies of the human and those relevant to the competencies of machines.

The allocation of sharable tasks is undertaken by a (digital) module, referred to as an *allocator* or *distributor*, the main functions of which consists of informing each agent of the allocation of tasks and preventing an agent from acting upon tasks that have been allocated to another. Furthermore, this allocator will indicate, particularly to the human operator, the method in which a task will be completed by the automatic system. This functionality can be handled either by an automatic system or by the operator himself [MIL 88b, KAM 89]:

– In *explicit* allocation, the allocation function is managed by an operator through a specialized interface. Therefore, this is what carries out the allocations in accordance with its own criteria.

– In *implicit* allocation, the allocation function is handled by an automatic system that carries out allocations using algorithms and criteria defined by the designer. These criteria may require the measuring or online estimation of different parameters of the human–machine system, for example the human workload, the performance obtained from the agents, errors and others.

The implicit mode has the advantage of relieving the operator of the task of allocation. Nevertheless, it is more complex to implement than the explicit mode, since the choice of algorithm and the criteria require serious study in order to succeed in the integration of the human operator within the human–machine system. Furthermore, decisional conflicts can appear between the operator and the allocator, if this operator is not in agreement with the allocation of tasks.

An intermediary mode can nevertheless be installed, which lets us take advantage of the two modes, explicit and implicit. This intermediary mode has been referred to as *explicit assisted mode* and consists of carrying out an initial allocation of tasks, but by giving the operator the possibility of modifying those allocations if they consider it necessary [CRE 02].

Different modes of DTA have been implemented and evaluated in the field of air traffic control. An experimental platform integrating a realistic traffic simulator has facilitated the realization of evaluations with professional controllers. Before describing the main results of these studies, a short introduction to air traffic control is given.

8.3.2. *Air traffic control*

French air traffic control is a public service in charge of the security of aircraft, the regulation of traffic and the economics of flights that travel across the territory. Security is, of course, the main constraint to be respected. It consists of detecting the situations, referred to as air conflicts where there is a risk of collision between airplanes. Separation standards set the minimal distances within a plane (six nautical miles) and within an altitude (1,000 feet) for the airplanes. The studies presented more specifically concern the control "en route" in which the air space is divided into sectors. Each sector is monitored by two controllers who have the same qualifications but not the same tasks, as seen in Figure 8.4 [DEB 92].

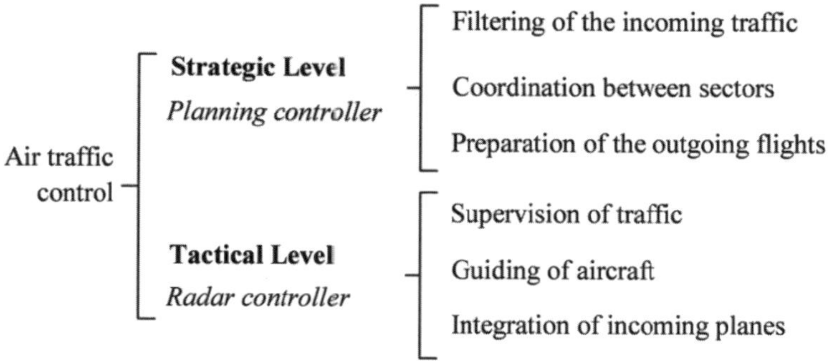

Figure 8.4. *Main tasks for air traffic controllers*

The first controller has a tactical role and is referred to as *radar controller* (RC). The RC ensures the supervision of the air traffic in order to detect conflicts between vessels and to resolve them with guidance. For this, the controllers modify the trajectory of one aircraft, or more, by sending an instruction for the direction, level or speed via radio to the pilot(s). The second controller has a strategic role and is referred to as *planning controller* (PC). The PC must notably predict (and regulate) the future load of traffic the RC will have to manage. The PC then negotiates with the PCs of the adjacent sectors over the input and output levels of the aircraft. The PC therefore assures a filtering of the traffic; however, filtering has certain limits as the number of airplanes can increase indefinitely. Furthermore, changing the level of the airplanes remains a lesser performing solution than that of conflict resolution through modification of the direction, unless of course the airplane must descend for landing.

In order to tackle the increase in traffic, a possible solution consists of supplying an active support to the RC in order to increase the control capacity of the tactical level.

8.3.3. *First studies: SPECTRA projects*

8.3.3.1. *The SPECTRA V1project*

The first study here is the SPECTRA VI project [DEB 92], which evaluates a dynamic allocation of tasks between the RC and an assistance tool called SAINTEX. This system is capable of resolving simple conflicts between two airplanes (referred to as binary conflicts or duels). These conflicts correspond to sharable tasks. SAINTEX has its own resolution strategy and executes this task completely, from detection to the sending of instructions to different airplane pilots [ANG 92] (in this case, to the simulators that the platform implements). In order to avoid decisional sharable conflicts between SAINTEX and the RC, the binary conflicts can only be considered sharable when no other aircraft can interfere with them, in other words when there is no other flight at the same level within a 40 nautical mile radius.

Two allocation modes have been evaluated in this project: *explicit* and *implicit*. For the *explicit mode*, the RC himself assigns some sharable conflicts to SAINTEX through intermediary interfaces, essentially electronic stripping that contains the flight plans of the vessels traveling in the sector. These conflicts are highlighted by a specific color, with the flight plans showing the order of the diversions computed by SAINTEX, which will be applied if the conflict is assigned to it, all within a limit date. The conflicts are assigned by default to the RC, who can modify the assignment within this limit date, provided the RC does not act directly on the aircraft involved.

The second mode is the *implicit mode*, in which the allocation of sharable conflicts is directed by an automatic system. This system ensures the online evaluation of the RC workload and when this load is higher than a specific value (obtained experimentally), any new sharable conflict is allocated to SAINTEX. In the opposite case, they are allocated to the RC in order to maintain a sufficient workload, so that competencies do not degrade in the long term. No *a posteriori* modification of these allocations is carried out – referred to as *non-preemptive* allocation [DEB 92] – to prevent the RC being disrupted, for example by assigning them a new task at the last minute. The workload estimator is based on an aggregation of the weights assigned to each of the tasks to be undertaken by the RC (flight surveillance, integration of a flight, transfer to the next sector, level change to be completed, diversion and return to the track, etc.). This estimator has been validated throughout these experimentations with an online subjective estimation by the RC of their workload [DEB 92, VAN 94].

The traffic scenarios used for these experiments involve a significant number of aircrafts in order to evaluate the use of these cooperation modes and generate around 20 conflicts an hour of different types (sharable and non-sharable). Three modes have been evaluated: *unassisted* mode, *explicit mode* and *implicit mode*, with two classes of controllers: professional controllers from the Air Traffic Control Center of Athis-Mons (near Orly Airport) and engineers from CENA, with experience in the control room, but still considered novices. In order to compare the different experimentations, different criteria have been used:

– The global workload evaluated *a posteriori* through the TLX method (*task load index*) [HAR 88].

– A performance criterion based on the fuel consumption of aircraft.

– A performance criterion based on security, by counting the number of near-misses, in other words violations of the minimal distance constraints between vessels.

An experimental protocol has been implemented to compare three assistance modes: *unassisted mode* where the controller is alone when faced with the traffic and will be used as a reference for the two other modes, the controller assistance mode through *implicit dynamic allocation* and the assistance mode through *explicit allocation*. The results of these experiments seem to be in favor of explicit and implicit cooperation modes compared to the unassisted control, due to the reduction in the global workload. Based on the performance criteria, the implicit mode gives better results. Nevertheless, this experimentation highlights human confidence problems toward the assistance system [DEB 02, HOC 02]. In the *implicit mode*, the professional controllers had better results than the novice controllers; the allocation of tasks not being modifiable by the controllers, they therefore experienced imposed allocations. However, in the *explicit mode*, the novice controllers obtained better results than the professional controllers, since they were more inclined to allocate conflicts to SAINTEX. The questionnaires that were completed at the end of the experiment corroborate these results: the professional controller preferred the explicit mode since they can regain control of conflicts from SAINTEX when the resolution computed by this system does not satisfy them. Furthermore, the implicit mode poses the problem of conflict between the authority and the responsibility of the controller (see Chapter 9 of this book). Nevertheless, the explicit mode brings with it a heavier workload for controllers, due to the strategic activity of managing two sources of computation that are the RC himself and SAINTEX.

8.3.3.2. *The SPECTRA V2 project*

These results have therefore led to a more precise examination of the organization of air traffic control [LEM 96, HOC 99]. In the second study, referred to as SPECTRA V2, the dynamic allocation of sharable conflicts has been preserved on a tactical level, between the RC and the SAINTEX; however, the role of allocation has shifted to a more strategic level, one managed by the PC [CRE 02]. The aim of this approach is to preserve the benefits of the implicit mode – avoiding an additional workload for the RC due to the management of the allocation – all the while letting the controllers modify the allocation of tasks as they see fit, in other words avoiding the conflict between authority and responsibility.

One of the tasks of the PC consists of predicting the future workload of the tactical level, and when the traffic becomes too high, the PC must "filter" the traffic. Giving this controller the job of allocator is therefore coherent since the PC must continuously evaluate the potential resources of the tactical level. Nevertheless, since our studies are within a high-traffic environment, a new "strategic" assistance referred to as allocation planning has been conceived in order to support the PCs in the allocation of tasks at the tactical level between the RC and the SAINTEX [CRE 96]. The resulting mode is referred to as the *explicit assisted mode*.

Allocation planning is composed of three main modules:

– The first module ensures the prediction of the set of tasks that will need to be completed at the tactical level and characterizes these tasks according to four pieces of information: a start time, an end time, an execution time and an indicator of potential for the allocation of binary conflicts.

– The second module predicts the RC workload. Using the tasks and their characterization carried out by the previous module, this module verifies the feasibility of their scheduling. Indeed, it is not possible to predict the real scheduling that will be carried out by the RC. This module therefore only computes the periods where the scheduling will not be possible and which correspond to periods of overload.

– Finally, the last module, based on periods of overload computed by the second module, ensures the allocation of sharable conflicts to SAINTEX.

In order to justify the allocation propositions proposed by allocation planning, an interface provides the PC with the detected binary tasks and conflicts, as well as the periods of computed overload. This information is presented in terms of two approaches: time (over 30 minutes of prediction) and the level of flights.

Two modes, as well as an unassisted reference situation, have been evaluated: the *pure explicit mode* directed by the RC and the PC and the *explicit assisted mode*, in which planning allocation ensures the initial allocation of sharable conflicts, but leaving the PC, who can modify them, with the *right to veto*.

In addition to the usual performance criteria, a cognitive analysis of the activities of controllers has been carried out [LEM 96, HOC 98]. The best results have been obtained in the *explicit assisted mode*, where the performance obtained is better and where the cooperation between the two controllers seems more efficient. Their activities are indeed better planned and much more reactive to events than in the unassisted reference situation. Furthermore, since the PC conducts the allocation of tasks with allocation planning, the PC participates much more in the decision-making process by providing suggestions for solutions to the RC.

Another important result comes from the interfaces proposed to the controllers. These allow the controller to position different color symbols on the strips and radar view in order to note information that seems important to them, for example aerial conflicts. Cognitive analysis of controllers' activities shows that these interfaces make up a *reduced CWS*: the two controllers make use of it in order to characterize the different situations and this allows them to cooperate without necessarily communicating. This reduced CWS does not however let them include all of the results from their activities, such as the information taken from a diagnosis or a decision that has been made.

The comparison between explicit and explicit assisted modes has nevertheless shown certain problems in the latter mode. Since in this model the RCs cannot act directly on the allocation of conflicts, they have a tendency to rely on the choices made by the PCs and to reduce monitoring of the conflicts allocated to SAINTEX. This phenomenon of complacency [SMI 97] is much less prominent in the pure explicit mode where the RC can modify allocations.

Finally, it seems that in certain situations, the resolution strategy of SAINTEX can disturb the controllers, with their strategies being potentially different. Indeed, the competencies of SAINTEX are limited to the resolution of binary conflicts (two airplanes considered as independent of the rest of the traffic); however, the controller can very well "introduce" to this initial problem a third airplane to resolve another conflict more easily. The representation of problems and tasks that stem from this, in this case the aerial conflicts and their resolutions, is not always the same. The inverse case could also appear: SAINTEX detects a problem, whereas in the opinion of controllers there are none, since they have the intention to act in a certain way over the flight, for example by making it land with the unknown intention and undetected by SAINTEX. Finally, it seems that the dynamic allocation of tasks, even though not originally designed for this purpose, also incorporates a debative type of cooperation since the controllers verify what SAINTEX does and can refuse to use it if they are not in agreement. Figure 8.5 shows those cooperative activities that are actually used in this experiment.

New research has therefore attempted to resolve these different problems.

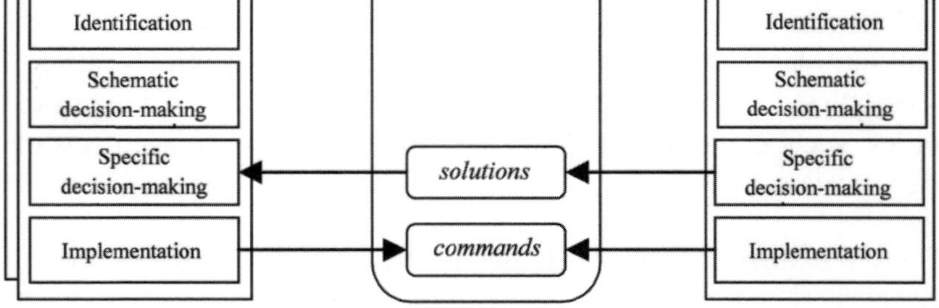

Figure 8.5. *Cooperative activities between the controllers and the SAINTEX within the SPECTRA V2 project*

8.3.4. *The AMANDA project*

In the AMANDA project (*Automation and Man–Machine Delegation of Action*) two principles are studied [DEB 98, DEB 02, DEB 10]:

– The principle of delegation in which the controllers delegate a task to a tool called STAR (French acronym for tactical resolution support system).

– The principle of CWS which must allow the two controllers and STAR to cooperate and share the same representation of the problems, in other words, the conflicts.

As is the case for SPECTRA projects, the controllers are assisted by a tool in line with an augmentative type of cooperative form; however, STAR does not have any competencies that aim to define the problems or strategies to resolve them. Nevertheless, STAR, using a resolution strategy defined by the controllers, looks to define a solution (a new trajectory) to resolve the aerial conflict. This cooperation is also of the integrative type. If controllers are in accordance with the solution proposed by STAR, they can delegate the corresponding tasks or rather jobs (surveillance and implementation). If they are not in agreement, they can modify the strategy or complete the task themselves. The cooperation is therefore debative. If, using the strategy defined by the controllers, STAR is not capable of finding a trajectory due to another airplane (referred to as an interfering plane since it spurs a new conflict with one of the planes from the initial problem), STAR will incorporate this new airplane into the problem. Thus, there is also a debative cooperation, this time over the definition of the problem. The controllers can then initiate a sort of negotiation with the machine, either by adding a new strategy incorporating the interfering airplane or by modifying the first strategy.

These cooperative strategies have been constructed around a CWS, which is shared here between the twc controllers, radar and planning, and STAR. Each agent therefore provides information over this medium depending on the activity/jobs defined, and can verify those provided by the others (mutual control), see Figure 8.6.

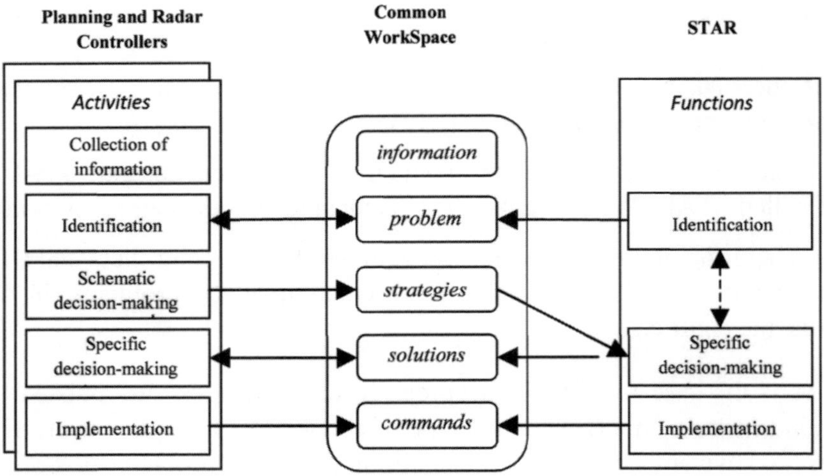

Figure 8.6. *AMANDA: CWS allocation of activities/functions between agents and cooperative activities*

The contents of the CWS have been defined based on a slightly particular experimentation, in which the future STAR system has been replaced by a second human RC. These two controllers have worked on the same traffic in which the airplanes were distributed between them in a static manner in order to compel them to cooperate [DEB 98, CAR 99, GUI 03a, GUI 03b]. The cognitive analysis of activities carried out has confirmed the high number of exchanges between the two RCs, which allowed them to construct and maintain a COFOR containing:

– the notion of a problem (which will replace the notion of aerial conflict), which involves two or more aircraft: for AMANDA, the problems are defined by clusters containing the airplanes directly involved in a problem and the interfering airplanes;

– the notion of strategy: for AMANDA, it is presented in the form of one or more directives for resolving a problem, for example "AFR123 PASS BEHIND BAW628";

– the notion of solution: for AMANDA, a solution is a plan, in other words a trajectory containing a beacon and directory instructions or the level to be assigned to aircrafts. These are plans delegated by the controllers to STAR for implementation later.

These principles have been evaluated with four pairs of professional controllers from the French Control Center of Bordeaux [DEB 09, GUI 04a, GUI 04b]. The principle of task delegation has received a positive reaction from controllers of different rankings, depending on their experience. Indeed, controllers with more experience appreciated the system more than those with less (less than 10 years). AMANDA has favored the anticipation of problems by the controllers, which is a positive aspect in terms of security. Nevertheless, certain RCs have anticipated too many problems, which has caused an unnecessary increase in their workload. The CWS has allowed controllers to share a COFOR with the assistance tool STAR, which favors the maintenance of situation awareness (SA) and has also provided the aerial controllers with an efficient cooperation medium between each other.

The following section presents another area of application of human–machine cooperation in which, this time, the level of automation is greater than in air traffic control.

8.4. Application to the process of nuclear combustibles reprocessing

8.4.1. *Introduction*

Represented schematically in Figure 8.7, the process studied involved the reprocessing of nuclear combustibles, the characteristics of which are described in a thesis by Leyval [LEY 91]. It is a continuous and heavily automated process. It is made up of two pulsating columns aligned "top and tail" and their power supply, the primary function of which is to separate the residue from the product of fission, from the initial phase containing uranium and plutonium, mixed together when used in nuclear power plants (see also Chapter 6).

A pulsating column receives an "organic" phase from the bottom and an "aqueous" phase from the top. The aqueous phase is an acidic dissolution of waste from nuclear combustion. It contains uranium, plutonium and the product of fission. The columns are pulsated by air following a variable frequency, which creates a chemical emulsion between the aqueous phase that drops and the organic phase that rises. Due to this emulsion, the organic phase carries uranium and plutonium, which are thus separated from the products of fission. This extraction is done in such a way that very little uranium and plutonium escapes the aqueous phase. Some of the produce from the fission is therefore taken with the organic phase. At the column's exit, the aqueous phase at the bottom contains the residue and the organic phase at the top contains the uranium and plutonium. The first column of the process is for the extraction of uranium and plutonium, whereas the second is used to cleanse the organic phase of the remains from the produce of fission after extraction. The structure of the pulsating columns contains three types of supply system: the danaids, the air-lifts and the dosing wheels, which, as a common aspect, are all based on very simple concepts. The first column is supplied by danaids providing the constant flow through the regulation of levels. The transfer between the two columns passes through a stamp pot that lets it eliminate the flow disruptions in the first column and, therefore, avoids the propagation to the second. The fission produce recovered at the output of the second column is reintroduced into the first column. They are therefore reincorporated into the cycle in order to be partly eliminated by extraction. All liquid flows, for security reasons, are achieved through the principle of air-lift. This process might be considered complex in part due to its looping nature and the strictly nonlinear physical phenomena that were in play. Furthermore, the very high-time constants will lead to the triggering of late traditional alarms. Due to this, human operators must be very vigilant and detect the appearance of a malfunction. The integration of an advanced supervision system is therefore justified.

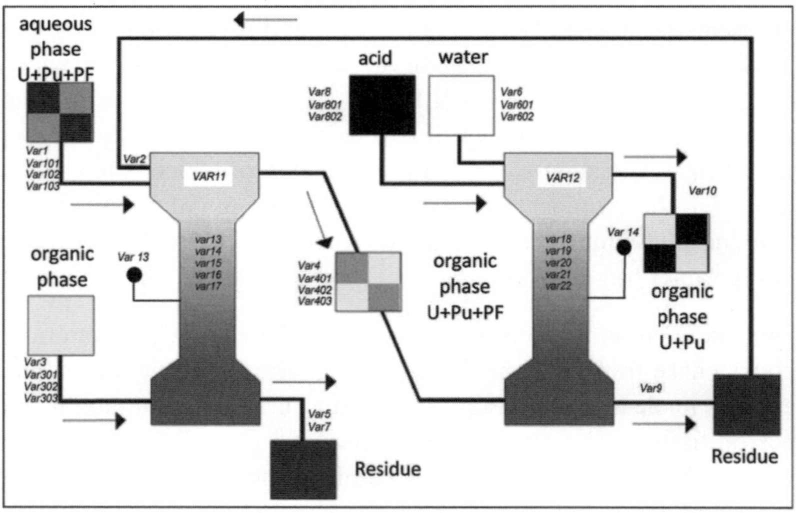

Figure 8.7. *The structure of pulsating columns*

8.4.2. *Human supervision tasks*

In this chapter, only the management aspect of the supervision is considered. In this context, the human supervision tasks are found at the intermediary level between strategic jobs (planning for example) and the tactical functions linked to the control system of the process. The human operator consistently receives continuous (measurements) and discrete information (state and alarms) from the supervision system. However, the operator transmits instructions to it intermittently. Supervision tasks require an extensive knowledge on the part of the operator and can be grouped into three classes, a classification adapted from [ROU 83] and completed:

– Management tasks, including the handling of failures. The operator can act in different ways over the process, either by modifying instructions (fine tuning, change in output, etc.) or by intervening during specific function modes such as the startup or the shutdown of the system for example. In this case, the operator is often required to apply predefined processes. On the other hand, following a malfunction, the operator must correct or compensate for the fault. These tasks highly depend on the nature of the diagnosed malfunction. In the case of a hardware failure or breakage, the operator, if possible, can launch backup or emergency systems in order to isolate the faulty sub-system, or continue to manage the process in degraded mode operation or, as a last resort, decide to shut down the system for repairs. In any case, they must plan a sequence of actions and execute them.

– Control tasks and ones for monitoring the normal operation of the system are referred to as surveillance tasks. The operator monitors the process, filters the available data and interprets it in order to detect a malfunction and/or to optimize the output. These tasks are subject to the establishment of the state of the process and require that the operator predicts the future evolution of the process while taking into account the planned actions. During this activity, the supervision operator finds himself exclusively in the position of observing the process operation. Indeed, the latter monitors the process by scanning the set of information available, without carrying out the slightest action that would modify the state. The supervision of the system also allows the operator to rethink their mental model of the state of the system in order to make the appropriate decisions when the system begins to operate abnormally or when an intervention is required on their part. This task therefore firstly involves vision; however, other senses such as hearing can also be simulated by signals (alarms for example). The supervision task is therefore characterized by an absence of actions over the process from the operator. However, this human operator must at all times be capable of characterizing the state of the system since the surveillance can be finished once a fault has been detected (see Figure 8.8). If the supervision operator detects it with an alarm, then this is a "passive detection". The reliability of the emitted alarms is therefore essential. On the other hand, if this detection is a result of the operator's observation of the process, then this is an "active detection" [RIE 01b]. Usually, the designer of supervision applications thinks only of improving the alarms and because of this, only the "integrative" human–machine cooperation form is considered. However, if the designer starts from the principle that operators have the competency and the supervision tools for cooperating with the operators and facilitating their decision-making, then the design approach becomes very different. Supervision tools, according to our approach, must make "debative" human–machine cooperation a priority. For this, the tools must be designed in order to improve, among other things, the active detection. Indeed, the operator is in a way outside the control–command loop since they do not receive enough information concerning the process. In other terms, SA [END 98]) is either facilitated very little or not at all.

– Diagnosis tasks, after having detected a malfunction either through alarms (passive detection) or by observation of the evolution of certain variables (active detection), the supervision operator must then diagnose the process. This diagnosis consists of determining the primary cause of the malfunction but also its effects and consequences on the process. Firstly, the operator must verify the validity of the malfunction, since the industrial system is controlled through several sensors of which some are used for the closed loop control. The operator must therefore determine if the malfunction is simply a disruption that will be countered by the regulation loops and will make the system capable of returning to a stable state. The capacity for abstraction and the fact that humans are capable of making decisions in partly familiar and/or uncertain situations explain why it is not possible to remove the operator from the supervision loop. Finally, it is certain that the earlier the detection of an abnormal event is, the easier it will be to counteract the negative effects on the process.

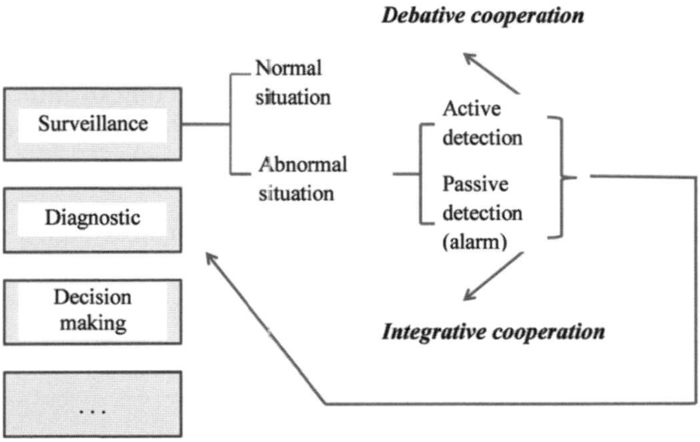

Figure 8.8. *From surveillance to diagnosis*

The supervision operator manages, in other words controls, commands, monitors and diagnoses, the system using graphical tools that we can classify in the following way:

– *Command views* facilitate the supervision of the operation of the command side and are able to react (regulation views, tuning views, modification of instructions, switch to automatic/manual of the control loops, etc.).

– *Synoptic views* represent the physical structure of the system. The term "P&ID diagram" (pipes and instrumentation devices diagram) is also used to refer to supervision where the measured variables are represented. Often, the command views are integrated with synoptic views

– *Tendency views* display the real evolution of one or more measured variables in real time.

– Finally, *alarm views* present the operator with a list of alarm descriptions and/or states classified according to criteria proper to each system (timeline, alarm type, geographical zone targeted by the alarm, etc.). Alarms are most often set off depending on the state of a sensor or even by a simple comparison of a value with a fixed threshold. The alarms therefore characterize a past event that leads to well-known problems of oscillation, or "Christmas trees", or of pertinence preventing a quick recognition of the state of the process.

Figure 8.9 shows the interactions between a human operator and a current supervision system. As has been shown, cooperation is only integrative and the operator must carry out by themselves the collecting of information and decision-making. This approach is not efficient since active detection is not taken into account by the designers of supervision applications.

Therefore, a design method for the supervision of systems adapted to humans is proposed.

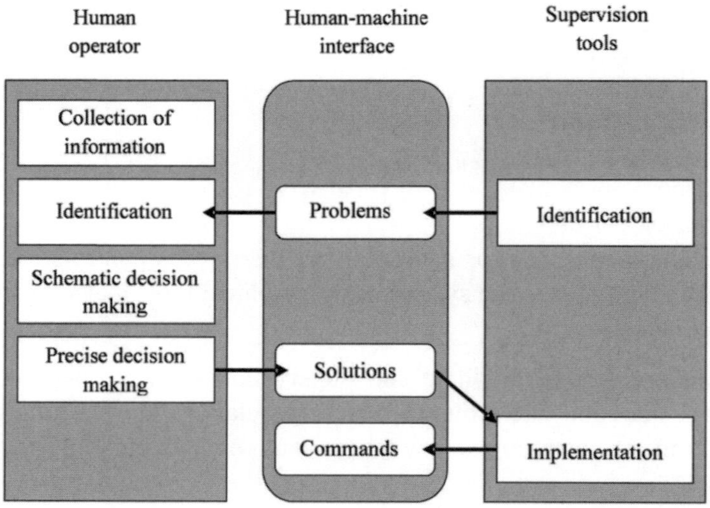

Figure 8.9. *Interaction between a human operator and a current supervision system*

8.4.3. *Design methodology of supervision systems adapted to humans*

Figure 8.10 proposes, in the form of a structural analysis and design technique (SADT) actigram, the design steps for supervision tools adapted to humans [RIE 01b]. The design of supervision tools is based on a preliminary analysis of the process that facilitates the formalization of the available knowledge of the system. The studied systems being of a large size, the analysis tools of complex systems are in this sense, interesting. The obtained models provide different points of view of the system (functional, dysfunctional, structural, topographical, behavioral, etc.) which can then be used, on the one hand, for the design of the automatic supervision and diagnosis system and, on the other hand, for the design of supervision tools. In order to be adapted to humans, the supervision tool must be founded on the design principles starting with an analysis of the tasks of the human operator and the knowledge of his decisional approach with a view to defining the efficient human–machine cooperation modes and integrating effective supervision and diagnosis algorithms. Finally, the evaluation step is essential and must contribute to the feedback experience.

The following section addresses the possible technical improvements for supervision systems.

8.4.4. *Improvement of the supervision and diagnosis system*

Research in this field is very active. Indeed, researchers in different fields (notably in control theory and computer science) propose original algorithms for the generation and processing of alarms. In this case, it is exclusively passive detection that will be improved.

In this chapter, the following definitions are adopted:

– *Detection*: this is the decision that consists in automatically determining whether the system has a fault or not. A fault is different from a breakdown, since the process continues to run but in a downgraded mode in the former case.

– *Isolation*: this term (as in diagnosis) has multiple definitions. We assume here that isolation corresponds to determining the source of the fault, and more precisely determining the measured variable, which is the source of the fault.

– *Identification*: this term consists of characterizing the fault by its cause.

Algorithms facilitating the completion of these three decision functions are often grouped together under the term fault detection and isolation. The terms surveillance system, alarm system or diagnosis system are often used to describe the systems incorporating these algorithms. The algorithms must enable an operator or an automatic system to make the right decisions and take the most appropriate actions. It is interesting to note that operators are not taken into account when devising algorithms. However, based on the realistic principle that it is not possible to completely remove the operator from the supervision loop, the direct consequence of this is that even if the regulation loops are very robust to disturbances and the detection, isolation and identification algorithms are very efficient, the human operator remains the final decision-maker. Consequently, the algorithms must also be considered as tools for human operators (see Figure 8.11).

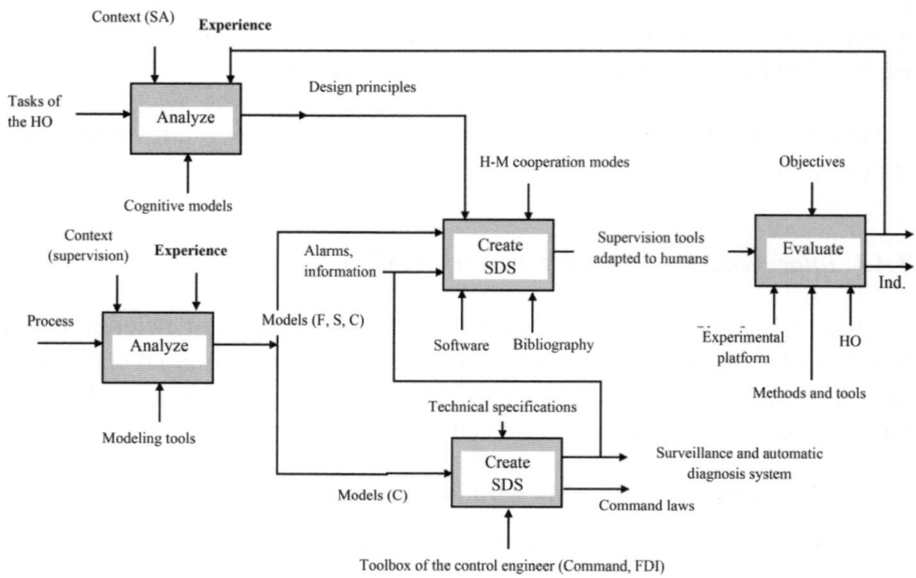

Figure 8.10. *Design steps for supervision tools adapted to the operator*

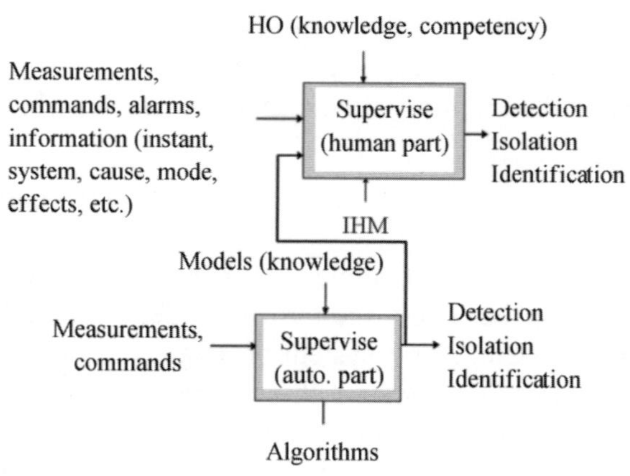

Figure 8.11. *Interactions between the human operator and algorithms*

The control and diagnosis community distinguishes between internal and external methods [GHE 97]. External methods are used when no analytical model is directly available for the process and, in this case, the technical issues of artificial intelligence are used most often to reproduce, in a simplified manner, human reasoning modes on the one hand, and to capitalize on human knowledge in the form of expertise and feedback experience on the other hand [DUB 97].

However, internal methods are used when knowledge of the system is sufficiently extensive, thus facilitating the creation of mathematical models during normal and/or unexpected operation (modeling of faults) of the system. The main diagnosis method using a model of normal operation consists of performing analytical redundancy [MAZ 96]. In fact, this is a generalization of material redundancy. An analytical redundancy is deduced from a behavioral model by using the available measured variables. Different techniques are possible. One involves estimating the parameters of the model online from the inputs and outputs of the system. These are methods founded on parameter estimation. In theory, the difference between the theoretical model and the observed model is representative of the faults. Another possibility involves comparing the values of a theoretical model with those obtained from the process. In principle, to avoid false alarms, the theoretical model of the normal operation is referred to as the "generation of residuals". When the values of the residuals are equal to zero, the process functions normally. In the opposite case, the difference can come from faults or the presence of noise, which needs to be eliminated by filtering the measurements. The values of these residuals in fact make up a signature or a vector identifying the fault. After having carried out the detection step, the signature is then compared with the theoretical signatures with a view to characterizing the fault [FRA 95].

The following sections detail the diagnosis method that we have used for the structure of the pulsating columns: approximate reasoning [EVS 00].

8.4.5. *Approximate reasoning*

Approximate reasoning is a diagnosis method developed by the GIPSA-Lab of Grenoble. It is very interesting, since it is based on, among others, a model in the form of a causal graph of the system – the variables are nodes and the transfer functions between variables are the arcs; see Chapter 6 of this book – which facilitates original and efficient human–machine communication. Figure 8.12 presents the principle of approximate reasoning. The idea is to use an analytical model for detection and, in addition, a causal model for isolation. Furthermore, the analysis of residuals is carried out not only according to amplitudes, but also according to variations using fuzzy logic.

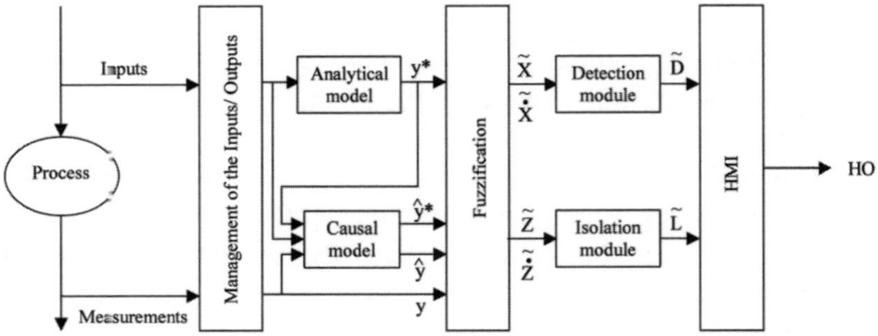

Figure 8.12. *Principle of approximate reasoning according to Evsukoff, Gentil and Montmain [EVS 00]*

The following sections outline the detection and isolation functions.

8.4.5.1. *The detection function*

The principle used for detection, in other words for the establishment of an abnormal state of the process, is the same as that used by methods with state estimation-based models. The fundamental problem of detection is the choice of thresholds at which fault appears. The approach proposed by the GIPSA-Lab [EVS 00] consists of characterizing the residuals according to two attributes and applying a "fuzzification" algorithm to them.

The first attribute involves the r values of the residual during the last n instants, represented by the vector $v(t)$ the components of which are defined by:

$$v_i(t)=r(t-i)=y(t-i)-y^*(t-i) \qquad i=0 \ldots n-1 \tag{8.1}$$

The second attribute involves the variations of the residual during the last m instants, represented by the derived vector of $v(t)$, the components of which are defined by:

$$v_j(t)=r(t)-r(t-j) \qquad j=1 \ldots m \tag{8.2}$$

The time windows are chosen *a priori*. The time window n is chosen in terms of the persistence of the fault. The time window m is chosen to improve the robustness of the decision since this corresponds to multiple estimations of the derivative in terms of the different time windows and thus allows us to manage the problems created by the noise in the measurements. At each instant, the attributes of the residual on each variable are recalculated for all of the $M = n*m$ combinations. The following idea involves blurring (fuzzying) the attributes of the residuals by introducing linguistic variables. Thus, Evsukoff, Gentil and Montmain [EVS 00] have defined five fuzzy partitions to describe the values of the residual: very negative, negative, zero, positive, very positive (*NN*, *N*, *Z*, *P*, *PP*) and three fuzzy partitions to describe the variations of the residual: negative, zero and positive (*N*, *Z*, *P*). The fuzzification of the attributes of the residual in each time window leads to the following descriptions:

$$\tilde{V}_i(t) = \left[NN\left(v_i(t)\right), N\left(v_i(t)\right), Z\left(v_i(t)\right), P\left(v_i(t)\right), PP\left(v_i(t)\right)\right] \qquad [8.3]$$

$$\tilde{\dot{V}}_j(t) = \left[N\left(\dot{v}_j(t)\right), Z\left(\dot{v}_j(t)\right), P\left(\dot{v}_j(t)\right)\right] \qquad [8.4]$$

At each instant, for each residual x, the membership function (μ) to each fuzzy sub-set is calculated, for each of the M combinations. The result is a matrix denoted as $\tilde{W}$ of M rows and 15 columns. Afterward, conclusions can be made by considering the rules represented in Table 8.1. We can see "logically" that the state "absolutely" normal (OK = 1, AL = 0) corresponds to a value and a zero tendency of the residual overall. Similarly, a residual with a very positive amplitude with a high tendency to diverge leads to an alarm (OK = 0, AL = 1).

			XP	
		N	*Z*	*P*
	NN	0/OK + 1/AL	0/OK + 1/AL	0.2/OK + 0.8/AL
	N	0/OK + 1/AL	0.4/OK + 0.6/AL	0.6/0K + 0.4/AL
X	*Z*	0.8/OK + 0.2/AL	1/OK + 0/AL	0.8/OK + 0.2/AL
	P	0.6/OK + 0.4/AL	0.4/OK + 0.6/AL	0/OK + 1/AL
	PP	0.2/OK + 0.8/AL	0/OK + 1/AL	0/OK+1/AL

Table 8.1. *Table of rules*

In matrix form, the rules are in the form of a matrix $\tilde{\phi}$ of fifteen rows (all of the possible states) and two columns (normal: OK and alert: AL). The partial conclusions (denoted by $\tilde{S}$ in matrix form) on the state of the variable are therefore obtained by computing:

$$\tilde{S} = \tilde{W}.\tilde{\phi} \qquad [8.5]$$

It then helps to define the engagement precondition of the alarm for the considered variable. For this, aggregation is done in the following way:

$$\text{D-(AL)} = \min(S_1(\text{AL}),\ldots,S_M(\text{AL})) \qquad [8.6]$$

$$\text{D-(OK)} = \max(S_1(\text{OK}),\ldots,S_M(\text{OK})) \qquad [8.7]$$

$$\text{D+(AL)} = \max(S_1(\text{AL}),\ldots,S_M(\text{AL})) \qquad [8.8]$$

$$\text{D+(OK)} = \min(S_1(\text{OK}),\ldots,S_M(\text{OK})) \qquad [8.9]$$

If D-(AL) – D-(OK) > ε_1, then the alarm is activated; this is then a "robust" conclusion to false alarms.

If D+(OK) – D+(AL) > ε_2, then the alarm is stopped. The conclusion is therefore "sensitive" to returning to the normal state of the variable.

The following section outlines the isolation function.

8.4.5.2. *The isolation function*

Once detection is achieved, the causal graph is used to isolate the source of the alarms, in other words the variable for which the behavior, be it considered as an alarm or not yet as an alarm, has led to the appearance of alarms over other variables. It is indeed possible that a variable having not varied much compared to its model and is therefore considered as normal, leads to heavy variations of other variables. The same situation arises when we have delays or high-time constants. The effects, visualized through alarms, therefore appear before the cause. The isolation algorithm is founded on a computation of the local residuals $\Delta\delta(t)$ in the following way:

$$\Delta\delta(t) = \delta(t) - \hat{\delta}(t) = \left(y(t) - y^*(t)\right) - \left(\hat{y}(t) - \hat{y}^*(t)\right) \qquad [8.10]$$

Where $\hat{y}$ is the predicted value of the variable y using the causal graph and the measurements taken from the measurements upstream (at the nodes), and $\hat{y}^*$ is the predicted value of the variable y using the causal graph and the values of the analytical model for the nodes directly upstream in the causal graph.

It is possible to describe the equation above in the following form:

$$\Delta\delta(t) = \delta(t) - \hat{\delta}(t) = \left(y(t) - \hat{y}(t)\right) - \left(y^*(t) - \hat{y}^*(t)\right) = y(t) - \hat{y}(t) - \Delta\tilde{y}^*(t) \quad [8.11]$$

If the residual is close to zero, this means that the detected disturbance is only due to the propagation of the faults upstream; in the inverse case, then it is a local fault [EVS 00].

This method requires the identification of reliable dynamic models that are difficult to obtain in real large-scale industrial systems. Furthermore, only passive detection is improved. Because of this, if the alarms are well defined and the operator must act on the system exclusively when the automatic part does not work, then the traditional problems of workload increase and loss of vigilance due to long periods of inactivity can appear. We rediscover the "ironies of automation" here, condemned at the time by Baindbridge [BAI 83] and the importance of cognitive characteristics of the human operator in the design of supervision tools [RIE 01a].

8.4.6. *The use of cognitive principles in the design of supervision tools*

The use of cognitive principles in the design of supervision tools leads us to define the following design concepts:

– Improving the perception of the state of the process by favoring active detection.

– Providing supervision views adapted to different supervision tasks.

– Favoring debative human–machine cooperation.

– Creating and providing the human operator with summarized or high-level information based on an implementation "adapted to humans" of the algorithms developed for supervision and diagnosis.

These four principles are now outlined and debated.

The models from cognitive psychology applied to the supervision of industrial processes [AMA 99] allow us to highlight the importance of SA. This is what guides decision-making. The ideal is therefore to provide the operator with the necessary and sufficient information to perceive the state of the system. Too often, the control engineer does not consider this fundamental aspect during the design of supervision tools. Improving the quality of alarms is not enough. Every aspect must be taken care of in order to guarantee an active behavior from the operator. It seems that a debative human–machine cooperation is the way the agents (human and artificial) exchange information with a view to improving decision-making, and is thus the most well adapted to systems with a high level of automation. Currently, supervision tools do not favor the collection of information. Operators must include all of the measurements from the system themselves before making a decision. Improving alarms does not solve the problem of information collection. The problem therefore remains in the design of human–machine systems and interfaces offering a detailed perception of the system state to the operator.

In order to reach this objective, supervision tools must provide the operator with multiple viewpoints (functional, dysfunctional, topographical, behavioral) of the system [JOU 01]. Supervision tasks (control, surveillance and diagnosis) have objectives that require a radically different information make-up and structure. Faced with the difficulty of proposing interfaces that adapt automatically to the context and therefore the state of the system, the most consistent solution consists of providing the operator with different viewpoints of the system. In particular, supervision tasks require views that provide the operator with a global and digitized overview of the process, whereas diagnosis tasks, due to the cognitive processes of the activity, require a more multi-point and hierarchical view of the process [LAM 98b]. Consequently, an advanced supervision system (ASS) must include what is referred to as "advanced" views dedicated to surveillance and other views dedicated to diagnosis and offered according to different viewpoints.

Today, in control rooms, the information contained in overviews is computed on the physical structure of the industrial system. Yet, in the last 10 years, researchers have explored presentation forms more adapted to the cognitive capacities of humans. "Ecological" interface and "direct perception interface" concepts have caught our attention. Based on the principles of ecological psychology promoted by Gibson [GIB 79], the concept of an ecological interface appeared toward the end of the 1980s, under the impetus of Vicente and Rasmussen [VIC 90] who proposed a theoretical approach to the design of interfaces for human–machine systems, based on the cognitive taxonomy of the decision-making process of the human operator as defined by Rasmussen [VIC 92, VIC 95]. According to Vicente and Rasmussen, in the design of interfaces, two aspects highlighted by cognitive psychology must be taken into account. The first one is linked to the work environment characterized by a certain complexity or variety that dictates the information content and the structure of the future interface. The second is related to the cognitive characteristics and perceptive properties of the human operator who will influence the choice of the information representation.

Thus, in the proposed ecological interface design (EID) process, the specification of the future interface information content is directed by the goals of the system studied and is correctly translated through the definition of different models of the system. Choosing the way in which to present the information involves Rasmussen's cognitive taxonomy (skill-based behavior, SBB; rule-based behavior, RBB, that is procedural; or knowledge-based behavior, KBB). These concepts are the result of numerous studies carried out on the cognitive behavior of humans applied to engineering and stemming from a multilevel hierarchical representation of knowledge. Additionally, the design of an ecological interface requires the creation of functional and structural models over several levels of abstraction (one model per level), without omitting the construction of links between these two types of models. A sufficiently complete model is therefore obtained to facilitate the identification of the content and the structure of the interface. It must be noted that the design phase of this type of interface must also facilitate the consideration of the ergonomy of the presentation, by suggesting the format of the information to be displayed.

The main purpose of an ecological interface resides in a representation of information favoring SBB and RBB from the supervision operator, thus resulting in an increase in performance. The EID process is fairly abstract however. The method for defining models remains at the discretion of the designers. Even the passing of models to interfaces is not very explicit. However, we remember that ecological interfaces aim to favor a direct perception of events and of the state of the system, thus relieving the operator of implementation work and the interpretation of information resulting from the process.

Among the proposed ecological interfaces, the mass data display (MDD) interface concept appeared in the 1980s. Instead of aggregating or filtering information, Elzer *et al.* [ELZ 95] purposefully chose to do the opposite by presenting the human operator with a maximum of data, which is of particular interest to them. Indeed, the MDD allows the human operator to quickly recognize and evaluate the changes in a graphical context [BOU 98]. This representation technique in fact provides the human operator with a global view of the entirety of the process, and the MDD concept translates the behavior of the process in a symbolic and graphical way. For example, a variable is represented by a bar, the position of which changes depending on its state. Since the MDD interface is adapted to human surveillance and fault detection tasks, its objective is to allow human operators to prioritize anticipation. Furthermore, the minimalist character of its representation facilitates its application in the frame of large-scale industrial systems.

MDD interfaces for existing complex processes have already been dealt with. For example, in the context of thermal power plants, the principle to be retained consists of representing the set of variables making up this process through small bars, with the angle relative to the horizontal plane and the thickness varying proportionally (see Figure 8.13) [BOU 98]. This type of interface is therefore valid since the three following conditions are fulfilled: the process is located around an operation point, the permanent plant state has been attained and the nominal values are known. Experiments were carried out in order to evaluate the performances of MDD interfaces compared to traditional interfaces on the one hand, and compared to views on tendency on the other hand [BOU 98]. Having been carried out with 10 subjects, these experiments have shown the efficiency of the MDD interfaces for the detection of faults. On the other hand, the performances of these seem more limited for diagnosis. Indeed, compared to traditional interfaces, MDD interfaces significantly shorten the detection time of abnormal events of the process through human operators. However, the error rate over the results on human expertise, when mentioning malfunctions, is quite a bit higher. Researchers explain this effect by recalling the initial vocation of this interface designed to alert human operators to abnormal events, and, in any case, to formulate reliable expertise [BOU 98]. Tendency views are therefore much more adapted.

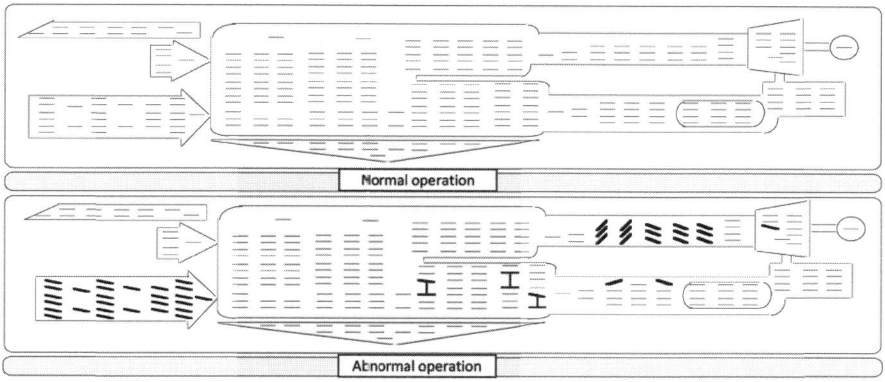

Figure 8.13. *Example of a mass data display*

The last proposed principle consists of defining summarized information, also referred to as high-level information. It is advisable to detail the different possible ways of processing the information. Measurements from the plant can undergo three types of processing: filtering, aggregation and synthesis [LAM 98c]:

– Filtering consists of choosing within a set E of measured variables, relative to a given object, a representative *a priori* sub-set E'.

– Aggregation, starting from a set E of variables measured in the same way, consists of finding a single variable denoted by E' which carries additional information. As an example, verification of a hydraulic balance allows us to define aggregated information.

– Synthesis consists of generating one or more artificial variables from measurements and a deep knowledge of the system. It therefore characterizes the passage from measurement to information or from digital to symbolic items. In other words, synthesis, if it is carried out by humans and if it is possible, requires a high cognitive effort. Currently, only filtering and aggregation of information are used in supervision viewpoints. Synthesis can be seen in two different ways:

- Through the use of dynamic models of the system operation (non-measured variable estimators, advanced alarms, values and tendencies of residuals, etc.).

- Through advanced processing over the available data (identification of signals, segmentations, tendency curves analysis, etc.).

A piece of synthetic information is either discrete (DSI) or continuous (CSI). Alarms from an advanced surveillance system set off through a comparison between the measurement and the dynamic model of the process are examples of DSIs. The availability of a dynamic model of the system within the surveillance system, by definition of the CSI, offers a big alternative to the exclusive presentation of measurements to the human operator. It is clear that the establishment of a CSI could sensibly change the nature of the "control", generally based on measurements and alarms, and therefore favors a more active role for the human operator by facilitating, among other things, cooperation with tools providing high-level information.

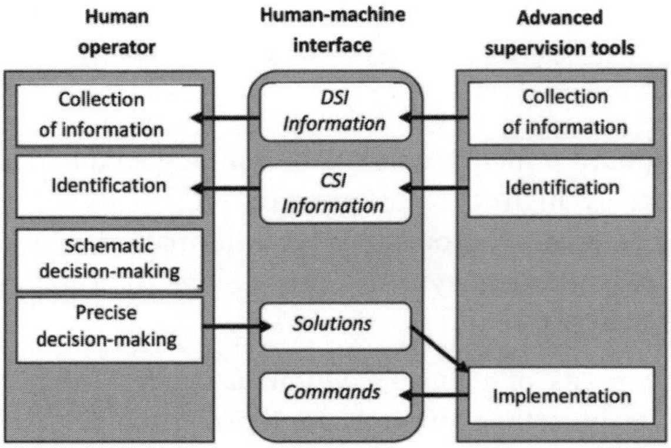

Figure 8.14. *Interactions between the human operator and supervision tools adapted to humans*

Nevertheless, to guarantee acceptability of the system by human operators, it is helpful to equip them with explanatory capacities. The objective is not to provide the operator with automatic explanations. It is rather a research topic in its own respect. The aim is to give the operator the means to reason over a model adapted to his diagnosis tasks on the one hand, and to be able to verify the information from supervision tools on the other hand. In other words, an ASS must provide the operator with information they can understand and verify. Furthermore, verification time of the information must be shorter than the time it would have taken for the human operator had they searched for the information themselves.

Figure 8.14 shows the interactions between the human operator and the supervision tools adapted to humans. The CSIs favor the collection of information and therefore an active behavior of the operator. Inversely, DSIs improve passive detection.

8.4.7. *An example of an advanced supervision system (ASS)*

Within the frame of research for the improvement of supervision systems for heavily automated and continuous processes, three labs, the LAMIH (Laboratory of Industrial and Human Automation Mechanics and Computer Science), the GIPSA-Lab (which was at the time called the Grenoble Laboratory of Automation) and the LIA (Laboratory for advanced informatics) of the CEA (Commissariat à l'Energie Atomique) at Marcoule, have participated in a common project for the last two years. This project was completed through experiments led by professional supervision operators. The three labs intervened at different levels according to their specialty. The LAMIH designed, created and evaluated a supervision system by incorporating supervision and diagnosis algorithms conceived by the GIPSA-Lab. The LIA collaborated at different stages of the project and provided expert knowledge of the process of implementation as well as the professional operators required for the evaluation.

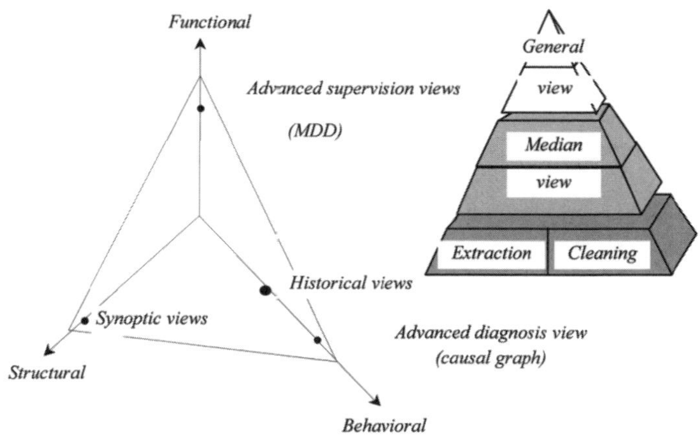

Figure 8.15. *Viewpoints and levels of abstraction*

The developed ASS incorporates the set of propositions made in the previous section, notably: the integration of surveillance and diagnosis algorithms based on approximate reasoning, tools used for each type of task (surveillance and diagnosis), a representation of the process according to different viewpoints, high-level information of explanatory capacities creating a debative cooperation between the operator and the supervision tools. Consequently, different human–machine interfaces representing the process according to different viewpoints and at different levels of abstraction ("whole–parts" decomposition) have been developed (Figure 8.15).

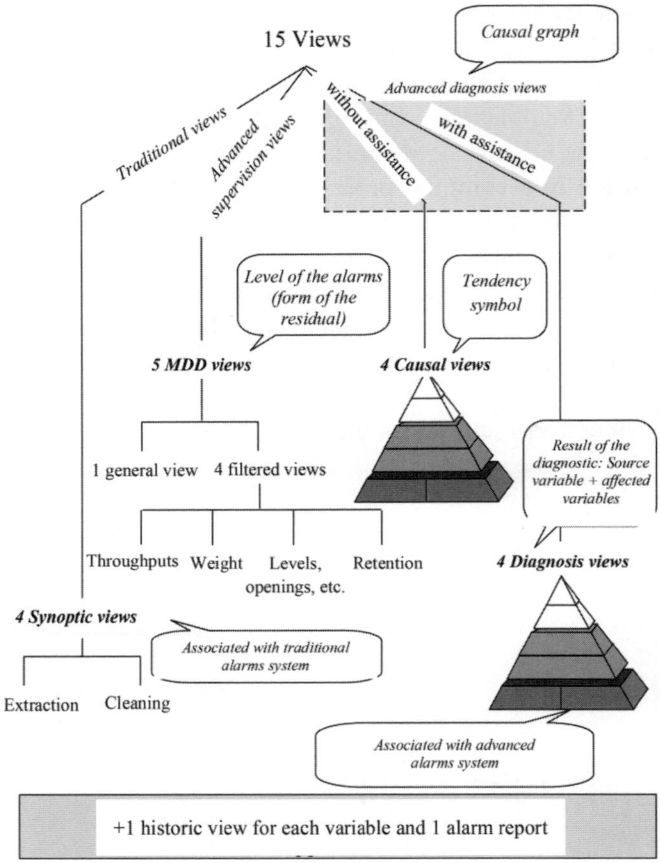

Figure 8.16. *Architecture of an advanced human-adapted supervision system (AHASS)*

The ASS produced [LAM 98c, RIE 01b] has two distinct parts: a traditional part proposing the views and tools that already exist in the industrial world and an advanced part that characterizes the CWS and favors the cooperation between the operator and the algorithms. The logical application has 17 views, positioned according to the different viewpoints (functional, structural and behavioral) across the process (Figure 8.16): one supervision view of the MDD type, two overviews that split the cleaning and extraction parts of the process, four filtered views similar to the MDD view where certain symbolic variables are replaced by their digital values, four causal views of the process depending on different levels of abstraction, four diagnosis views computed over the causal views, one historic view made available for each variable and one view of the alarm's report.

The application was designed to address the need to optimize its flexibility. Thus, it is possible to test breakdown scenarios according to different contexts, in other words, by only giving access to certain pieces of information. The steps of the operator, depending on the tools they have available to them, can therefore be compared, which helps to determine the qualities or drawbacks of these tools in relation to certain situations. A configuration file facilitates the authorization or prevention of access to certain views, to histories and alarm messages produced by one or both of the supervision systems (advanced or traditional depending on the threshold). The advanced supervision tools provide specific summarized information for surveillance and diagnosis. These have been defined and worked using the approximated reasoning algorithms shown previously.

8.4.7.1. *Synthetic information for supervision*

From an information point of view, the rules presented intrinsically contain original *a priori* information that is pertinent for the human operator: the form of the residual.

		$\dot{X}$		
		N	***Z***	***P***
X	***NN***	vsv	sv	sv
	N	sv	susp	susp
	Z	susp	norm	susp
	P	susp	susp	Sv
	PP	sv	sv	vsv

Table 8.2. *Table of rules*

It is possible to associate a particular state with each combination. We have chosen to define four states: very severe (vsv), severe (sv), suspect (susp) and normal (norm), see Table 8.2. More precisely, for each state (vsv, sv, susp, norm), a membership function is computed using the following product aggregation functions and the sum aggregation functions:

$$\mu_{(X \in \text{A AND } Y \in \text{A})} = \mu_{\text{A}}(X) * \mu_{\text{A}}(Y) \quad [8.12]$$

$$\mu_{(X \in \text{A OR } Y \in \text{A})} = \mu_{\text{A}}(X) + \mu_{\text{A}}(Y) - \mu_{\text{A}}(X) * \mu_{\text{A}}(Y) \quad [8.13]$$

The states "normal", "suspect", "severe" and "very severe" are then defined by combining the different membership functions. As an example, the "very severe" state corresponds to the following formula:

$$\mu_{tgr} = \mu(((X \in \text{NN}) \text{ AND } (\dot{X} \in \text{N})) \text{ OR } ((X \in \text{PP}) \text{ AND } (\dot{X} \in \text{P})) \quad [8.14]$$

In other words, the "very severe" state characterizes the fact that the residual is very negative and that it continues to drop. Furthermore, at each instant, the membership function at each primitive state is known. Thus, it is possible to propose artificial variables incorporating the set of results concerning the primitive states. A variable named "gravity" has thus been defined: it is denoted as μ_{al}, a value between 0 and 1, and is calculated at each sampling period by aggregating the membership function at each state [EVS 98, GEN 97, LAM 99, RIE 01a, RIE 01b]:

$$\mu_{al} = 1/6 *[(1 - \text{sgn}\,\mu_{gr})(1 - \text{sgn}\,\mu_{tgr})(1 - \mu_{norm} + \mu_{susp}) +$$

$$\text{sgn}\,\mu_{gr}(1 - \text{sgn}\,\mu_{tgr})(4 - \mu_{susp} + \mu_{gr}) + \text{sgn}\,\mu_{tgr}(5 - \mu_{gr} + \mu_{tgr})] \quad [8.15]$$

where *sgn* represents the sign function (0 if $\mu = 0$, 1 otherwise) and μ the indices of the membership functions of the fuzzy sub-sets severe, very severe, normal and suspect. The notion of "gravity" is uniquely tied with the tendency of the residual. The value 0 signifies that the membership function to the "normal" state is equal to 1. Conversely, a value of 1 signifies that the membership function to the "very severe" state is equal to 1. This variable is continuous and, due to this, must favor active detection, on the condition that it be well interpreted by the operator, particularly during slow evolutions of the process. It is absolutely possible to define other indicators, in terms of the searched objects and the type of surveillance algorithm being used. Incidentally, this is a future research topic that we have found to be very interesting.

The joint use of two functional analysis methods, the functional tree and SADT [LAM 98a, LAM 98b, LAM 98c] enables us to highlight the contents and the organization of advanced supervision views (general view and filtered views), based on the concept of MDD.

Figure 8.17 shows how the general MDD view is based on functional organization [2]. Each variable is represented by a rectangle [1], the color[2] of which varies according to the state of the variable. In order to represent the matter flow, the arrows link some of these blocks together [3]. The colors of the variables vary from green (normal state) to red (abnormal state) in a very progressive manner (palette of 32 shades of color). Each color is associated with an index of severity typical of the tendency of the residual. Concretely, green signifies that the variable is in a completely normal state (zero or near-zero value of the residual over the time window), whereas the color red corresponds to a very abnormal state (residual with a very high membership value to the very severe state). The color palette allows us to translate, in a gradual fashion, the set of intermediary states existing between normality and abnormality. It is evident that this representation is well adapted to abnormal and slow evolutions of the process.

The operator can also obtain the history of a variable by clicking on the rectangle representing it or information on the variable by clicking on its name [4].

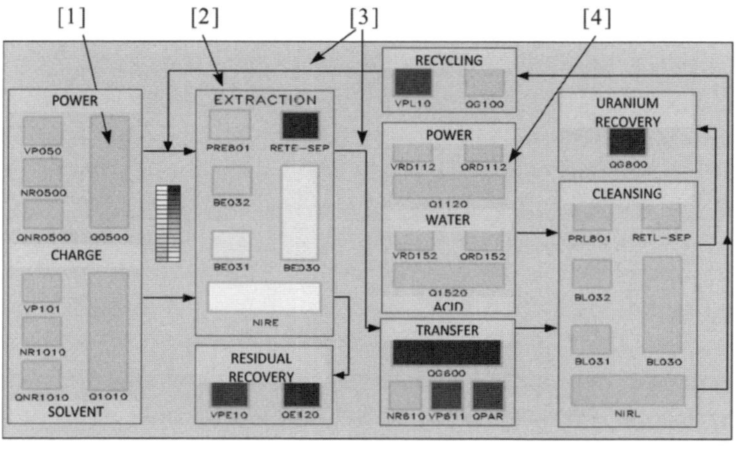

Figure 8.17. *General MDD view*

2 In figure 8.17, the panels are represented at different levels of gray, however the real interface is in color.

8.4.7.2. *Synthetic information for diagnosis*

The isolation function used in approximate reasoning gives reason to use a causal graph as a model. It is therefore possible to justify the results of the isolation using this graph, a medium that is easily understandable by the human operator.

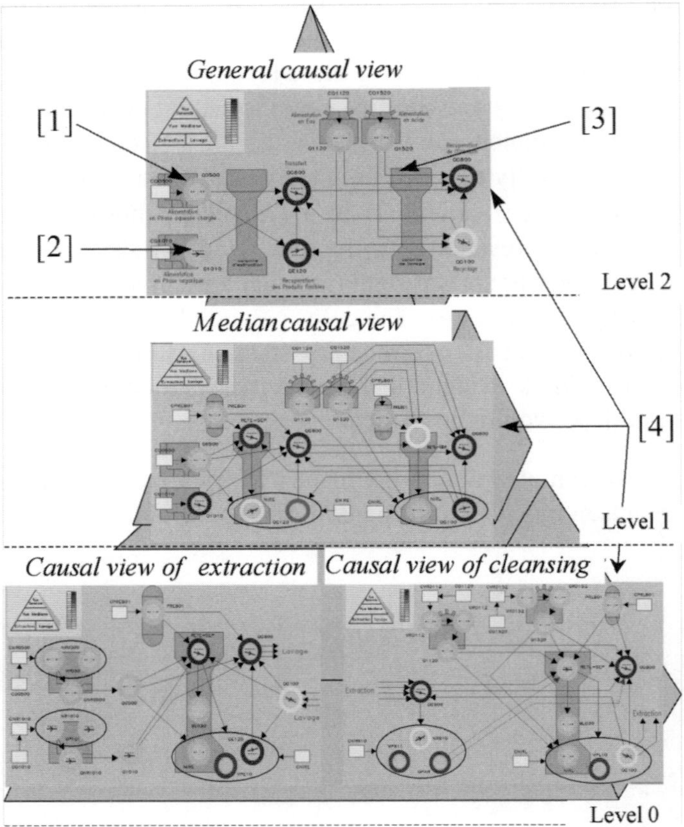

Figure 8.18. *Causal views*[3]

Nevertheless, in order to guarantee an active behavior of the operator and take into account the importance of SA, the ideal would be to provide the operator with the means to carry out and validate their own diagnosis. Human–machine interfaces for isolation assistance must therefore be created in order to support the decision-making process of the human operator. To

3 The views are in color following the same codes as in Figure 8.17.

this effect, it seems that the form of the residual is a pertinent piece of synthetic information for the diagnosis. However, the problem with graphical representation is apparent given the number of variables. A solution has been provided in the context of nuclear combustible reprocessing.

Presented in Figure 8.18, advanced diagnosis views, also referred to as causal views, are composed of causal graphs, the nodes [1] that contain the tendency symbol [2] representing the variables and the arcs [3], and the interactions between them. The organization of these views is in the form of a pyramid representing three levels of abstraction [4]. At the highest level, the operator has a general view of the process representing the flow variables. At the intermediate level, the median view provides them, on top of flow variables, with the variables reflecting the performances of the columns. At the lowest level, two detailed views of the process allow the operator to monitor the whole set of system variables. The definition of the variables at each level of abstraction is based on the use of multilevel flow modeling (MFM) [LAM 98a, LAM 98b, LAM 98c, LIN 11a, LIN 11b].

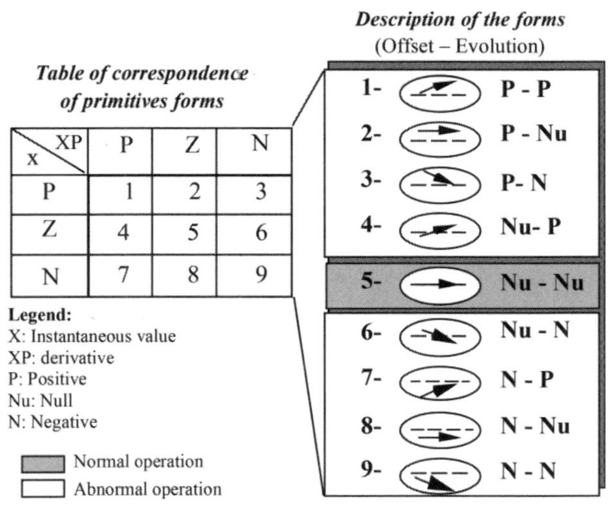

Figure 8.19. *Table of primitive states*

A node consists of a variable color that in fact takes the same principle as the MDD view. At the center of this disk is the tendency symbol [2], which informs the operator of the form of the distance (residual) between the real value and the variable (measurement) and its theoretical value during the normal operation (model).

Indeed, nine primitive states have been retained (see Figure 8.19) over the fifteen initial ones used in the detection and isolation algorithms. A symbol has been assigned to each primitive state. This is made up of an arrow corresponding to the tendency of the residual relative to the reference model, represented by a dotted horizontal line. State 5 corresponds to a normal situation. States 2, 3, 4, 6, 7 and 8 correspond to suspect situations, whereas states 1 and 9 correspond to a severe situation.

We note that the set-values are represented by rectangles. Finally, the regulation loops are characterized by an oval containing the set of variables (nodes) involved in the loop (command and measurement). This representation facilitates a considerable aggregation of the causal graph and favors its readability, while clearly indicating the contribution from the command part. As with other views, a click on the node gives the history of the corresponding variable. Furthermore, a click on the name of the variable gives information on its type, localization in the process, etc. Information on the arc linking the variables can also be obtained by clicking on it. Finally, through a simple click, it is possible, after the appearance of alarms, to display the result on the graph, which is possible due to isolation. Thus, the variables at the source of the problem are displayed as well as the propagation through the graph.

The human operator can himself easily verify the information provided by the system. The same model (causal graph) is used by the algorithms in order to present the results and provide support to the operator's reasoning. This is therefore in some ways a CWS created to facilitate debative human–machine cooperation.

All of the proposed concepts have been evaluated at the CEA of Marcoule (France) by professional operators (six experts and a novice). Without detailing these experimentations, which is not the aim of this chapter, the following results were obtained. The interest of the operators with the new concepts of ASS has been proved. Globally, the obtained results tend to prove the utility and usability of the developed ASS. On a more detailed level, the results from objective data (indicators of the usage of the different viewpoints for example) or subjective data (questionnaires) show that the subjects have appreciated and assimilated the inherent principles with the following concepts:

– A categorization of the supervision views into specific views (surveillance and diagnosis).

– The use of an MDD representation, particularly appreciated by operators for its capacity of providing, in a functional manner, the set of variables of the process in a symbolic form.

– The choice of a hierarchical organization for advanced diagnosis views or views without assistance which, even if the proposed hierarchy is not the best, proposes additional viewpoints of the process, used by operators at different stages of their diagnosis.

– The use of a dynamic representation of the system behavior with a causal graph, which, in addition to enabling the representation of the results of the isolation–support system, provides support to the operator for reasoning.

– The concept of high-level information based on residuals, be it the "alarm level" on the advanced supervision views or the tendency symbol qualified by some as a shortcut to the history (evolution over time of the residual).

The most interesting result of the surveillance task relates to detection time. Active detection has been considerably improved through advanced supervision views (MDD view with the "gravity" as the synthetic information). All of the subjects have detected an abnormal operation of slow evolutions in less than five minutes, whereas the traditional supervision system set off the alarm after 26 minutes and the advanced system set it off after 9 minutes. Furthermore, the median causal view has been used in the frame of the supervision of the process and the detection of faults. With the MDD, these views have been qualified by operators as "watch keeping views", in other words views in which the most representative variables of the operation of the process appear, which ensured the reliability of the supervision.

With diagnosis, the causal graph has shown to be a representation of the behavior of the system, which is appreciated by operators, seconded by representative information of the evolution of the state of the residual, with the tendency symbol particularly praised. This symbol has the advantage of synthetic information susceptible to considerably improving future supervision views. Nevertheless, in the hierarchical organization provided, the most abstract view, the general view, is of little interest for operators and must therefore be removed. For the diagnosis tools, this is useful during moments of uncertainty over the origin of the fault. Our approach to diagnosis assistance is interesting to operators as it lets them be in control of their actions while allowing them to verify hypotheses. Thus, the boundary between assistance and assistantship is never crossed. It therefore seems that debative human–machine cooperation is adapted to systems showing a high level of automation since it has been accepted, understood and used by human operators.

8.5. Conclusion

The supervision of automated systems is an activity in which the human operator is present and has an essential role despite the increase in automation, since they remain responsible for the security of the system. Consequently, the design of supervision systems requires a different approach to automation, one which is more systematic than analytical. This has brought us to consider original automation solutions based around human–machine cooperation concepts. In this chapter, we have shown how it is possible to favor it and illustrated our point with two applications of supervision systems design.

The first application was air traffic control, which is (today) considered as a process with a low level of automation. In different projects, the concept of DTA has been implemented and evaluated with the aim of modifying the level of automation online, such that it adapts to the needs of human operators. For this, the contents of a CWS adapted to the field of air traffic have been defined in order to favor cooperative activities between the controllers on the one hand and an assistance tool for solving aerial conflicts on the other hand. The latter, in order to avoid any decision-making conflict with the operators, is particularly capable of taking into account the strategies provided by the controllers for determining the solutions to aerial conflicts.

The second application deals with a process with a high level of automation: the reprocessing of nuclear combustibles. The main idea in this case is to favor active detection by the establishment of a debative human–

machine cooperation. For this, the idea is to provide synthetic information, implement advanced detection and diagnosis algorithms and favor the detection of abnormal operation by the human operator. In other words, everything is done to allow the operator to have SA in harmony with the real state of the process. Experiments with expert operators have made the concepts proposed in this ASS partly validated.

In both cases, "human engineering" has led to original solutions in the field of automation. All of the principles of human–machine cooperation detailed in this chapter can be applied to automated systems where the human has a supervision role. For example, numerous applications can be considered in the field of driving (low level of automation) or in the management of large networks (high level of automation).

8.6. Acronyms

AHASS: *Advanced human-adapted supervision systems*

AMANDA: *Project's name: Automation and Man–Machine Delegation of Action*

CTD: *Cognitive task design*

CWS: *Common work space*

DTA: *Dynamic tasks allocation*

HMC: *Human–machine cooperation*

LOA: *Level of automation*

MDD: *Mass data display*

P&ID: *Piping and instrumentation diagram*

PC: *Planning controller*

PLAF: *Module's name*: *Planning of Allocation*

RC: *Radar controller*

SAINTEX: *Module's name*: *in reference to the famous pilot (and writer) – Saint-Exupéry* (*Vol de Nuit*, 1931)

SPECTRA: *Project's name*: *Experimental Dynamic Task Allocation System for Air Traffic Control*

STAR: *Module's name*: Tactical System for Assistance to Resolution

8.7. Bibliography

[AMA 99] AMAT-NEGRI A-L., Simpatico: a cognitive model of the pilot – from concepts to specifications, PhD Thesis, University of Valenciennes et Hainaut-Cambrésis, France, 1999.

[ANG 92] ANGERAND L., LE JEANNIC H., Bilan du projet SAINTEX, Report R92009, Centre d'études de la navigation aérienne (CENA), France, 1992.

[BAI 83] BAINBRIDGE L., "Ironies of automation", *Automatica*, vol. 19, no. 6, pp. 775–779, 1983.

[BEN 92] BENTLEY R., RODDEN T., SAWYER P., *et al.*, "An architecture for tailoring cooperative multi-user display", *Proceedings of CSCW'92*, Toronto, Canada, November 1992.

[BOU 98] BOUSSOFFARA B., ELZER P., "About human pattern matching and time information in S&C of large technical systems", *Proceedings of the 17th European Annual Conference on Human Decision Making and Manual Control*, Valenciennes, France, pp. 161–166, 1998.

[CAR 99] CARLIER X., HOC J.-M., "Role of a common frame of reference in cognitive cooperation: sharing tasks in air-traffic-control", in HOC J.M., MILLOT P., HOLLNAGEL E., CACIABUE P.C., (eds.), *Proceedings of the 7th European Conference on Cognitive Science Approach to Process Control CSAPC'99*, Presses Universitaires de Valenciennes, p. 212, 1999.

[CRE 96] CREVITS I., DEBERNARD S., MILLOT P., "Application criteria of dynamic task allocation principles", *5th International Conference on Human Aspects of Advanced Manufacturing: Agility and Hybrid Automation*, Maui, HI, August, 1996.

[CRE 02] CREVITS I., DEBERNARD S., DENECKER P., "Model building for air-traffic controllers workload regulation", *European Journal of Operational Research*, vol. 136, no. 2, pp. 324–332, 2002.

[DEB 92] DEBERNARD S., VANDERHAEGEN F., MILLOT P., "An experimental investigation of dynamic task allocation between air traffic controller and AI system", *Proceedings of the 5th IFAC/IFIP/IFORS/IEA Symposium on Analysis, Design and Evaluation of Man-Machine Systems*, La Haye, The Netherlands, pp. 9–11, June 1992.

[DEB 98] DEBERNARD S., CREVITS I., CARLIER X., *et al.*, Projet AMANDA: rapport final de la première phase, no. CENA/NR98-185, Convention de Recherche LAMIH/CENA 96/C007, September 1998.

[DEB 01] DEBERNARD S., HOC J.-M., "Designing dynamic human-machine task allocation in air traffic control: lessons drawn from a multidisciplinary collaboration", in SMITH M.J., SALVENDY G., HARRIS D., KOUBEK R. (eds.), *Usability Evaluation and Interface Design: Cognitive Engineering, Intelligent Agents and Virtual Reality*, vol. 1, Lawrence Erlbaum, London, pp. 1440–1444, 2001.

[DEB 02] DEBERNARD S., CATHELAIN S., CREVITS I., *et al.*, "AMANDA Project: delegation of tasks in the air traffic control domain", in BLAY-FORNARINO M., PINNA-DERY A-M., SCHMIDT K., ZARATE P., (eds.), *Cooperative Systems design*, IOS Press, Lansdale, pp. 173–190, 2002.

[DEB 03] DEBERNARD S., HOC J.-M., "Conception de la répartition dynamique d'activités entre opérateur humain et machine dans le contrôle de trafic aérien. Leçons tirées d'une collaboration pluridisciplinaire", *JESA, Analyse et commande des systèmes dynamiques*, vol. 37, no. 2, pp. 187–212, 2003.

[DEB 08] DEBERNARD S., ANNEBICQUE D., CREVITS I., *et al.*, "Vers une modélisation des systèmes coopératifs", *Actes de la Conférence internationale francophone d'automatique, CIFA 2008*, Bucarest, Romania, September 2008.

[DEB 09] DEBERNARD S., GUIOST B., POULAIN T., *et al.*, "Integrating human factors in the design of intelligent systems: an example in air traffic control", *International Journal of Intelligent Systems Technologies and Applications*, vol. 7, no. 2, pp. 205–226, 2009.

[DEB 10] DEBERNARD S., POULAIN T., MILLOT P., "A common work space for supporting human-human and human-machine cooperation in ATC", *HCI-Aero 2010*, Cape Canaveral, FL, November, 2010.

[DEC 94] DECORTIS F., PAVARD B., "Communication et coopération: de la théorie des actes de langage à l'approche ethnométhodologique", in PAVARD B., (ed.), *Systèmes Coopératifs de la modélisation à la conception*, Octarès, Toulouse, pp. 21, 50, 1994.

[DUB 97] DUBUISSON B., "Détection et diagnostic de pannes sur processus", *Techniques de l'Ingénieur, Traités Mesures et Contrôles*, vol. R.8, pp. 75–97, 1997.

[ELZ 95] ELZER P., BEUTHEL C., BOUSSOFFARA B., *et al.*, "Advantages of mass data display in process S & C", *IFAC Symposium on Analysis, Design and Evaluation of Man Machine Systems*, Boston, MA, pp. 439–444, 1995.

[END 98] ENDSLEY M., "Design and evaluation for situation awareness enhancement", *Proceedings of the Human Factors Society: 32nd Annual meeting*, Santa Monica, CA, pp. 97–101, 1988.

[EVS 00] EVSUKOFF A., GENTIL S., MONTMAIN J., "Fuzzy reasoning in co-operative supervision systems", *Control Engineering Practice*, vol. 8, no. 4, pp. 389–407, 2000.

[FRA 95] FRANK P.M., "Fault diagnosis in dynamic systems using analytical and knowledge-based redundancy: a survey and some new results", *Automatica*, vol. 26, no. 3, pp. 459–474, 1995.

[GAR 91] GARLAND D.J., "Automated systems: the human factor", NATO ASI series, vol. F73, *Automation and Systems Issues in Air Traffic Control*, pp. 209–215, 1991.

[GEN 97] GENTIL S., RIERA B., LAMBERT M., *et al.*, Filtrage d'alarmes en temps réel, Rapport final Convention DRET Scientifique et Technique, no. 93.34.098.00.470.75.0, France, May 1997.

[GHE 97] GHETIE M.A., Vers le concept de redondance algorithmique pour le diagnostic; le projet DIALOGS, Doctoral Thesis, University Henri Poincaré-Nancy 1, France, 1997.

[GIB 79] GIBSON J.J., *The Ecological Approach to Visual Perception*, Houghton-Mifflin, Boston, MA, 1979.

[GRI 99] GRISLIN E., MILLOT P., "Specifying artificial cooperative agents through a synthesis of several models of cooperation", in HOC J.M., MILLOT P., HOLLNAGEL E., CACIABUE P.C., (eds.), *Proceedings of the 7th European Conference on Cognitive Science Approach to Process Control CSAPC'99*, p. 212, Presses Universitaires de Valenciennes, Valenciennes, 1999.

[GUI 03a] GUIOST B., DEBERNARD S., POULAIN T., *et al.*, "Evaluation of human-human cooperation to design a support tool for air-traffic controller", *22nd European Conference on Human Decision Making and Manual Control, Proceedings of EAM 2003*, University of Linköping, Sweden, pp. 167–174, 2–4 June 2003.

[GUI 03b] GUIOST B., DEBERNARD S., MILLOT P., "Definition of a common work space", *10th International Conference of Human–Computer Interaction, Proceedings of HCI International 2003*, Crete, Greece, vol. 3, pp. 442–446, 2003.

[GUI 04a] GUIOST B., DEBERNARD S., POULAIN T., *et al.*, "Evaluation of a new ATC support system based on human machine task delegation", *IFAC – HMS 2004*, Atlanta, GA, 7–9 September 2004.

[GUI 04b] GUIOST B., DEBERNARD S., POULAIN T., *et al.*, "Supporting air-traffic controllers by delegating tasks", *Proceedings IEEE-SMC*, La Haye, The Netherlands, pp. 164–169, September 2004.

[HAR 88] HART S.G., STAVELAND L.E., "Development of NASA-TLX (task load index): results of empirical and theoretical research", in HANCOCK P.A., MESHKATI N. (eds.), *Human Mental Workload*, Elsevier Science Publishers, Amsterdam, pp. 139–183, 1988

[HOC 96] HOC J.-M., *Supervision et contrôle de processus: la cognition en situation dynamique*, Presses Universitaires de Grenoble, Grenoble, 1996.

[HOC 98] HOC J.-M., LEMOINE M -P., "Cognitive evaluation of human-human and human-machine cooperation modes in air traffic control", *The International Journal of Aviation Psychology*, vol. 8, no. 1, pp. 1–32, 1998.

[HOC 99] HOC J.-M., MILLOT P. (ed.), "Human-Machine Cooperation", *Numéro Spécial de la Revue Le Travail Humain*, vol. 62, no. 3, p. 110, 1999.

[HOC 00] HOC J.-M., "From human-machines interaction to human-machine cooperation", *Ergonomics*, vol. 43, no. 7, pp. 833–843, 2000.

[HOC 01] HOC J.-M., "Towards a cognitive approach to human-machine cooperation in dynamic situations", *International Journal of Human-Computer Studies*, vol. 54, pp. 509–540, 2001.

[HOC 02] HOC J.-M., DEBERNARD S., *Respective Demands of Task and Function Allocation on Human-Machine Co-operation Design : A Psychological Approach*, Taylor & Francis, London, vol. 14, no. 4, pp. 283–295, 2002.

[HOL 00] HOLLNAGEL E., "Principles for modeling function allocation", *International Journal of Human-Computer Studies*, vol. 52, pp. 253–265, 2000.

[JON 97] JONES P., JASEK C., "Intelligent support for activity management (ISAM): an architecture to support distributed supervisory control", *IEEE Systems, Man and Cybertenics, Part A: Systems and Humans*, vol. 27, no. 3, 1997.

[JOU 01] JOUGLET D., MILLOT P., "Performance improvement of technical diagnosis provided by human-machine cooperation", *Proceedings of the IFAC/IFIP/IFORS/IEA Conference Analysis Design and Evaluation of Human Machine Systems*, Kassel, Germany, pp. 379–384, September 2001.

[KAM 89] KAMOUN A., DEBERNARD S., MILLOT P., "Comparison between two dynamic task allocations", *2nd European meeting on Cognitive Science Approaches to Process Control CSAPC 89*, Siena, Italy, October 1989.

[KAR 95] KARSENTY L., BREZILLON P.J., "Cooperative problem solving and explanation", *Expert Systems with Applications*, vol. 8, no. 4, pp. 445–462, 1995.

[KAR 00] KARSENTY L., "Cooperative work: the role of explanation in creating a shared problem representation", *Le Travail Humain*, vol. 63, no. 4, pp. 289–309, 2000.

[LAM 98a] LAMBERT M., RIERA B., MARTEL G., "Design of a new supervision system involving new man-machine interactions", *Proceedings of the IEEE SMC IMACS Conference on Computational Engineering in Systems Applications, CESA 98*, Hammamet, Tunisia, April 1998.

[LAM 98b] LAMBERT M., RIERA B., MARTEL, G., "Analysis, design and evaluation of a new supervision system for the control of a nuclear fuel reprocessing system", *Proceedings of the Seventh IFAC/IFIP/IFORS/IEA Symposium on Analysis, Design and Evaluation of Man-Machine Systems*, Kyoto, Japan, pp. 455–460, September 1998.

[LAM 98c] LAMBERT M., RIERA B., MARTEL G., "Design of a new human-machine system for the supervision of a highly automated continuous process", *Proceedings of the IEEE International Conference on Systems, Man and Cybernetics*, vol. 2, San Diego, CA, pp. 1290–1295, October 1998.

[LEM 96] LEMOINE M.-P., DEBERNARD S., CREVITS I., *et al.*, "Cooperation between humans and machines: first results of an experimentation of a multi-level cooperative organisation in air traffic control", *Computer Supported Cooperative Work: The journal of Collaborative Computing*, vol. 5, no. 2–3, pp. 299–321, 1996.

[LEM 98] LEMOINE M.P., Coopération Hommes-Machines dans les procédés complexes: modèles techniques et cognitifs pour le Contrôle de Trafic Aérien, Thèse de doctorat, University of Valenciennes et Hainaut-Cambrésis, France, 1998.

[LER 96] LEROUX M., "Méthodologie de spécification d'interface guidée par un modèle cognitif : application à la spécification des futures aides à la décision pour les contrôleurs aériens", *8e Journées d'Interaction Homme-Machine*, Grenoble, France, 1996.

[LEY 91] LEYVAL L., Raisonnement causal pour la simulation de procédés industriels continus, Doctoral Thesis, University of Grenoble, France, 1991.

[LIN 11a] LIND M., "An introduction to multilevel flow modeling", *International Journal of Nuclear Safety and Simulation*, vol. 2, no. 1, March 2011.

[LIN 11b] LIND M., "Control functions in MFM: basic principles", *International Journal of Nuclear Safety and Simulation*, vol. 2, no. 2, June 2011.

[MAZ 96] MAZZOUR E.H., HERAUD N., ALFONSI M., "Utilisation des relations de redondance analytique pour la tolérance de défaillances de capteurs", *RAIRO-APII, JESA*, vol. 30, no. 9, pp. 1273–1288, 1996.

[MIL 88a] MILLOT P., *Supervision des procédés automatisés et ergonomie*, Hermès, Paris, 1988.

[MIL 88b] MILLOT P., KAMOUN A., "An implicit method for dynamic task allocation between man and computer in supervision posts of automated processes", *Proceedings of IFAC/IFIP/IEA/IFORS Conference on Analysis Design and Evaluation of Man Machine Systems*, Oulu, Finland, pp. 77–82, June 1988.

[MIL 97] MILLOT P., HOC J.-M., "Human-machine cooperation: metaphor or possible reality?", *ECCS97, Workshop on Human-Machine Cooperation*, Manchester, UK, April 1997.

[MIL 98] MILLOT P., "Concepts and limits for human machine cooperation", *IEEE SMC CESA 98, Computational Engineering in System Application Conference*, Hammamet, Tunisia, April 1998.

[MUN 01a] MUNDUTEGUY C., Reconnaissance d'intention et prédiction d'action pour la gestion des interactions en environnement dynamique, Doctoral Thesis, Conservatoire national des arts et métiers, Paris, July 2001.

[MUN 01b] MUNDUTEGUY C., "Des représentations singulières d'une situation d'interaction aux prédictions d'action convergentes", *Actes des journées en psychologie ergonomique, EPIQUE 2001*, Nantes, 29–30, October 2001.

[PAC 02] PACAUX-LEMOINE M.-P., DEBERNARD S., "Common work space for human-machine cooperation in air traffic control", *Control Engineering Practice*, vol. 10, no. 5, pp. 571–576, 2002.

[PAC 11] PACAUX-LEMOINE M.-P., DEBERNARD S., GODIN A., *et al.*, "Levels of automation and human-machine cooperation: application to human-robot interaction", *18th IFAC World Congress*, Milan, Italy, August 2011.

[PAR 00] PARASURAMAN R., SHERIDAN T.B, WICKENS C.D., "Model for types and levels of human interaction with automation", *IEEE Transactions on Systems, Man and Cybernetics Part A: Systems and Humans*, vol. 30, vol. 3, pp. 286–297, 2000.

[PIA 65] PIAGET J., *Etudes sociologiques*, Droz, Geneva, 1965.

[RAS 83] RASMUSSEN J., "Skill, rules, and knowledge; signals, signs and symbols, and other distinctions in human performance models", *IEEE Transaction on Systems, Man and Cybernetics*, vol. 15, no. 3, pp. 257–266, 1983.

[RAS 85] RASMUSSEN J., "The role of hierarchical knowledge representation in decision making and system management", *IEEE Transaction on Systems, Man and Cybernetics*, vol. 15, no. 2, 1985.

[RIE 01a] RIERA B., Contribution à la conception d'outils de supervision adaptés à l'homme, Habilitation à diriger des recherches, University of Valenciennes et du Hainaut-Cambrésis, 2001.

[RIE 01b] RIERA B., "Specifications, design and evaluation of an advanced human adapted supervisory system", *Cognition, Technology and Work*, vol. 3, pp. 53–65, 2001.

[ROU 83] ROUSE W.B., "Models of human problem solving: detection, diagnosis and compensation for systems failure", *Automatica*, vol. 19, no. 6, 1983.

[ROY 94] ROYER V., "Partage de croyances : condition nécessaire pour un système coopératif?", in PAVARD B., (ed.), *Systèmes Coopératifs de la modélisation à la conception*, Octarès, Toulouse, pp. 253–270, 1994.

[SCH 91] SCHMIDT K., "Cooperative work: a conceptual framewok", in RASMUSSEN J., BREHMER B., LEPLAT J. (ed.), *Distributed Decision-making : Cognitive Models for Cooperative Work*, John Wiley and Sons, Chichester, pp. 75–110, 1991.

[SMI 97] SMITH P.J., MCCOY E., LAYTON C., "Brittleness in the design of cooperative problem solving-systems: the effects on user performance", *IEEE Transactions on Systems, Man and Cybernetics. Part A: Systems and Humans*, vol. 27, pp. 360–371, 1997.

[TAH 05] TAHBOUB K.A., "A novel human-machine interaction architecture-intention recognition approach", *16th IFAC World Congress*, Prague, Czech Republic, 4–8 July 2005.

[VAN 94] VANDERHAEGEN F., CREVITS I., DEBERNARD S., *et al.*, "Human-machine cooperation: toward and activity regulation assistance for different air-traffic control levels", *International Journal of Human-Computer Interaction*, vol. 6, pp. 65–104, 1994.

[VIC 90] VICENTE K.J., RASMUSSEN J., "The ecology of human-machine systems: mediating 'direct perception' in complex works domains", *Ecological Psychology*, vol. 2, pp. 207–250, 1990.

[VIC 92] VICENTE K.J., RASMUSSEN J., "Ecological interface design: theoretical foundations", *IEEE Transactions on Systems, Man and Cybernetics, SMC 22*, vol. 4, pp. 589–606, 1992.

[VIC 95] VICENTE K.J., CHRISTOFFERSEN K., PEREKLITA A., "Supporting operator problem solving through ecological interface design", *IEEE Transactions on Systems, Man and Cybernetics*, vol. 25, no. 4, pp. 529–545, 1995.

[WAN 98] WANGERMANN J.P., STENGEL R.P., "Principled negotiation between intelligent agents: a model for air traffic management", *Artificial Intelligence in Engineering*, vol. 12, no. 3, pp. 177–187, July 1998.

[WAN 02] WANG H., LIAO S., LEJIAN LIAO L., "Modeling constraint-based negotiating agents", *Decision Support Systems*, vol. 33, no. 2, pp. 201–217, June 2002.

[ZOU 98] ZOUINAR M., "Modélisation des processus de constitution du contexte partagé pour l'analyse et la conception des systèmes coopératifs", *Actes de la conférence Ingénierie des connaissances 98 (IC'98)*, Pont-à-Mousson, France, 13–15 May 1998.

9

Human–Machine Cooperation and Situation Awareness

9.1. Introduction

In critical systems, decision-making calls upon the cognitive functions necessary for the supervision, management and recovery of faults. Human operators involved in the operation and management of this type of system can have two types of role: a positive role because of their motivation and capacity for dealing with imprecise and unexpected situations, as well as a negative role due to their propensity to commit errors. From the negative point of view, the human is a problem and designers attempt to control it, sometimes by defining procedures to rigorously follow in order to avoid hazardous behavior [ZHA 04]. This constitutes a certain form of automation of the human operator.

From the positive point of view, humans must remain free in their behavior and capable of coming up with new solutions when they are faced with problems unknown to them. Naturally, they can commit errors; however, they are also capable of detecting and correcting them and especially learning from their errors. Due to this, the designer of the human–machine system is faced with a dilemma: how to combine the two roles, "a completely predictable *proceduralized* behavior" and an "innovative behavior where the human is aware of the situation" and deals with unknown situations.

Chapter written by Patrick MILLOT and Marie-Pierre PACAUX-LEMOINE.

Placing humans into an organization that allows for the improvement of situation awareness (SA) seems to be a promising approach to respond to this dilemma. We have defined SA in Chapter 1 in reference to Endsley, as the capacity of a human for developing an internal representation of the current situation, of its environment, and for predicting the probable future states of this environment [END 95a]. SA is defined by three components: the perception of events in space time (SA1), the understanding of their significance (SA2) and the projection of their state in the near future (SA3).

Since human activities are generally developed in groups, current research is naturally oriented toward the understanding and modeling of collective SA. Furthermore, it involves understanding the collective management of SA to promote it among a group of humans and artificial agents [SAL 95, SAL 08, SHU 05].

At the same time, works on cooperative organizations including cognitive agents, humans or machines have been developed in the last two decades and evaluated in different fields: air traffic control, the supervision of large systems such as nuclear power plants, telecommunication networks [MIL 11] or even the management of reconnaissance robots, or cooperation inside the cockpit of a fighter aircraft [PAC 11]. These works have shown an increase in the capacity to solve problems and a decrease in the workload (WL) when the tasks are carried out by cooperative teams. We deduce in this final chapter that *collective SA* can be improved by such a cooperative organization. We firstly show the similarity between human cooperation and collective SA, and then show how the concept of a *common work space (CWS)*, which helps to mediate the cooperation between agents, can also serve to improve collective SA.

9.2. Collective situation awareness

This research topic is in full expansion; several studies are being carried out in distributed systems as diverse as road traffic, the development of troops or air traffic. New ideas are emerging for evaluating and maintaining the level of human SA through adaptive and cooperative automation. We use the term *collective SA* to designate the different variants that we find in the literature related to the SA acquired through several humans or artificial agents:

– Team-SA: Salas and his colleagues suggest as a definition "an understanding of the situation shared between the members of a team"

[SAL 95]; Shu and Furuta [SHU 05] rely on the model of Endsley [END 95a] to propose a combination of *individual* SA and *mutual* SA (SA that the cooperative entity has built using the activities of others); and Salmon and his colleagues [SAL 08] define *team*-SA as the combination of the SA of members of the team (*team member* SA), its *shared*-SA and the *combined*-SA of the whole team.

– *Shared*-SA: according to [END 95a], this refers to the level of recovery of SA elements common to the members of a team and the SA required by each member to fulfill their responsibilities.

– *Distributed*-SA: Stanton and his colleagues [STA 06] suggest that the knowledge relative to SA is distributed between the human and artificial agents; the property, the use and sharing of this knowledge are dynamic and depend on the task and its associated goals. Also, the agents could have different levels of SA for the same situation; these levels of SA can overlap and become complementary such that the SA deficiencies of one agent can be compensated by the SA of another. Each agent therefore plays an important part in the development and sustainability of the SA of other agents.

Salmon *et al.* [SAL 08] conclude that teams need to share data to a high degree of SA, such as the significance of SA, the goal elements and the projected states. The performance of the team therefore depends on shared goals, the interdependency of the actions of the members of the team and the distribution of work between them (see Figure 9.1).

Similar to the definitions of SA previously mentioned in Chapter 1, the lack of consistency in the definitions of team-SA reflects a lack of convergence of the foundations of the models of the phenomenon. Nevertheless, this conveys the importance of the phenomenon. This recalls the rapidly growing appearance of definitions of WL and related evaluation methods that dominated the literature in the 1980s. However, Salmon's team-SA model seems very similar to the human–machine cooperation model that we developed in human engineering [MIL 11, PAC 11] and presented in Chapters 7, 8 and 9 of this book. We will therefore analyze the pertinence of transposing this model for team-SA.

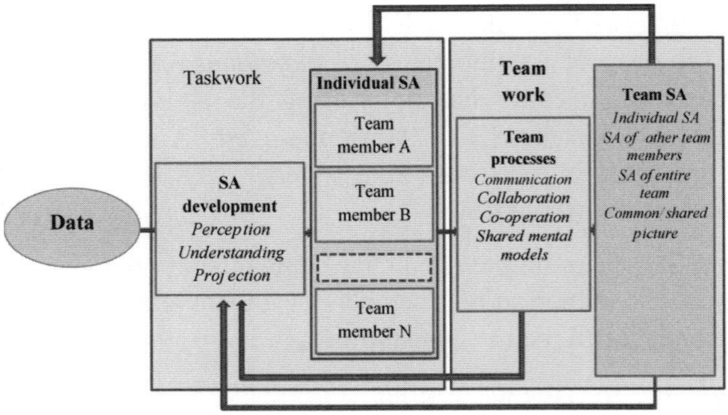

Figure 9.1. *Model of team-SA adapted from Salmon et al. [SAL 08]*

9.3. Structural approaches of human–machine cooperation

The theoretical concepts of human–machine cooperation have resulted in a collection of studies and exchanges between multidisciplinary researchers conducted over more than 20 years in different fields of application. The presentation that we give in this chapter naturally borrows concepts from Chapters 7 and 8 in order to put it within the context of the human-centered design steps.

Historically, everything started with the idea of bringing flexible support to the human operator, in other words providing them with this support whenever they feel they need it and then ceasing this support when the need is no longer there. This idea for responding to the "Bainbridge[1] syndrome" aimed for an adaptive level of automation, with "additional automation during overloaded situations" and less automation during less loaded situations, thus allowing for the human to remain present and active within the loop. A first approach consists of a *dynamic allocation of tasks* – and therefore of the *authority* – between the human and the support tool [MIL 88a, MIL 88b]. A second approach suggested supporting the operator through a tool that provides advice but with the tool connected according to a hierarchical structure, in other words it had no authority [MIL 89].

1 Lisanne Bainbridge suggested that an increase in the level of automation would lead to a decrease in vigilance and human expertise and therefore, the capacities of the operator to efficiently solve a problem, thus paradoxically weakening the performance of the overall system rather than improving it.

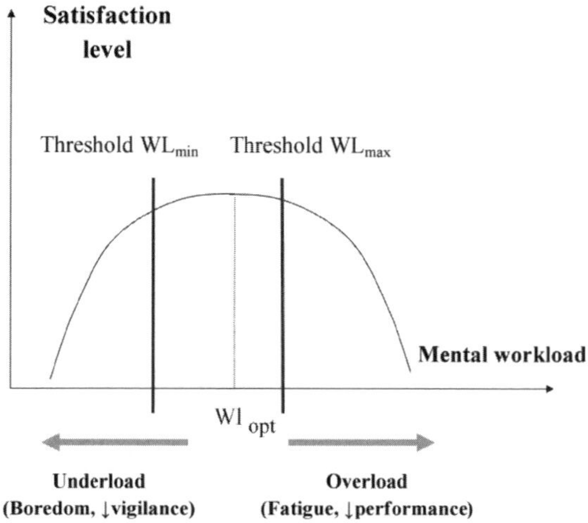

Figure 9.2. *Range of control of the mental workload (WL) [MIL 88a]*

9.3.1. *Dynamic allocation of tasks: horizontal cooperation structure*

The first approach toward showing the feasibility[2] of a dynamic allocation of tasks between a single human operator and a single agent machine was based on optimal control theory in order to optimize the global performances of the supervised process, while regulating the human WL [MIL 88a, MIL 88b]. The principle consists in iteratively allocating additional tasks to one of the two decision-makers – human or machine – that contributes toward the improvement of the performance of the process, until one of the human WL limits (WL_{min} or WL_{max}) is reached (see Figure 9.2). In other words, if the human is more "efficient" than the machine, the number of his tasks is iteratively increased, but without surpassing the WL_{max} threshold. Conversely, if the machine is more efficient, additional tasks are allocated to it until the WL_{min} human under load limit is reached. The structure of the system is represented in Figure 9.3. Two questions then arise:

– How does this dynamic task allocator between the two decision-makers work? Is it human or is it an algorithm? When the task allocator is controlled by the human operator according to their own estimation of their WL, we refer to *implicit allocation*. The algorithm is therefore of the optimal control

2 The idea was proposed by Greenstein [GRE 81], however, without experimentally showing its feasibility.

type; in other words, the *sharable tasks* can be completed interchangeably by the operator and the machine (see also Chapter 1 of this book).

– If it is an algorithm, how does it assess the WL of the human decision-maker? The WL is assessed in real time independently of the operator, according to a model that calculates the ratio between the required execution time and the available time for completing the task and has been experimentally validated using a modified Cooper–Harper scale [MIL 88a, MIL 88b] (see also Chapter 1).

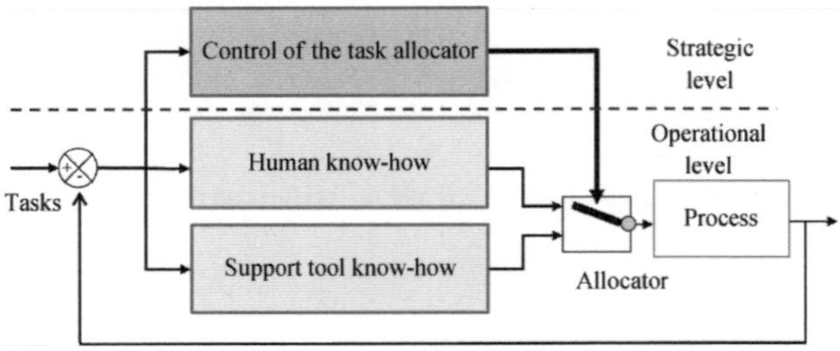

Figure 9.3. *Structure of the human–machine dynamic allocation of tasks*

The allocation method has been tested on a laboratory simulator involving 96 independent variables that are randomly disturbed. The performance is evaluated as the sum of absolute errors of the disturbed variables. The variables are allocated between a human operator and a second "decision-maker". The human must detect the disturbed variables and compensate for them by modifying the set-values. The second decision-maker consists of several proportional integral derivative (PID) regulators, which play the same role. Implicit and explicit dynamic allocations between human and PID were carried out with 10 different human operators and the results have shown that there exists an allocation of the variables to be corrected between the two which optimizes the performance of the process, under the WL_{min} and WL_{max} constraints. The human WL is regulated to be between these thresholds.

9.3.2. *Vertical structure for cooperation*

In a *vertical structure* – or even hierarchical – the operator is accountable for all of the variables of the process and he must, if necessary, call upon the decision-making support tool that will provide advice (see Figure 9.4).

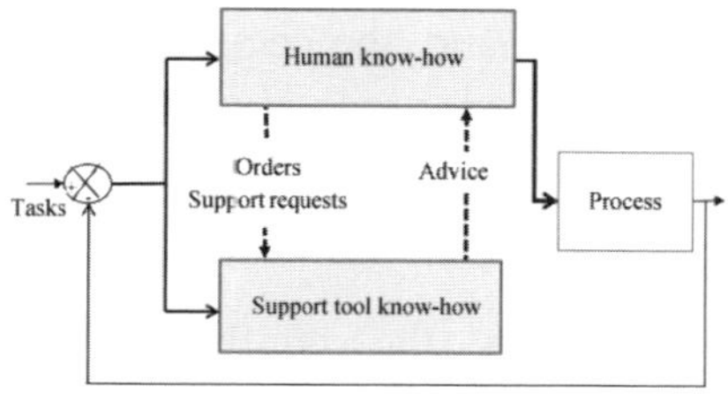

Figure 9.4. *Vertical structure for cooperation*

This second principle has been developed to insert a support system providing the human with diagnosis and action advice in a supervision post of a simulated nuclear process [MIL 89]. When a complex problem is being solved, due to a process malfunction, the human operator can have three types of reaction when faced with the action advice proposed by the support system:

1) He does not have a solution and will therefore be tempted to apply the one proposed by the system.

2) His own solution coincides with that of the support system and he is therefore comfortable with his own decision.

3) His own solution differs from that of the support system, and in this case, a conflict appears between the two decision-makers.

The design steps that we have defined therefore consist in letting the operator of this role resolve the conflict between him and the support system. We have therefore analyzed, during the design of the support system – and its interfaces – what could have been the steps toward the resolution of the conflict led by the operator, and deduced which contributions the support system can provide to aid the operator and, due to this, the corresponding human–machine interfaces.

The method consists in keeping the decisional approaches of each decision-maker parallel (see Figure 9.5), simulating the potential conflicts the operator could have and providing them with preventative measures. The steps of the operator are modeled by the Rasmussen ladder [RAS 83] in four stages: detection of events, situation evaluation (observation of information and identification of the system state), decision-making (definition of a strategy and of the tasks) and action. The reasoning of the support system is supposed to be decomposable into three stages; the same as those for the human model, but on the one hand, the decision-making stage is reduced to the choice of a pre-established process in a base of rules (rule-based behavior) and on the other hand, the action is absent since it is carried out by the operator.

The operator only detects the conflict after the work of each decision-maker has been completed. Due to this, the consensus must be found by "back-tracking" through the resolution steps of the problem. For example, different action processes can come either from different diagnoses, or from a diagnosis having led to several possible solutions. The diagnosis is therefore a possible point of consensus. The support system must be able to provide its diagnosis to the operator for him to be able to manage the conflict. Furthermore, different diagnoses can originate from an omission of certain variables by one of the decision-makers during the evaluation of the situation. Because of this, the set of considered variables is a second possible point of consensus which the system can help to find by displaying the data used to make its diagnosis.

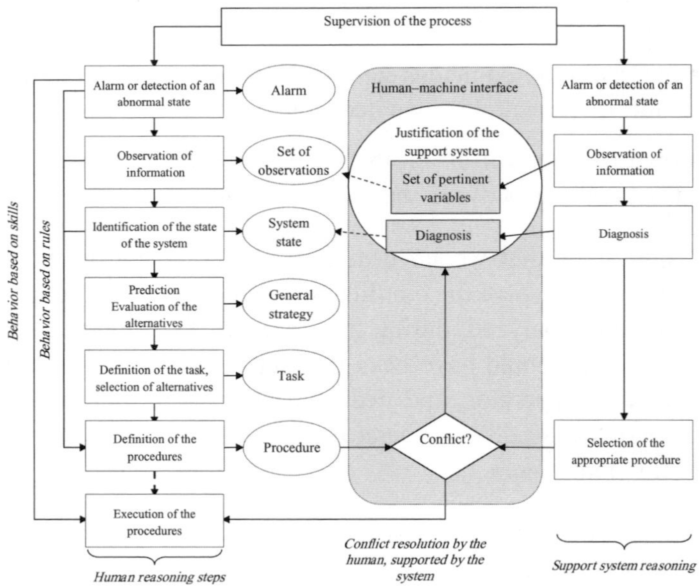

Figure 9.5. *Overview of the steps for the resolution of decisional conflicts between the operator and the support system*

The interface of the support system has therefore been created to provide the operator on demand with, on the one hand, the diagnosis carried out by the support system and, on the other hand, the variables that the latter has considered in its reasoning. This information will be used to facilitate the steps for the resolution of conflicts by the operator.

9.3.3. *Multilevel structure for the dynamic allocation of tasks*

In order to understand the two *implicit* and *explicit* dynamic allocation principles in more depth, a wider study has been carried out in the field of air traffic control, using a realistic simulator with professional controllers in collaboration with the CENA (French acronym for Center for the Study of Air Navigation[3]). The studies of the dynamic allocation of air traffic control tasks have been outlined in Chapter 8. We provide only a quick summary here to allow the reader to understand the additional interpretations that we will make.

The aims of "en route" air traffic control consist of supervising and controlling the traffic in order to avoid collisions between airplanes. The air space is divided into geographical sectors controlled by two controllers. A *radar controller* (RC) has a tactical role and supervises the traffic from a radar screen and communicates with the pilots. The main role of the RC is to preventively detect aerial conflicts when the airplanes run the risk of crossing each other at a distance less than the minimal separation standards that could lead to a collision. The RC must also resolve the conflict by ordering the pilot of one of the planes to change their trajectory. The second controller has a strategic role and is referred to as the *planning controller* (PC). The PC coordinates the transfers of airplanes coming from adjacent sectors or going out toward them. The PC also anticipates the density of the traffic and regulates the WL of the RC. The PC can also directly assist the RC in cases of overloaded traffic by executing certain tactical tasks.

In a first series of experiments[4] carried out by professional air traffic controllers, the two principles of implicit and explicit task allocation have been implemented between the RC and an automatic aerial conflict detection

3 Formed from DTIA "Direction technique et industrielle de l'aéronautique" belonging to the DGAC "Direction générale de l'aviation civile".

4 This initial study is known as SPECTRA V1, *Système de partage expérimental des tâches de contrôle de trafic aérien version 1*, (Experimental system for task allocation in air traffic control).

and resolution system referred to as SAINTEX[5]. The allocation of aerial conflicts is based on the respective know-how (KH) of the RC and SAINTEX in order to regulate the WL of the RC. For SAINTEX, the KH is limited to conflicts between the two simplest airplanes. For the RC, this KH is not limited, but its capacities are limited by his WL. The experiments have shown better performance of the traffic (safety and overall fuel consumption) as well as better regulation of the WL of the RC in the implicit allocation mode [DEB 92, MIL 93, VAN 94]. Nevertheless, dedicated questionnaires submitted to the RCs at the end of the experiment have revealed a *refusal* on their part to accept this type of organization in a real-life situation, despite the improvement in performance and the safety levels achieved. The main reasons for this refusal lie with the loss of control by the RC of the tasks allocated to SAINTEX in implicit mode, for which the human RCs nevertheless remain legally accountable.

Due to this, in a second set of experiments[6], we have introduced the PC at the strategic level to manage the allocation of tasks between the RC and SAINTEX. This is a multilevel organization in which the RC at the tactical level can negotiate and cooperate with the PC at the strategic level, in order to resolve (or prevent) allocation conflicts between the RC and SAINTEX. The PC was helped to this effect by a planning support module referred to as planning of allocations (PLAF) and which was integrated according to a hierarchical structure [CRE 94, MIL 95] (see Figure 9.6). From a structural point of view, the dynamic task allocation at the operational level (horizontal CR-SAINTEX structure) was controlled by the task allocator commanded by the PC at the strategic level according to a vertical PC–PLAF cooperation.

Another mode of cooperation between the PC and PLAF referred to as "assisted explicit mode" has also been introduced, according to which PLAF proposed a pre-allocation of conflicts to either the RC or the SAINTEX, with the possibility of modification by the PC, thus rejoining level 5 or 6 of Sheridan's levels of automation (see Chapter 1).

The results from the experiments have therefore shown good acceptance levels for the organization by the controllers while maintaining a good performance [HOC 98, LEM 96]. Another important result is that the success of dynamic allocation can depend not only on technical criteria, such as

5 The SAINTEX name is chosen in reference to the famous French aviator and poet Antoine de Saint-Exupery author of "Le petit prince".
6 This second study is referred to as SPECTRA V2.

performance and safety, but also on human criteria related to organization such as the balance between *authority* and *responsibility*[7] as well as *legal and social acceptance* as we have shown in Chapter 1. In this second study, authority has been given to the holders of responsibility.

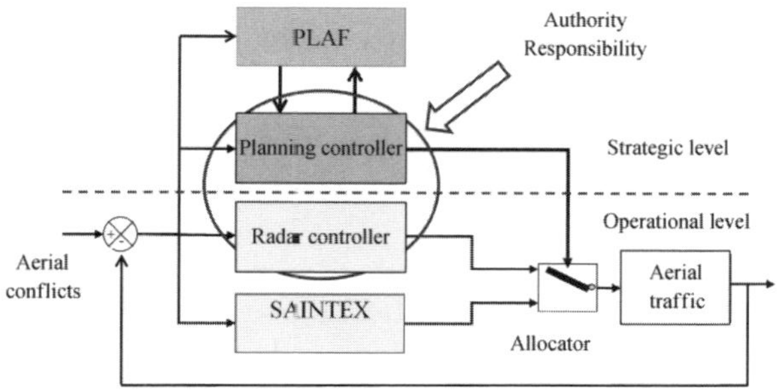

Figure 9.6. *Multilevel organization for the dynamic allocation of tasks*

It is undeniable that these studies have formed the first steps toward a human–machine cooperation in the sense that they have surpassed the purely technical point of view of the control engineer to highlight typically human *decisional conflict* problems with the organization or the machine. These conflicts are linked to the roles of humans in the system and result in *automation levels* and the trading *of authority* and *responsibility* with the machine. This structural approach has provided structural answers to cooperation questions. However, we have quickly noticed that these structural answers were incomplete and must be complemented by functional approaches to the cooperation principles and the respective qualities that are asked of humans and machines to this effect. They therefore also led to the definition of move advanced theoretical foundations that we now look at.

9.4. Human–machine cooperation: a functional approach

9.4.1. *Cooperative agents, forms of cooperation*

The generic term "agent" is used to designate humans as well as machines. Each *cooperative agent* shows two attributes: a *know-how* (KH) with its capacities for controlling the process or executing the task (problem

7 Translator's note: here we do not make the distinction between the terms *responsibility*, which refers to organizational aspects, and *accountability*, which is more related to legal issues.

solving, perception of information, decision-making, actions, etc.) and a *know-how-to-cooperate* (KHC) allowing it to cooperate with other agents involved in the task. The KHC is an important concept stemming from the definition given by Millot and Hoc [MIL 97] and adapted from Piaget [PIA 77]: "two agents cooperate if they respect two conditions: 1) each pursues its goals while managing interferences with the goals, resources and the processes of the other agent; and 2) each attempts to manage interferences such that it facilitates the individual activities of the other and/or the common task where it exists. The symmetry of this definition can only be partly respected".

KHC groups two classes of capacities: those related to the *management of interferences* (MI, detection, creation and resolution of interferences) between the goals of agents – which can be seen as activities of *coordination* [MIL 98a] – and those that help to *facilitate the goals* (FG) of other agents (*coordination, FG*). Interferences can be positive (facilitators) for example, in mutual control where an agent mentally carries out the tasks of the other in order to allow for a different approach and thus improving the performance. It can also be negative, for example during an access conflict at an office printer.

A first question involves the methods for determining the makeup of a cooperative team and its operation during the execution of a task.

A model developed by Millot [MIL 98a] and intended to formalize the foundations for the construction of the KHC of an agent outlines the three steps followed by agents.

1) Before cooperation, they carry out two activities: they coordinate their respective KHs relating to the task and prepare the cooperation by deciding a common – or shared – plan to achieve the goals. To this effect, they decompose and share – if possible – the task to be executed in a cooperative manner, according to their respective KHs and their availability (WL).

2) They work cooperatively on the same task or on their respective sub-tasks while resolving possible conflicts (coordination or management of interferences; we will see the importance of the forms of cooperation later).

3) They must group the partial solutions relative to the sub-tasks if they exist. These three steps must also be seen as three episodes for constructing the cooperative team and then executing the task in cooperation.

Another, more functional point of view has been proposed by Hoc [HOC 01] and includes three stages (see also Chapter 7 of this book). At the highest level, referred to as *meta-cooperation*, the agent constructs the cooperative team he/it wants to establish according to multiple models: the model of the task, the model of the agents with a KH related to the task (and the performance he/it expects from other agents) and the model of himself/itself as well as communication codes. The second level is referred to as *cooperation in planning* where the agents develop (while maintaining) a COmmon Frame Of Reference, denoted as COFOR, which is the common part of their respective *occurring representations of the situation*. The COFOR allows them to construct common or complementary plans to solve problems. The third level is referred to as *cooperation in action*, which consists of detecting and managing interferences during the execution of the task. Hoc's point of view is more directed toward "mechanisms", whereas the previous one is more useful for defining the architecture of a software agent.

Other cooperation strategies can be chosen depending on the context, for example Woods *et al.* [WOO 04] propose for a hierarchical human–robot organization, a *direction based on a plan* – direction following the intentions of a "head" for which the cooperative goal is to adapt the plan in terms of the disturbances and opportunities – and a *direction based on constraints*.

It is equally important to answer a second question: how will the agents cooperate – operate together – on the task: Will they break it down or share the sub-tasks? Will they collectively execute it and will they compare their respective results, etc. A response given by Schmidt [SCH 91] outlines three *forms of cooperation*:

1) *Augmentative*: when the agents have a similar KH and the task can be broken down into similar and independent sub-tasks.

2) *Debative*: when the agents have a similar KH but the task is not broken down; they each execute it and compare their results; conflicts can then be produced and the KHC must contribute toward resolving them *via* the COFOR which helps to provide explanations

3) *Integrative*: when they have different but complementary KHs and the task can be broken down into different and complementary sub-tasks.

These three forms of cooperation exist in *human organizations* and are incidentally often combined. An example of an *augmentative form* can be observed at the counter of a bank: when the queue is too long, the head of the agency can decide to open a second counter and the line is split into two, thus reducing the load of the first counter. An example of a *debative form* exists in the mutual control established between the pilot and the co-pilot of an airplane. Integrative forms exist in the coordination of the trades intervening in the construction of a house.

Grislin and Millot [GRI 99] have shown that these three forms are generic for all forms of cooperation. In other words, if cooperation exists, it must be under one of these forms, otherwise it does not exist. This is interesting for the creation of the KHC of a general artificial agent (and for human–agent cooperation interfaces), which could therefore group the union of KHCs relative to each form together.

We can summarize the main attributes of an agent in the following way: KH regroups its capacities in carrying out the task and KHC regroups its capacity to cooperate with other agents. We can further break down KH and KHC into two parts each: one external part dedicated to exchanges with the exterior, and an internal part steered toward the processing of information and decision-making: problem solving, modeling of the attributes of other cooperating agents in accordance with the models proposed in section 9.4.1 (see Figure 9.7).

9.4.2. *Organization and cooperation*

With the goal of a concrete implementation, the agents must be integrated into an organization. We therefore rediscover two natural generic structures empirically defined in section 9.3 [MIL 95]:

– In the *vertical structure*, an agent at the highest level has the authority; the second at the lower level is limited to giving advice. Interferences such as conflicts are managed by the KHC, with the superior agent acting as the *coordinator*, the other being *benevolent* and its KH being limited to the capacity of *facilitation of goals* of the other. The agents' KH therefore completes the structure (see Figure 9.8). In the example of the resolution of the decisional conflict between the supervision officer of the nuclear power plant and the diagnosis support tool presented in section 9.3 (see Figure 9.5), the steps established for the preventive resolution of conflicts and equipping the system with the corresponding interfaces from its creation was of a *debative* form. There was no real cooperation since the agent did not have a model of the human and did not anticipate the needs of the human on its own; the operator managed the KH on its own. On the other hand, these steps strongly foreshadowed cooperation;

Know-how (KH)

- **Internal capacity** to solve problems (relative to the process):
  - Competencies (knowledge, rules, skills);
  - Reasoning capacities (inferences, etc).
- **External capacity** to:
  - Perceive information (of the process and the environment);
  - Act (on the process).

Know-how-to-cooperate (KHC)

- **Internal capacity** to:
  - Construct a model of other agents (KH and KHC) and of the COFOR;
  - Deduce the intentions of other agents;
  - Analyze the task, identify the forms and structures of cooperation;
  - Produce a common plan and share it according to the forms of cooperation;
  - Coordinate the other agents;
  - Accepting to respond to the demands of other agents
- **External capacity** to:
  - Communicate with the other agents (codes etc.).

Figure 9.7. *Attributes of a cooperative agent*

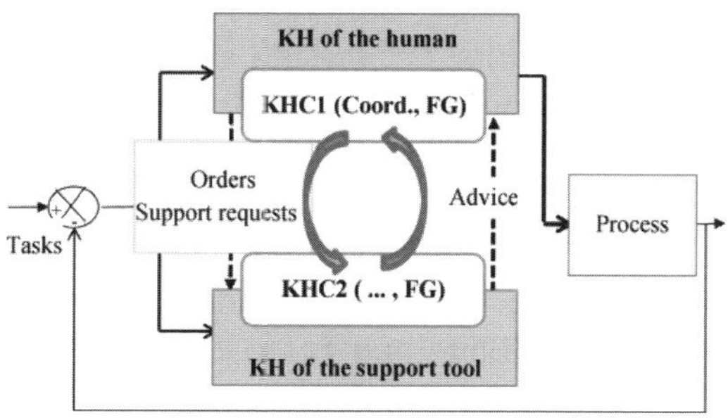

Figure 9.8. *Vertical structure of cooperative agents*

– In the *horizontal* structure, the two agents are at the same hierarchical level (from an operational point of view) and the tasks are shared between them *via* a task allocator located at a higher level (for example, at the strategic level, see Figure 9.9). The *augmentative and integrative* forms are well suited to this structure, as in the example of the bank counters, the task allocator is the head of the agency; whereas in the example of the house, the allocator is the architect.

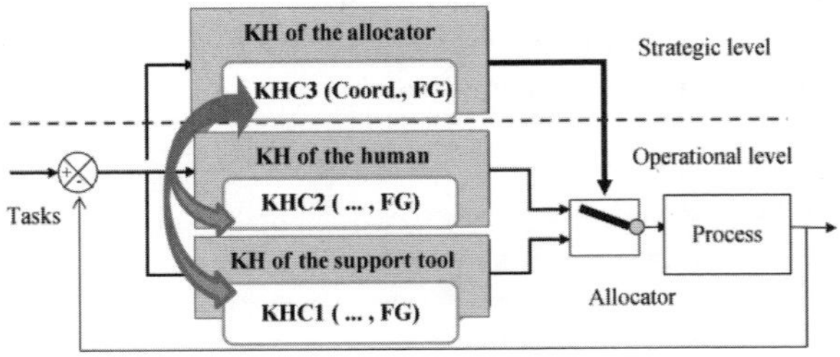

Figure 9.9. *Complete horizontal structure of cooperative agents*

The reader will find more details on the relations between the forms of cooperation and structures in [MIL 98b, MIL 11].

9.4.3. *Human factors activating or inhibiting cooperation*

The KHC is generally asymmetrical between the human and the machine – with the advantage going to the former – due to the difficulty in implementing the coordination capacities within a machine. The reader will find examples of this asymmetry in [MIL 98b].

We should also note another important class of factors influencing cooperation either as activators or, on the contrary, as inhibitors. They concern the way in which the agent, particularly human, here perceives:

1) his difficulties in carrying out the task and thus feels a *need to cooperate* with another agent to accomplish it ;

2) the difficulty he can feel when faced with another relative to *trust* and *self-confidence*, which we will interpret as the *willingness to cooperate.*

The *need to cooperate* appears when the agent feels overloaded and requires assistance. If his WL is below a certain threshold, then the agent does not feel any *need to cooperate* and feels entirely capable of carrying out the task on his own. If the *WL* increases, the agent will feel the *need to cooperate*, regardless of the overload that has spurred this cooperation, and finally, if the *WL* continually increases, the agent can be so overloaded that the execution of the task becomes impossible even if he is assisted and all cooperation is therefore felt as futile (see Figure 9.10).

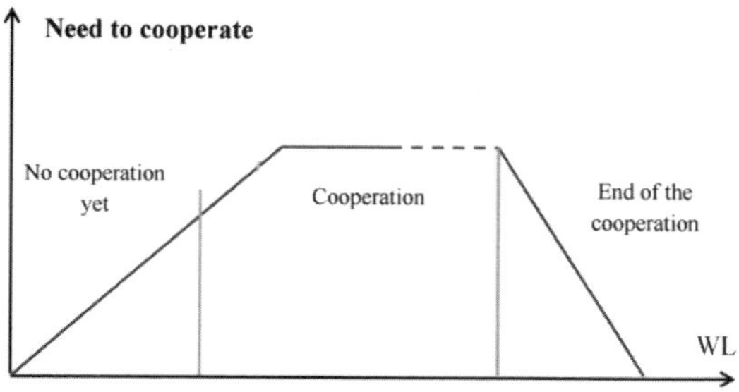

Figure 9.10. *The need to cooperate*

The *willingness to cooperate* of the human agent depends on the trust he has toward his collaborator and his self-confidence. It can be seen as an increasing function of *trust* and a decreasing function of *self-confidence*. Thus, the more the human has trust in his collaborator (human or not), the more he will want to cooperate with him. Conversely, the more self-confidence he has, the less he will want to cooperate. This is consistent with the study of Lee and Moray [LEE 92] of the allocation of functions between an operator and an automatic control system. Furthermore, Rajaonah *et al.* [RAJ 08] have shown that confidence in the interaction also plays a part.

We highlight the great importance of other human factors that influence the relation between the human and the automated agent:

– The position of the agent in a hierarchical organization and the necessity of an allocation of the authority which must be consistent with the completeness of the KH and KHC of the agent [MIL 11]. Moreover, the *allocation of authority* must be consistent with legal factors such as *responsibility* and *accountability* [FLE 12].

– The unacceptable effects of *overconfidence* [INA 13], which can lead to phenomena of *complacency* [BIL 96] or *negligence* [GOO 01] of the human and result in a drop in situation awareness (SA).

9.4.4. *Multilevel cooperative organization*

Activities can be broken down into three levels of abstraction: a *strategic level* corresponding to a global plan, a *tactical level* in which the plan is broken down into tasks and an *operational level* to carry out the actions that make up the tasks. Cooperation can follow the same decomposition, with the agents being able to take a place (and execute activities) on one or multiple levels depending on the completeness of their KH and KHC. As we have seen in Chapter 1, the human is more naturally involved in the tactical and strategic levels, whereas an artificial agent will more likely be placed at an operational level. Cooperation can therefore be established within a level or between levels [HOC 98] (see Figure 9.11). In the example of the SPECTRA V2 project, the organization is made up of two levels with a human–machine cooperation at each level – between the RC and the SAINTEX at the operational level and between the PC and the PLAF at the level above – as well as an inter-level cooperation between the PC and the RC (see Figure 9.6) [CRE 93, LEM 96].

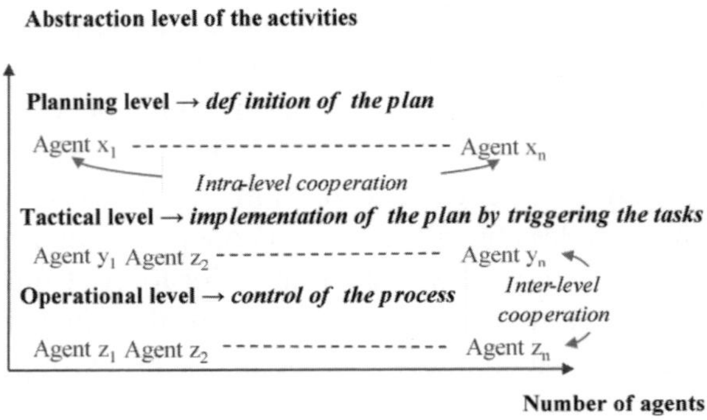

Figure 9.11. *Multilevel cooperation*

9.4.5. *Common work space (CWS)*

We have seen in section 9.4.1 that the agents construct and maintain a COFOR to solve problems and make decisions in a cooperative manner. One way to support cooperation and maintain the COFOR consists in facilitating the perception and understanding of the state of the process as well as the behaviors and the points of view of the agents. We notice a high convergence with Endsley's SA levels SA1 and SA2. Such a support medium is referred to as a CWS, and it has been evaluated in three fields of application: the cockpit of a fighter aircraft, reconnaissance robotics and air traffic control.

9.4.5.1. *Example in the cockpit of a fighter aircraft*

The first study takes place in a two-seater fighter aircraft simulator. The aim is to analyze the cooperative activities between the pilot and the weapons system officer, in order to establish a model of this cooperation.

This model has therefore been used to construct a CWS intended for supporting the cooperation in degraded situations. Consequently, this model has been used as a reference for imagining future cooperation between the pilot and a drone or between the pilot and another operator placed in an AWACS[8].

The construction of the CWS begins with the pilot and the navigator acquainting themselves with the mission file that has been developed by the command center (see Figure 9.12). This file, referred to as the "red file", groups together the set of information collected on the theatre of operations, notably threats, the objectives to be known or to destroy, the radio frequencies to be used and the entrance/exit timeslots into the enemy territory. Each member of the team reads the file to share their knowledge of the situation with their collaborator and to create the flight plan, which becomes the initial CWS (see Figures 9.13 and 9.14).

Figure 9.12. *Two collaborators reading the "red file"*

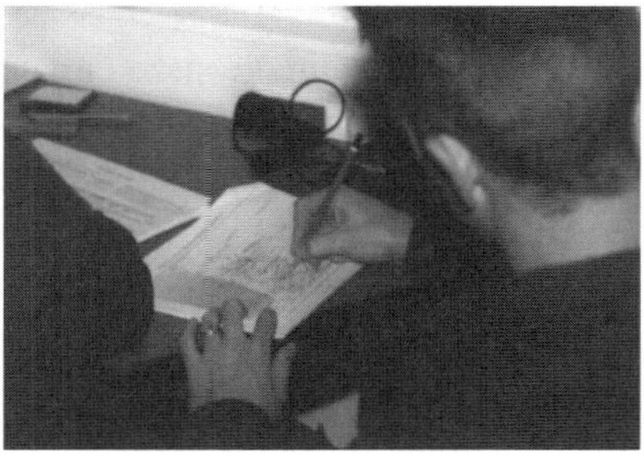

Figure 9.13. *The two collaborators sharing their knowledge of the situation*

8 AWACS: *Airborne warning and control system.*

The CWS is therefore created in reference to the tactical situation, in other words a view of the flight plan completed thereafter with the information pertaining to enemy positions (circles in Figure 9.14) [PAC 02]. Mission preparation software has been created in order to allow the members of the team to take advantage of their annotations on the representation of the tactical situation (SITAC: interface available at the stations of the two operators). The annotations concern individual tasks as well as the cooperative tasks. This software was therefore also an assistance for the planning of cooperation, in the sense that it encourages operators to clarify and adjust their cooperative tasks. This is *cooperation in planning*. An example of an annotation is given in Figure 9.14.

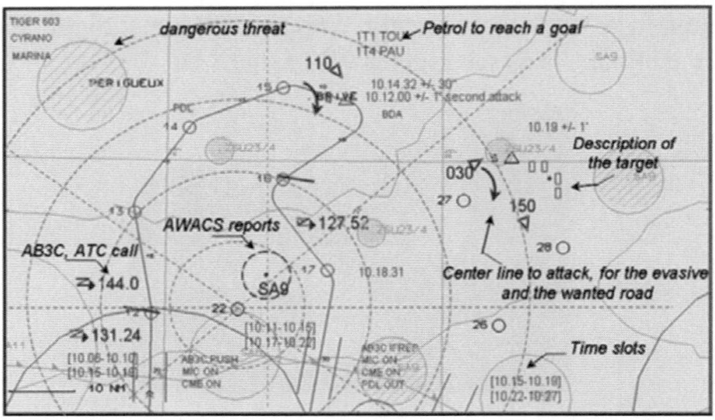

Figure 9.14. *Flight plan resulting from cooperation*

During the mission, the tactical situation is enhanced with information pertaining to the individual and cooperative activities of each operator (see Figure 9.15). The CWS is therefore the plan outlined during the preparation of the mission and then enhanced during the mission execution. At each step, each operator is capable of obtaining information on the current activity or the intentions of the other by interpreting the CWS. The study has further shown that this type of cooperation support tool prevents the operators from mutually interrupting each other, particularly if they are fully invested in their own tasks. A second important result is that the cooperation is maintained within the team even when the operators are no longer in a position to communicate with each other over the radio for safety reasons.

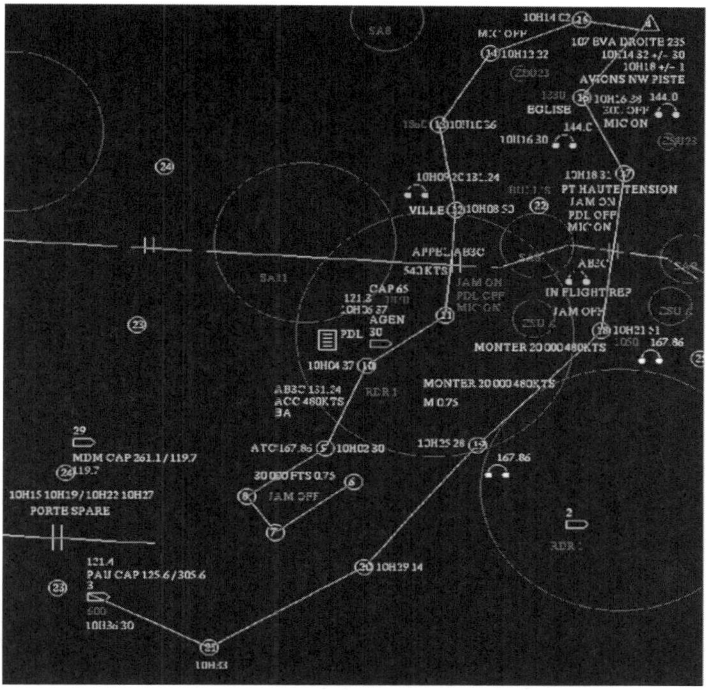

Figure 9.15. *The CWS is the annotated plan acting as a support medium for cooperation*

9.4.5.2. *Example in reconnaissance robotics*

The second study aims to establish a CWS for maintaining the cooperation between a human operator and a partially autonomous reconnaissance robot. During the reconnaissance mission, the operator is positioned at a distance from a target zone and sends the robot toward this target zone [PAC 11]. The robot can move on its own and transmit information to the operator. This allows the operator to dedicate more time to observations impossible for the robot that does not have all of the objects in its database. Four modes of autonomy of the robot have been defined: in M0 mode, the control is completely manual; conversely in M3, the robot chooses and carries out its movement algorithm autonomously. In the modes M1 and M2, the robot must ask for the operator's permission: in M1, it must wait for the response of the operator before moving and in M2, it only waits for a limited amount of time before it can act autonomously if the operator has not given a response.

Operators have been asked to prepare a mission according to the information available to them on the competencies of the robot following a theoretical and practical briefing, and on the theatre of operations based on a file they have been given. They must construct a mission plan according to the reconnaissance objectives and the constraints associated with the terrain, notably obstacles and zones with disruptive communication with the robot due to masking.

The mission plan is described by goals. Each chain of goals is associated with a task, itself associated with a mode of autonomy (from M0 to M3). The sensorimotor behavior of the robot is also described (follow-up of a border of forest, follow-up of a seamark[9]), as well as the methods of perception of the robot for carrying out the reconnaissance. The analysis of mission preparations has highlighted that the operators have a bad representation of the competencies of the robot, notably in terms of the management of communication (KHC).

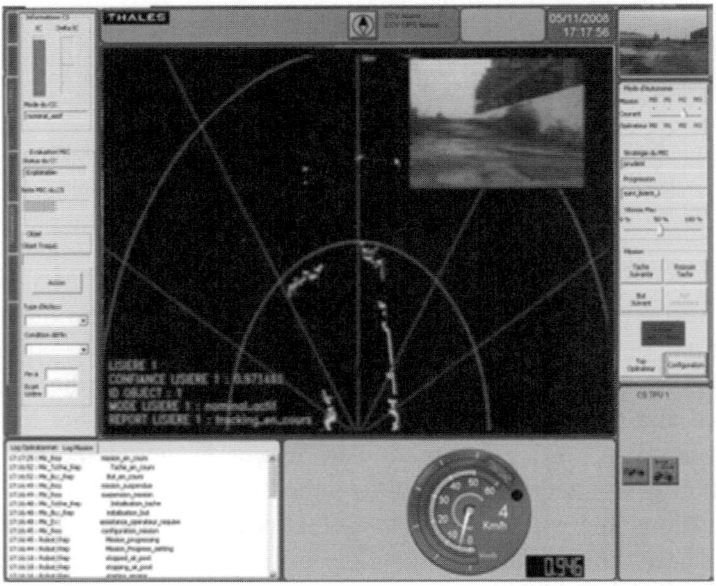

Figure 9.16. *CWS used for cooperation in robot reconnaissance*

9 Translator's note: the French word *amer* denotes a mark that can be seen from everywhere on the sea in order to guide ships. Robotics extends the word to marks that guide robots.

A CWS has been defined to inform the operator of the objectives of the robot (geographical points), the analysis of the information it has carried out and the movement algorithm it has chosen in autonomous mode (see Figure 9.16). This provides information to the operator for the next objective of the plan. The CWS is a screen on which the robot displays the result from its sensors, a video radar view of the scene notably showing the targets and obstacles. The right side recalls the current goals and task as well as the mode of autonomy of the robot. The left side shows a detailed auto-evaluation of the robot in its capacity for mobility. The bottom part gives the history of the individual or cooperative tasks and the speed of the robot. The CWS has greatly facilitated the cooperative operation of the robot [PAC 11].

9.4.5.3. *Example in air traffic control*

The third project in human–machine cooperation in air traffic control follows the SPECTRA projects presented at the start of this chapter and is referred to as AMANDA[10]. The introduction of a CWS for maintaining the cooperation with a support tool for aerial control referred to as STAR[11] has been outlined in detail in Chapter 8 of this book, which we advise the reader to take a look at. We will simply say that the cooperation is established over the cognitive activities of the perception of information, the evaluation of the situation, schematic decision-making, precise decision-making and actions (in this sense comparable to functions rather than tasks) and that the three forms of Schmidt cooperation are possible [DEB 09]. The CWS is implemented as a graphical interface where the problems to be resolved are displayed – aerial conflicts in this case – as well as the evolution of their resolution over time following cooperation [DEB 10] (see Figure 9.17). The conclusions of the study outlined in Chapter 8 clearly indicate that AMANDA favors the anticipation of problems by controllers and that the CWS has allowed the PC and RC controllers to share a COFOR with STAR, which favors the maintenance of their SA and has provided them with an efficient cooperation medium between them. We are therefore tempted to conclude that after this study, the CWS favors not only the SA1 and SA2 levels but also SA3.

10 AMANDA: *Automation and Man–Machine Delegation of Action.*
11 STAR: French acronym for tactical conflict solving support system.

9.4.5.4. *Synthesis*

These human–machine cooperation principles highlight all of the necessary interactions between human agents and machines. The COFOR is at the center of the cooperation and the CWS is an interaction method created to maintain this cooperation. In the first study, the CWS is a mediation tool between two humans in a cockpit which supports the perception of information, the analysis of the situation and decision in the three forms of cooperation: integrative, augmentative and debative. In the second study, the form of cooperation was mostly integrative, with the robot providing the human with data he was not able to perceive and the human delegating partial control of movement to the robot. In the third study, the cooperation has been extended between the controllers and the STAR to the three forms of cooperation and four stages of problem solving (perception, analysis of the situation, decision and action).

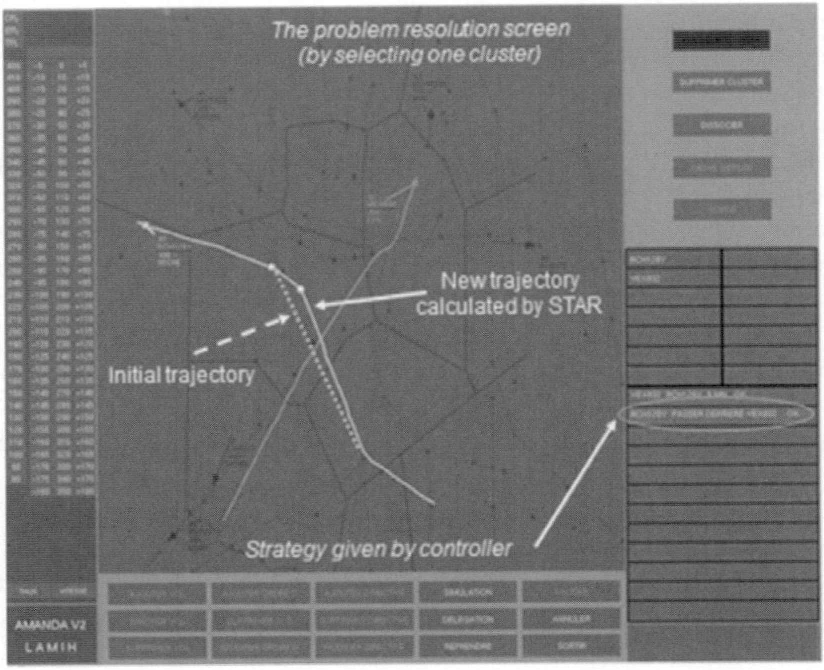

Figure 9.17. *Example of view displayed on the CWS: the "problem solving" view with the route deviating to the left and the list of problems on the right*

A generalization of such a CWS is shown in Figure 9.18 in which each agent can bring and take away information from each stage of the problem solving (two-directional arrows). Furthermore, they are capable – by cooperating – of adjusting the level of involvement (one-directional arrow) and authority (directed arcs) of each agent as suggested by Flemisch *et al.* [FLE 12].

The following section proposes and argues for a mutual enhancement of team-SA using a CWS.

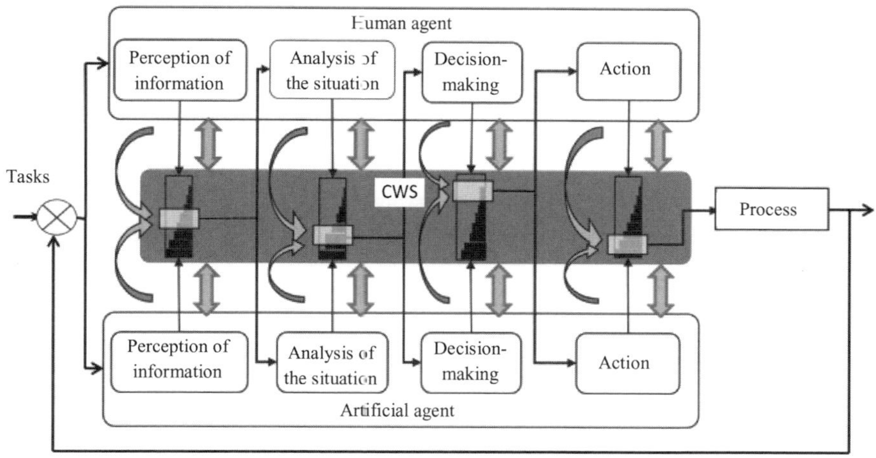

Figure 9.18. *CWS for mediating human–machine cooperation*

9.5. Common work space for team-SA

In the team-SA model in Figure 9.1, teamwork includes the “team processes” (communication, cooperation, shared mental model), which produce team-SA (individual SA, SA of other members of the team and SA of the entire team) [SAL 08]. We consider that the activities of the “team processes” are included in the cooperation activities described in section 9.4. In particular, the agents construct a COFOR and use it to cooperate, which creates an individual but also collective SA.

To this effect, a CWS should be a useful medium, as shown in Figure 9.19. The agents Ag_i have their own KH_i for executing their task and their KHC_i to construct a COFOR useful for cooperation and collective SA. Each agent constructs his/its model of the other (KH, KHC) *via* a direct or mediated communication, *via* the observation of the actions of the other agent and the intentions that he/it deduces and via the interferences. Each agent constructs a representation of this COFOR by combining its own occurring representation of the situation with an image of the representation of the other agents created through inference. In the same way, they construct their own team-SA.

A CWS could help agents analyze the intentions of other agents and outline their own SA as well as the SA of other agents and the team-SA. The CWS would then be an explicit external representation of the team-SA in the same way that it supports the steps for cooperation.

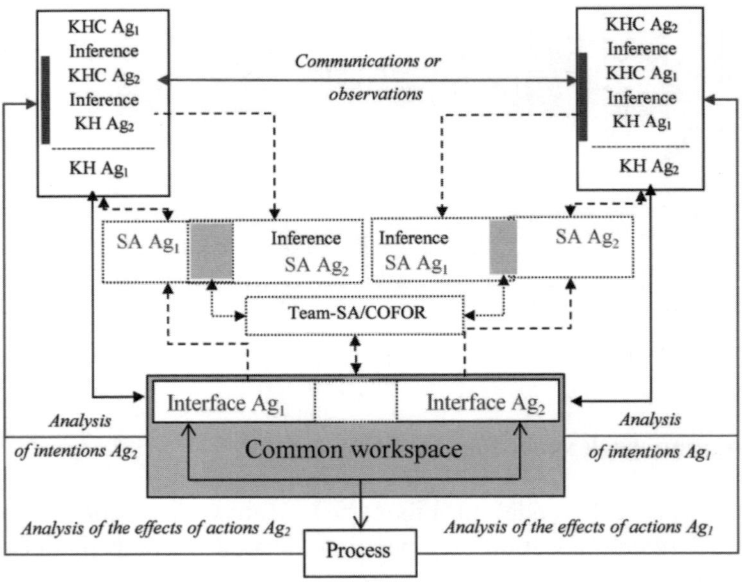

Figure 9.19. *Principle of the CWS for team-SA*

An example of a virtual camera (VC) system is proposed by Platt, Millot and Boy [PLA 13] to facilitate the operations and SA of an astronaut during interplanetary exploration. VC takes the form of a tablet given to the astronaut, which serves as a cooperation interface with the mission chief and the scientists (for example, the geologist) on Earth. These transmit strategic decisions to the astronaut relative to the mission, for example collecting samples of rock in a given area. With the astronaut being on the other planet, they can encounter access difficulties or dangers not yet discovered by humans. Additional information, useful for exploration and/or safety phases, can then be provided to the astronaut either by being available in memory or being acquired online (for example, via "Google Moon"). Furthermore, the tablet can also be a control interface for a space vehicle. This is therefore a CWS between the three distant human actors of the mission but also between the astronaut and the decision-making support tools installed in the vehicle. Several modes of human–human and human–machine cooperation are therefore established and seem promising for the improvement of the SA of the astronaut or even the SA of the team. An initial evaluation using the SAGAT method (see Chapter 1 and [END 95b]) gives encouraging results. Further tests will be carried out to confirm them.

9.6. Conclusion

This final chapter has placed human–machine cooperation within the human-centered design problem by recalling the purpose of SA, in other words the human capacity to perceive events in space time (SA1), to understand their significance (SA2) and to project their state in the near future (SA3). After addressing the concept of team-SA, we have outlined the evolution of approaches to cooperation, firstly structural, from the concept of dynamic task allocation between humans and decision support systems, as well as functional and multilevel in particular the principle of a CWS.

As interesting similarities exist between human–machine cooperation and team-SA, we have compared the two theories and proposed a CWS to enhance the team-SA model. Three examples showing the purpose of a CWS have been outlined and we have deduced a multi-agent architecture in which the CWS can help to construct and maintain a COFOR, as well as the team-SA. A preliminary validation has been shown in a spatial exploration application and will be continued.

The common use of the CWS in both human–human or human–machine cooperation and team-SA is particularly interesting in the sense that it offers a possible solution to the "designer's dilemma": allow the humans to remain capable of imagining solutions for new problems, while allowing them to limit and correct their errors, or even learn from their errors. This is a research perspective that the human engineering community must adapt by investing in the field of collective SA.

9.7. Bibliography

[BAI 83] BAINBRIDGE L., "Ironies of automation", *Automatica*, vol. 19, no. 6, pp. 775–779, 1983.

[BIL 96] BILLINGS C., "Human-centered aviation automation: principles and guidelines", *NASA Technical Memorandum 110381*, February 1996.

[CRE 93] CREVITS I., DEBERNARD S., VANDERHAEGEN F., *et al.*, "Multi-level cooperation in air-traffic control", *4th International Conference on Human-Machine Interaction and Artificial Intelligence in Aerospace*, Toulouse, France, 26–28 September 1993.

[DEB 92] DEBERNARD S., VANDERHAEGEN F., MILLOT P., "An experimental investigation of dynamic task allocation between air traffic controller and AI system", *Proceedings of the 5th IFAC/IFIP/IFORS/IEA Symposium on Analysis, Design and Evaluation of Man-Machine Systems*, La Haye, The Netherlands, pp. 9–11, June 1992.

[DEB 09] DEBERNARD S., GUIOST B., POULAIN T., *et al.*, "Integrating human factors in the design of intelligent systems: an example in air traffic control", *International Journal of Intelligent Systems Technologies and Applications*, vol. 7, no. 2, pp. 205–226, 2009.

[DEB 10] DEBERNARD S., POULAIN T., MILLOT P., "A common work space for supporting human-human and human-machine cooperation in ATC", *HCI Aero 2010, International Conference on Human Computer interaction in Aerospace Cape Canaveral*, November 2010.

[END 95a] ENDSLEY M., "Toward a theory of situation awareness in dynamic systems", *Human Factors*, vol. 37, no. 1, pp. 32–64, 1995.

[END 95b] ENDSLEY M., "Measurement of situation awareness in dynamic systems", *Human Factors*, vol. 37, pp. 65–84, 1995.

[FLE 12] FLEMISCH F., HEESEN M., HESSE T., *et al.*, "Towards a dynamic balance between humans and automation: authority, ability, responsibility and control in shared and cooperative control situation", *Cognition Technology & Work*, vol. 14, no. 14, pp. 3–18, 2012.

[GOO 01] GOODRICH M., OLSEN D., CRANDALL J., *et al.*, "Experiments in adjustable autonomy", *Proceedings of the IJCAI01 workshop on Autonomy, Delegation, and Control: Interacting with Autonomous Agents*, August 2001.

[GRE 81] GREENSTEIN J.S., REVESMAN M.E., "A Monte-Carlo simulation investigating means of human computer communication for dynamic task allocation", *IEEE Conference on Cybernetics and Society*, New York, USA, pp. 488–494, 1981.

[GRI 99] GRISLIN E., MILLOT P., "Specifying artificial cooperative agents through a synthesis of several models of cooperation", in HOC J.M., MILLOT P., HOLLNAGEL E., CACIABUE P.C. (eds.), *Proceedings of the 7th European Conference on Cognitive Science Approach to Process Control CSAPC'99*, Presses Universitaires de Valenciennes, Valenciennes, pp. 73–78, 1999.

[HOC 98] HOC J.M., LEMOINE M.P., "Cognitive evaluation of human-human and human-machine cooperation modes in Air Traffic control", *The International Journal of Aviation Psychology*, vol. 8, no. 1, pp. 1–32, 1998.

[HOC 01] HOC J.M., "Towards a cognitive approach to human-machine cooperation in dynamic situations", *International Journal of Human-Computer Studies*, vol. 54, no. 4, pp. 509–540, 2001.

[INA 13] INAGAKI T., ITOH M., "Human's overtrust in and overreliance on advanced driver assistance systems: A theoretical framework", *International Journal of Vehicular Technology*, p. 8, 2013. Avialable at http://dx.doi.org/10.1155/2013/951762.

[LEE 92] LEE J.D., MORAY N., "Trust, control strategies and allocation of function in human machine systems", *Ergonomics*, vol. 35, no. 10, pp. 1243–1270, 1992.

[LEM 96] LEMOINE M.-P., DEBERNARD S., CRÉVITS I., *et al.*, "Cooperation between humans and machines: first results of an experimentation of a multi-level cooperative organization in air traffic control", *Computer Supported Cooperative Work*, vol. 5, pp. 299–321, 1996.

[MIL 88a] MILLOT P., *Supervision des procédés automatisés et ergonomie*, Hermès, Paris, 1988.

[MIL 88b] MILLOT P., KAMOUN A., "An implicit method for dynamic task allocation between man and computer in supervision posts of automated processes", *Proceedings of IFAC/IFIP/IEA/IFORS Conference on Analysis Design and Evaluation of Man Machine Systems*, pp. 77–82, Oulu, Finland, June 1988.

[MIL 89] MILLOT P., TABORIN V., KAMOUN A., "Two approaches for man-computer cooperation in supervisory tasks", *Proceedings of IFAC/IFIP/IEA/IFORS Conference on Analysis Design and Evaluation of Man Machine Systems*, Xi-An, China, September 1989.

[MIL 93] MILLOT P., DEBERNARD S., "Men-machines cooperative organizations: methodological and practical attempts in air traffic control", *IEEE/SMC'93 Conference*, Le Touquet, France, 17–20 October 1993.

[MIL 95] MILLOT P., MANDIAU R., "Men-machine cooperative organizations: Formal and pragmatic implementation methods", in HOC J.M., CACCIABUE P.C., HOLLNAGEL E., (eds.), *Expertise and Technology: Cognition Computer Cooperation*, Lawrence Erlbraum Associates, Hillsdale, NJ, chapter 13, pp. 213–228, 1995.

[MIL 97] MILLOT P., HOC J.-M., "Human-machine cooperation: metaphor or possible reality?", *Proceedings of the 2nd European Conference on Cognitive Science ECCS'97*, pp. 165–174, Manchester, UK, 1997.

[MIL 98a] MILLOT P., "Concepts and limits for human machine cooperation", *IEEE SMC CESA 98, Computational Engineering in System Application Conference*, Hammamet, Tunisia, April 1998.

[MIL 98b] MILLOT P., LEMOINE M.P., "An attempt for generic concepts toward human-machine cooperation", *IEEE SMC'98 Conference*, San Diego, US, October 1998.

[MIL 11] MILLOT P., DEBERNARD S., VANDERHAEGEN F., "Authority and cooperation between humans and machines", in BOY G. (ed.), *Handbook for Human-Machine Interaction*, Ashgate, Wey Court East, 2011.

[PAC 02a] PACAUX-LEMOINE M.-P., LOISELET A., "A common work space to support cooperation in the cockpit of a two seater fighter aircraft", in BLAY-FORNARINO M., PINNADERY A.M., SCHMIDT K., *et al.*, (eds.), *Cooperative Systems Design: A Challenge of Mobility Age*, IOS Press, Amsterdam, pp. 157–172, 2002.

[PAC 02b] PACAUX-LEMOINE M.-P., DEBERNARD S., "A common work space to support the air traffic control", *Control Engineering Practice, A Journal of IFAC*, vol. 10, pp. 571–576, 2002.

[PAC 11] PACAUX M.P., DEBERNARD S., GODIN A., *et al.*, "Levels of automation and human-machine cooperation: Application to human-robot interaction", *IFAC World Conference*, Milan, Italy, 2011.

[PIA 77] PIAGET J., *Etudes sociologiques*, 3rd. ed., Droz, Genève, 1977.

[PLA 13] PLATT D., MILLOT P., BOY G., "Virtual camera (VC) system to improve astronaut and mission operations situation awareness while exploring other planetary bodies", *IFAC/IFIP/IEA/IFORS Conference on Analysis Design and Evaluation of Human Machine Systems*, Las Vegas, USA, August 2013.

[RAJ 08] RAJAONAH B., TRICOT N., ANCEAUX F., *et al.*, "The role of intervening variables in driver-ACC cooperation", *International Journal of Human Computer Studies*, vol. 66, no. 3, pp. 185–197, 2008.

[RAS 83] RASMUSSEN J., "Skills, rules and knowledge; signals, signs, and symbols, and other distinctions in human performance models", *IEEE Transactions on Systems, Man and Cybernetics*, SMC13, no. 3, May–June 1983.

[SAL 95] SALAS E., PRINCE C., BAKER P.D., *et al.*, "Situation awareness in team performance", *Human Factors*, vol. 37, no. 1, pp. 123–126, 1995.

[SAL 08] SALMON P., STANTON N., WALKER G., *et al.*, "What really is going on? Review of situation awareness models for individuals and teams", *Theoretical Issues in Ergonomics Science*, vol. 9, no. 4, pp. 297–323, 2008.

[SCH 91] SCHMIDT K., "Cooperative work: a conceptual framework", in RASMUSSEN J., BREHMER B., LEPLAT J., (eds.), *Distributed Decision Making: Cognitive Models for Cooperative Work*, John Wiley and Sons, New York, pp. 75–110, 1991.

[SHU 05] SHU Y., FURUTA K., "An inference method of team situation awareness based on mutual awareness", *Cognition Technology & Work*, vol. 7, pp. 272–287, 2005.

[STA 06] STANTON N.A., STEWART R., HARRIS D., *et al.*, "Distributed situation awareness in dynamic systems: theoretical development and application of an ergonomic methodology", *Ergonomics*, vol. 49, pp. 1288–1311, 2006.

[TAY 90] TAYLOR R.M., "Situational awareness rating technique (SART): the development of a tool for aircrew systems design", *Situational Awareness in Aerospace Operations (AGARD-CP-478), NATOAGARD*, Neuilly-sur-Seine, France, pp. 3/1–3/17, 1990.

[VAN 94] VANDERHAEGEN F., CREVITS I., DEBERNARD S., *et al.*, "Human-machine cooperative: toward an activity regulation assistance for different air traffic control levels", *International Journal of Human Computer Interaction*, vol. 6, no. 1, pp. 65–104, 1994.

[WOO 04] WOODS D., TITTLE J., FEIL M., *et al.*, "Envisioning human-robot coordination in future operations", *IEEE Transactions on Systems, Man and Cybernetics*, vol. 34, pp. 210–219, May 2004.

[ZHA 04] ZHANG Z., POLET P., VANDERHAEGEN F., *et al.*, "Artificial neural network for violation analysis", *Reliability Engineering and System Safety*, vol. 84, no. 1, pp. 3–18, 2004.

Conclusion

In 1992, one of the French pioneers in automation presided over the CNRS evaluation committee of the laboratory which would become LAMIH at Valenciennes and concluded the evaluation with this sentence: "our field not only requires control scientists, it is lacking in 'practitioners in automation' […] and your work focusing on humans in the loop will contribute to this"[1]. Throughout this book, we show that even with increasingly elaborate technology, humans and especially teams of humans all have their place within systems. *Human-centered design* such as we have described it in the Introduction is the result of the conjunction of knowledge, experience and fields stemming from not only engineering and cognitive sciences, but also from social, economic and legal sciences. We have mentioned the multilevel flow modeling approach by Morten Lind who treats all of these subject areas as different points of view in an attempt to model large systems. The AUTOS pyramid by Guy Boy is a complementary approach for defining the five areas to take into account in the process of human-centered design.

This book aims to be methodological but also practical, as illustrated by the examples of applications in complex real-life systems. We have mentioned transport systems (aerial, automobiles, railways), production systems (nuclear energy, nuclear waste reprocessing industry) and the

---

Conclusion written by Patrick MILLOT.

1 "Notre discipline n'a pas seulement besoin d'automaticiers elle manque d'automatiseurs… et vos travaux sur l'humain dans la boucle y contribuent". Translator's note: this was a pun intended between the French words *automaticien* and *automatiseur*.

networks that are now systems of systems, such as road or air traffic control. The human is increasingly less isolated in the loop, as he works within a team or a crew. He is no longer just a decision-maker, but also a collaborator, of course, with not only other humans in his team but also increasingly more with tools equipped with decision-making capacities and even actions that can substitute the human, but which he must supervise and be in charge of, which he remains responsible for, and sometimes accountable to.

The fact that the human remains responsible for the tool for which the organization replaces him is one of the paradoxes that the practitioners in automation will need to fix, by taking into account the societal environment and the *individual acceptability* so that the human will want to stay in the loop, the *social acceptability* so that he will want to cooperate with others, *legal acceptability* so that the *authority* remains consistent with the *responsibility* and *know-how* and finally the *economical acceptability* so that the challenges we must raise are easily accomplishable.

Can we escape what we refer to as the designer's dilemma? On the one hand, the human commits errors, and designers tend to impose procedures to follow up on him (and limit his degrees of freedom), but on the other hand, he is capable of innovating, inventing and solving new problems, so that he must absolutely be kept active and vigilant in the loop. Aside from the usual dogmatic responses that oppose the traditionalists of *militant ergonomy* to the *excessive* control engineers, partisans of "everything without humans" but who have not yet found the means to fulfill their ambitions, we have explored the answers of the scientific community following two areas of research: one referred to as *human reliability*, which attempts to understand the mechanics of human errors (and accidents) and to put in place barriers to prevent them, and the other which aims to help humans make good decisions by supporting them with *decision support systems* or helping them with action support tools with which he can cooperate. These two research areas have been presented in this book.

An intermediary area of research seems promising: allowing the human operator to remain *aware of the situation*, in other words, according to Endsley, to perceive events in space time (SA1), to understand their significance (SA2) and project their state in the near future (SA3). The four last chapters have established concepts intended to define and implement *human–machine cooperation* or to mediate *human–human cooperation*. It turns out that the *common work space*, a mediation tool for the *COmmon Frame Of Reference* essential to cooperation, is also used to favor team situation awareness or team-SA. Examples have shown the feasibility of this idea in piloting fighter jets, in human–robot cooperation and in air traffic control. This is a particularly promising perspective which opens up to the *human engineering* community, knowing that the international community of "human factors" is already invested in this area of research.

List of Authors

Pietro Carlo CACCIABUE
Polytechnic University of Milan
Italy

Christine CHAUVIN
CRPCC
University of Southern Brittany
Lorient
France

Serge DEBERNARD
LAMIH
University of Valenciennes
France

Bernard DUBUISSON
Heudiasyc
University of Technology of Compiègne
France

Hélène FONTAINE
IFSTTAR
Marne-la Vallée
France

Jean-Michel HOC
IRCCyN
Ecole Centrale
Nantes
France

Pierre LOSLEVER
LAMIH
University of Valenciennes
France

Gilles MALATERRE
IFSTTAR
Marne-la Vallée
France

Marine MILLOT
CEREMA
Direction Territoriale Méditerranée
Aix en Provence – Les Milles
France

Patrick MILLOT
LAMIH
University of Valenciennes
France

Jacky MONTMAIN
LGI2P
Ecole des Mines d'Alès
Nîmes
France

Marie-Pierre PACAUX-LEMOINE
LAMIH
University of Valenciennes
France

Jean-Christophe POPIEUL
LAMIH
University of Valenciennes
France

Thierry POULAIN
LAMIH
University of Valenciennes
France

Bernard RIERA
CReSTIC
University of Reims
France

Philippe SIMON
LAMIH
University of Valenciennes
France

Frédéric VANDERHAEGEN
LAMIH
University of Valenciennes
France

Peter WIERINGA
TU-Delft
Netherlands

Index

B

C

D

P

Q, R

S

T

U, V, W